29th Annual International Laser Physics Workshop (LPHYS'21)

Journal of Physics: Conference Series Volume 2249

Online
18 – 23 July 2021

ISBN: 978-1-7138-5503-3
ISSN: 1742-6588

Printed from e-media with permission by:

Curran Associates, Inc.
57 Morehouse Lane
Red Hook, NY 12571

Some format issues inherent in the e-media version may also appear in this print version.

Institute of Physics
Dirac House, Temple Back
Bristol BS1 6BE UK

Phone: 44 1 17 929 7481
Fax: 44 1 17 920 0979

techtracking@iop.org

Additional copies of this publication are available from:

Curran Associates, Inc.
57 Morehouse Lane
Red Hook, NY 12571 USA
Phone: 845-758-0400
Fax: 845-758-2633
Email: curran@proceedings.com
Web: www.proceedings.com

TABLE OF CONTENTS

Journal of Physics: Conference Series **2249** (2022) 011001 doi:10.1088/1742-6596/2249/1/011001

The twenty-ninth annual International Laser Physics Workshop (LPHYS'21) was held in an exclusively virtual format on Zoom from July 19 to July 23, 2021.

In 2020, due to the Covid-19 pandemic that paralyzed the entire world, the twenty-ninth Annual International Laser Physics Workshop could not be held in person in Lyon, France, and had to be postponed until 2021. This year the Workshop was carried out in an exclusively virtual format on Zoom. The LPHYS'21 Steering and Advisory&Program committees and the LPHYS'21 Management Team have worked hard to make this event possible.

We kept the conference virtual Zoom format as closely as possible to the standard in-person format of the previous years, albeit without popular extravagant welcome parties, sumptuous gala dinners, and many other in-person activities and friendly gatherings. However, the cordial atmosphere that had always ruled at our annual meetings was kept on.

The following is the list of the Proceedings Editors:

Vanderlei S. Bagnato
University of São Paulo, São Carlos, SP, Brazil
vander@if.sc.usp.br

Olga Kocharovskaya
Texas A&M University, College Station, TX, USA
kochar@physics.tamu.edu

Mikhail V. Fedorov
A.M. Prokhorov General Physics Institute, Russian Academy of Sciences, Moscow, Russia
fedorov@ran.gpi.ru

Igor Meglinski
College of Engineering and Physical Sciences, Aston University, Birmingham, UK
i.meglinski@aston.ac.uk

Sergei P. Kulik
Lomonosov Moscow State University Moscow, Russia
sergei.kulik@physics.msu.ru

Vladimir A. Makarov
International Laser Center, Moscow State University, Moscow, Russia
makarov@ilc.edu.ru

Antonino Di Piazza
Max Planck Institute for Nuclear Physics, Heidelberg, Germany
dipiazza@mpi-hd.mpg.de

Kirill A. Prokhorov
A.M. Prokhorov General Physics Institute, Moscow, Russia
cyrpro@gpi.ru

Vyacheslav I. Yukalov
Joint Institute for Nuclear Research, Dubna, Moscow Region, Russia
yukalov@theor.jinr.ru

Alexander Fedotov
National Research Nuclear University MEPhI, Moscow, Russia
symposium@theor.mephi.ru

Plenary speakers were world-renowned scientists:

Gérard Mourou
École Polytechnique, Palaiseau, France

Paul Corkum
University of Ottawa, Ottawa, Canada

Margaret Murnane
JILA, University of Colorado, Boulder, CO, USA

Henry Kapteyn
JILA, University of Colorado, Boulder, CO, USA

Jorge Rocca
Electrical and Computer Engineering, Colorado State University, Fort Collins, CO, USA

Marlan O. Scully
Texas A&M University, Princeton, NJ, USA
Baylor University, Waco, TX, USA

Mikhail Lukin
Harvard University, Cambridge, MA, USA

Gerd Leuchs
Max Planck Institute for the Science of Light, Erlangen, Germany

Philippe Grangier
Laboratoire Charles Fabry, Institut d'Optique Graduate School, Palaiseau, France

Phil Bucksbaum
Stanford University, Stanford, CA, USA

The Workshop is being organized by:

- A M Prokhorov General Physics Institute, Russian Academy of Sciences (RAS), Moscow, Russia

- Astro Ltd., Hong Kong

- Institute of Physics, London, UK

- International Laser Center, Moscow State University, Moscow, Russia

Sponsors:

- International Laser Center, Moscow State University, Moscow, Russia

- Astro Ltd., Hong Kong

- IOP Publishing Limited, Bristol, UK
- Quantum Technology Centre, Lomonosov Moscow State University, Moscow, Russia

Peer Review Statement

All papers published in this volume have been reviewed through processes administered by the Editors. Reviews were conducted by expert referees to the professional and scientific standards expected of a proceedings journal published by IOP Publishing.

- **Type of peer review:** Undefined
- **Conference submission management system:** Morressier
- **Number of submissions received:** 23
- **Number of submissions sent for review:** 22
- **Number of submissions accepted:** 22
- **Acceptance Rate (Submissions Accepted / Submissions Received × 100):** 95.7
- **Average number of reviews per paper:** 1.45
- **Total number of reviewers involved:** 32
- **Contact person for queries:**
  Name: Alexander Yevseyev
  Email: staffeditor@lasphys.com
  Affiliation: Astro Science Ltd - Director

Weakly Aligned Molecules: From Molecular Detectors to Room-Temperature Tunable Masers

Colin Roy

Center for Nonlinear Sciences and Department of Physics, University of North Texas, 1155 Union Circle #311427, Denton, Texas 76203, USA

Zorica Branković

University of Belgrade, Institute for Multidisciplinary Research, Kneza Višeslava 1a, 11030 Belgrade, Serbia

Yuri Rostovtsev

Center for Nonlinear Sciences and Department of Physics, University of North Texas, 1155 Union Circle #311427, Denton, Texas 76203, USA

E-mail: rost@unt.edu

Abstract.

We have developed and demonstrated a new mechanism of manipulation of population in molecular rotational levels in a weakly aligned molecules. The mechanism is based on an adiabatically changing electric field interacting with the molecules with dipole moments. Treating molecules as simple rotators, we have described their behavior using the density matrix taking into account the relaxation processes. We have considered the interaction of the weakly aligned molecules with a microwave field in a high finesse cavity. We have found that, on one hand, the population inversion can be reached in the ensemble of the weakly aligned molecules to be used for the maser operation at room temperature. On the another hand, we have found that the enhancement of the absorption can reach the theoretical limit and be used for gas sensing with high sensitivity and selectivity. Such sensors can efficiently analyze the multi-gas mixtures and be used for a huge range of applications – stretching from technology, sciences, control of environment, biology and medicine.

1. Introduction

The maser (microwave amplification by stimulated emission of radiation), the coherent source of the microwave radiation similar to the laser, was demonstrated in 1954 [1]. However, unlike lasers [2], the masers are much less widely used because in order to function, they must be cooled to temperatures close to absolute zero. In 2012, it was demonstrated that a maser could operate at room temperature using the organic molecule pentacene. However, it only produced short pulses of maser radiation. For continuous operation of the maser, the crystal would likely have melted [3]. Only recently, it was reported that a new maser is created that operates continuously [4].

Meanwhile, we have to underline here that the maser technology being used where sensitive detection of microwave radiation is essential. The masers could be used in a range of applications such as medical imaging and airport security scanning. They have more traditionally been used in deep space communication and radio astronomy. As well as medical imaging and airport security scanning, masers could play a pivotal role in improving sensors to remotely detect bombs, new technology for quantum computers, and might even improve space communication methods to potentially find life on other planets. Thus, the development tunable maser working at room temperature is very important.

On the other hand, to detect molecular gas dispersion or absorption or to create population inversion in molecular levels at room temperature is difficult for rotational frequencies because the population difference is very low at room temperature and probability to absorb photon is very low. Indeed, the probability for photon being absorbed is given by

$$P_{abs} = 1 - \exp\left[-\frac{\omega_{ab}\wp_{ab}^2 N_0 L}{c\hbar\Gamma_{ab}} n_{ab}\right] \quad (1)$$

where ω_{ab} is the frequency of transition, $\wp_{ab}$ is the dipole moment, Γ_{ab} is the width of transition, N_0 is the gas density, and n_{ab} is the population difference between corresponding levels. It is clear that low frequency ω_{ab} and small n_{ab} results in low absorption probability. But it would be desirable to enhance the probability of absorption for individual molecules to 100 %, so each molecule can absorb one photon. Then, this can be extremely useful to reach maximal absorption for gas detectors as well as creation of population inversion in a molecular gas for the maser operation.

In this paper, we show that developing techniques to control and to manipulate the populations in the rotational molecular levels can enhance the sensitivity of gas detection, and also it can open possibilities of creation population inversion that can be used for developing masers working at room temperature.

Applying a dc electric field aligns molecules along the field, and modifies the rotational energy levels of molecules [5, 6] (see Figure 1(a)). Aligned molecules can be viewed as simple pendulums that have the energy structure of a simple harmonic oscillator (see Figure 1(a)). Even a small electric field produces the weak alignment that can be take advantage for development various techniques of quantum control of rotational molecular states. The control is based on an adiabatically changing electric field that interacts with the rotational structure of the molecules with dipole moments. In particular, these new techniques are important for developing gas masers at room temperature as well as the developing of new gas sensors with higher selectivity and sensitivity [7, 8].

There are several physical mechanisms to change absorption that are related to the so-called quantum coherent effects. Quantum coherence effects, such as coherent population trapping [9] and electromagnetically induced transparency (EIT) [10, 11, 12], have been the focus of a broad range of research activity for the past two decades since they drastically change the optical properties of the media. In EIT, for example, absorption practically vanishes in both the CW and the pulsed regime [11]. A medium with an excited quantum coherence, phaseonium [10], can be used to make an ultra-dispersive prism [13] which will have several orders of magnitude greater angular spectral dispersion compared to a conventional one [14, 15, 16, 17, 18]. The importance and success of proper quantum engineering was demonstrated, in particular, for the pulse shaping in coherent Raman spectroscopy [19] that allows researchers to improve sensitivity. Here, we propose a new approach that results in the enhancement of the population difference between corresponding molecular levels and reaches the theoretical maximum of absorption.

The new approach to the problem is based on introducing mechanism based on an adiabatically changing electric field interacting with the rotational structure of the molecules with dipole moments. Normally the gases do not absorb radiation if optical density is very small. Contrary, using this approach, every gas molecule absorbs a photon. This can be reached by using an adiabatically changing electric field that is sweeping the resonant frequency through the frequency of the ac probe field resulting in complete population transfer in the molecular system. Together with the change of photon numbers, the phase change of the probe field occurs. Both changes can be used as signals for gas sensing. Using proper tailoring of the adiabatically changing electric field allows researchers to resonantly enhance the sensor signal reaching the theoretical limit.

2. Interaction with chirping adiabatic pulses

Let us consider the interaction of the adiabatically changing fields with a two-level system. For the two-level atom, the Hamiltonian is given by

$$\hat{H} = \hbar\Delta|a\rangle\langle a| + \hbar\Omega(|a\rangle\langle b| + |b\rangle\langle a|), \tag{2}$$

where a and b are the excited and ground levels of the atom, $\Delta = \omega_{ab} - \omega$ is the detuning between frequency of the field ω and the resonance frequency of the transition ω_{ab}, Ω is the coupling due to interaction with adiabatically changing field. The eignevalues are

$$\lambda_{\pm} = \omega_{\pm} = \frac{\Delta}{2} \pm \sqrt{\left(\frac{\Delta}{2}\right)^2 + \Omega^2} \tag{3}$$

and eigenvectors are given by

$$|\pm\rangle = \frac{\omega_{\pm}|a\rangle + \Omega|b\rangle}{\sqrt{\omega_{\pm}^2 + \Omega^2}} \tag{4}$$

Let us note here that for $\Omega = 0$ and $\Delta > 0$: $\omega_+ = \Delta$, $|+\rangle = |a\rangle$ and $\omega_- = 0$, $|-\rangle = |b\rangle$, and for $\Omega = 0$ and $\Delta < 0$: $\omega_+ = 0$, $|+\rangle = |b\rangle$ and $\omega_- = \Delta$, $|-\rangle = |a\rangle$. For $\Omega \neq 0$, adiabatically changing Δ from $-\infty$ to $+\infty$ leads to adiabatical changes $\omega_{\pm} \to \omega_{\mp}$ and $|\pm\rangle \to |\mp\rangle$, as

$$\omega_+(\Delta) = \begin{cases} 0, & \Delta \to -\infty \\ \Delta, & \Delta \to +\infty \end{cases} \quad \text{and } |+\rangle(\Delta) = \begin{cases} |b\rangle, & \Delta \to -\infty \\ |a\rangle, & \Delta \to +\infty \end{cases} \tag{5}$$

$$\omega_-(\Delta) = \begin{cases} \Delta, & \Delta \to -\infty \\ 0, & \Delta \to +\infty \end{cases} \quad \text{and } |-\rangle(\Delta) = \begin{cases} |a\rangle, & \Delta \to -\infty \\ |b\rangle, & \Delta \to +\infty \end{cases} \tag{6}$$

and the states $|a\rangle$ and $|b\rangle$ can be expressed in terms of $|\pm\rangle$ states as

$$|a\rangle = \frac{\sqrt{\omega_+^2 + \Omega^2}|+\rangle - \sqrt{\omega_-^2 + \Omega^2}|-\rangle}{2\sqrt{\left(\frac{\Delta}{2}\right)^2 + \Omega^2}} \text{ and } |b\rangle = \frac{-\omega_- \sqrt{\omega_+^2 + \Omega^2}|+\rangle + \omega_+ \sqrt{\omega_-^2 + \Omega^2}|-\rangle}{2\Omega\sqrt{\left(\frac{\Delta}{2}\right)^2 + \Omega^2}}. \tag{7}$$

Then, an arbitrary state $|\psi_i\rangle = A|a\rangle + B|b\rangle$ changes into $|\psi_f\rangle = Be^{i\phi_a}|a\rangle + Ae^{i\phi_b}|b\rangle$, performing swap the populations in the states. It is a remarkable because it can lead to the maximal absorption on one hand, and to creation of population inversion on the other. Below there are applications of this swap of populations to the gas sensing and to the maser operation.

3. Theoretical Model

For simplicity and to understand essential physics, we consider molecules with electric dipole moments $\vec{\wp}$, and the Hamiltonian in the dc electric field $\vec{\mathcal{E}}_0$ is given by

$$\hat{H} = B\hat{J}_z^2 - \vec{\wp} \cdot \vec{\mathcal{E}}_0 \tag{8}$$

Journal of Physics: Conference Series **2249** (2022) 012001 doi:10.1088/1742-6596/2249/1/012001

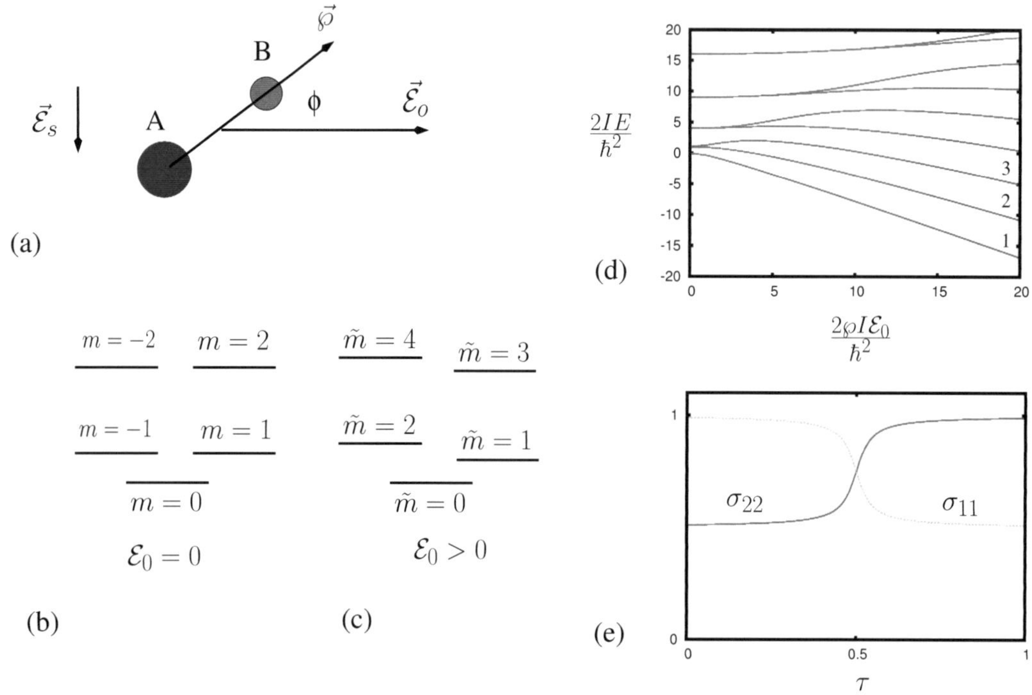

Figure 1. (a) Molecular rotator with electric dipole $\vec{\wp}$, where ϕ is the angle of rotation with respect to the electric field $\vec{\mathcal{E}}_0$, and the ac electric field $\vec{\mathcal{E}}_s$ orthogonal to the dc field that is tuned to the resonance with the molecules transitions; (b) the energy structure of the molecular rotator with no electric field ($\mathcal{E}_0 = 0$); (c) the energy structure of the molecular rotator with electric field ($\mathcal{E}_0 > 0$); (d) Molecular energy $2IE_{\tilde{m}}/\hbar^2$ dependence on $2I\wp\mathcal{E}_0/\hbar^2$. The numbers 1, 2, and 3 correspond to the states with $\tilde{m} = 0$, $\tilde{m} = 1$, and $\tilde{m} = 2$ correspondingly. (e) Dependence of the population in levels $\tilde{m} = 1$ and $\tilde{m} = 2$ on time τ.

Here, we can view symmetric molecules as simple rotators (see Figure 1(a)) to simplify our consideration [20] ($B = \hbar^2/2I$, I is the moment of inertia).

The first excited rotational states are split in the dc electric field due to Stark effect (see Figure 1(b,c)), and this energy splitting depends on $\mathcal{E}_0$

$$\frac{2I(E_2 - E_1)}{\hbar} = F\left(\frac{2\wp I\mathcal{E}_0}{\hbar}\right)$$

The splitting can be tuned to any frequency $\omega_0 = \dfrac{(E_2 - E_1)}{\hbar}$, and the transition can be used for maser operation and for spectroscopy as well. The strong (dc or slowly varying) electric field control the population and the splitting between levels, and the ac electric field orthogonal to the dc field is tuned to the resonance with the molecules transitions. To pump maser the ac pump field is used. Then, for maser operation and spectroscopy, the maser field and the probe field are ac fields with frequencies from RF to microwave range. The ac pump and maser and probe fields $\mathcal{E}_s$ can be created by the high finesse resonance circuits or high finesse microwave cavities. The equation for the Rabi frequency $\Omega_s = \wp\mathcal{E}_s/\hbar$ is given by

$$\dot{\Omega}_s = -i(\omega_0 - \omega_s)\Omega_s + i\Omega_a^2\rho_{ab} \tag{9}$$

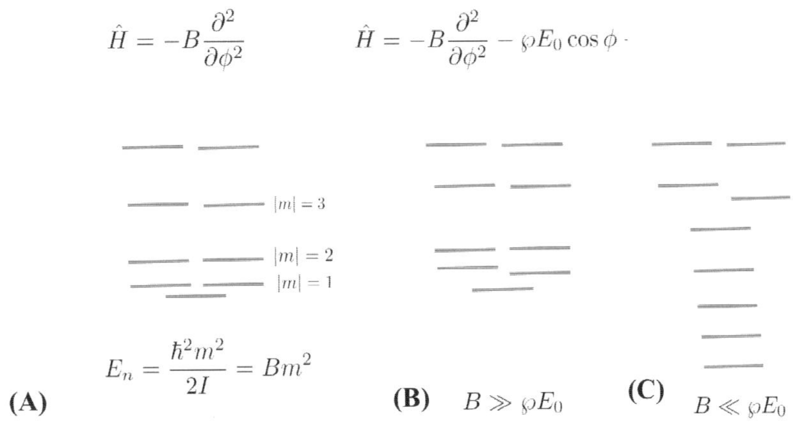

Figure 2. Molecular level structure in the different dc electric fields. (A) $\mathcal{E}_0 = 0$, no dc field, level structure of rotator; (B) $\mathcal{E}_0$ is weak ($B \gg \wp\mathcal{E}_0$), one can see small splitting of the low levels of rotator; (C) $\mathcal{E}_0$ is strong ($B \ll \wp\mathcal{E}_0$), one can see that the low levels of rotator look similar to the levels of the simple harmonic oscillator (see Eq. (15) and Eq. (16)).

where ω_s is the frequency of the resonance circuit, $\Omega_a^2 = \dfrac{2\pi\omega_0\wp^2 N}{\hbar}$ is the cooperative frequency, and ρ_{ab} is the molecular coherence between the molecular states involved, ρ_{ab} is the element of the density matrix ρ.

Taking into account relaxation in all processes described above is very important. For this purposes, we use the Liouville-von Neumann equation for the density matrix to describe the time evolution of the molecules. The decoherence of the individual three-level atomic system is taken into account using Lindblad superoperator formalism [21].

$$\frac{\partial\rho}{\partial t} = \frac{i}{\hbar}[\rho, H] - \hat{\Gamma}[\rho] \tag{10}$$

where $\hat{\Gamma}[\rho]$ is the relaxation matrix for all components of the density matrix ρ. The set of density matrix equations are given by

$$\dot{\rho}_{ab} = -(\Gamma_{ab} + i\Delta)\rho_{ab} + i n_{ab}\Omega_s \tag{11}$$

$$\dot{\rho}_{bb} = i(\rho_{ba}\Omega_s - \Omega_s^*\rho_{ab}) \tag{12}$$

$$\dot{\rho}_{aa} = i(\rho_{ab}\Omega_s^* - \Omega_s\rho_{ba}) \tag{13}$$

where $n_{ab} = \rho_{aa} - \rho_{bb}$, and $\Delta = \omega_{ab}(\Omega_0) - \omega_0$.

The modification of the energy structure leads also to redistribution of population in the rotational levels. Namely, the population difference between levels $\tilde{m} = 1$ and $\tilde{m} = 2$ (see in Figure 1(d) and even more clear see Figure 2(A), (B), (C)) is given by

$$N_2 - N_1 = \frac{\hbar^2 N_0}{2I k_B T Z} F\left(\frac{2\wp I\mathcal{E}_0}{\hbar^2}\right), \text{ where } Z = \sum_{\tilde{m}=0}^{\infty} e^{-\frac{E_{\tilde{m}}}{k_B T}} \tag{14}$$

where k_B is the Boltzmann constant, T is the room temperature, and N_0 is the molecular gas density. The population distribution for case of the electric field strong enough as shown

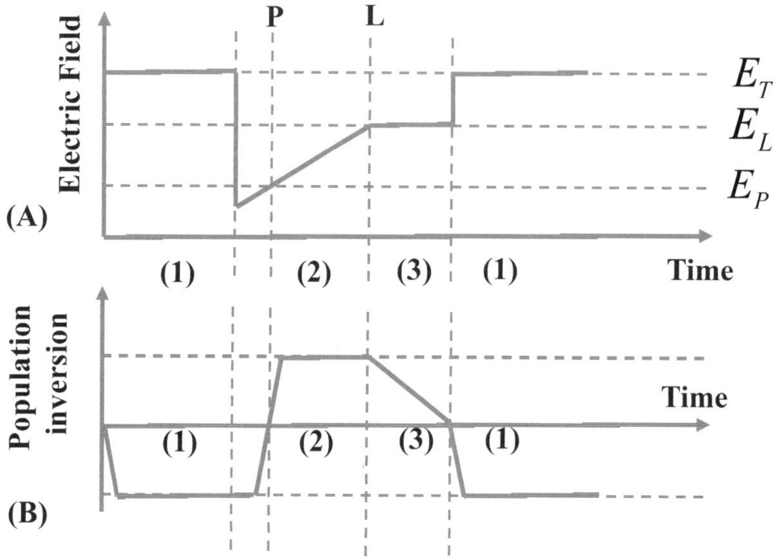

Figure 3. Time dependence of the adiabatically changing strong electric field (arbitrary units) $\mathcal{E}_0$ shown in (A). The corresponding population inversion between corresponding levels (arbitrary units) is shown in (B). Different time periods corresponds different phases of molecules in the cavity. (1) - the thermal pumping; (2) - adiabatic change of electric field to produce population inversion between corresponding levels due to some strong field (as shown in Figure 1(e)); (3) - reaching the resonance with maser transition, microwave amplification, depleting all population inversion and then returning to the pumping stage (1).

in Figure 2(C) will transfer to the same population difference in Figure 2(B) with adiabatic change of the electric field $\mathcal{E}_0$. We can see that this is similar to the magnetic cooling with the adiabatically changing magnetic field [22, 23, 24].

In our approach, we use a molecular gas with weakly aligned molecules in a dc electric field. For complete alignment, a much stronger dc electric (or strong ac electric fields) has to be applied. Indeed, for the total molecular alignment, we need the energy difference between molecular levels be larger than the thermal energy, i.e.,

$$\hbar\omega_{\mathcal{E}} \gg k_B T$$

where k_B is the Boltzman constant, and

$$\wp\mathcal{E}_0 \gg \left(\frac{k_B T}{\hbar B}\right)^2 \hbar B$$

that gives a simple estimation of the dc electric field to be of the order of $\mathcal{E}_0 \simeq 100$ MV/cm. We consider the very weak alignment of molecules in the dc field which leads to the resonance with the ac electric field. For the weak alignment, we need to have the dc electric field be sufficiently strong that the coupling is comparable with the rotational splitting, i.e.

$$\wp\mathcal{E}_0 \simeq \hbar B,$$

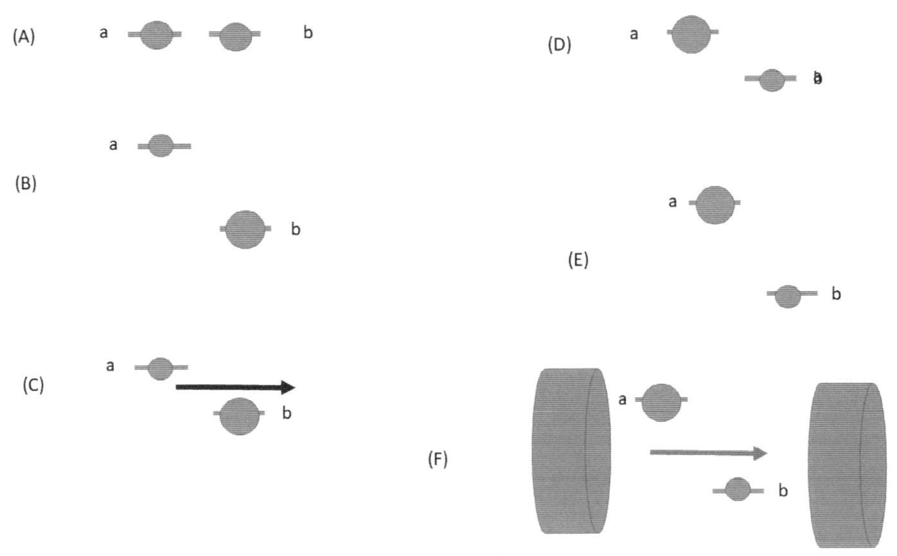

Figure 4. Plot shows all stages of population preparation for maser action. (A) Two selected levels from the levels of the molecular rotator that are chosen for serve as a maser transition; (B) Applying a dc field that splits levels, and it creates population difference by thermal equilibrium at room temperature; (C) The splitting tuned to be in resonance with low frequency field to swap population between levels; (D) The swap of population occurs between levels a and b; (E) The transition is tuned by adiabatic field to the maser cavity frequency; (F) The molecular population inverted levels tuned to cavity frequency and in the cavity produce maser radiation.

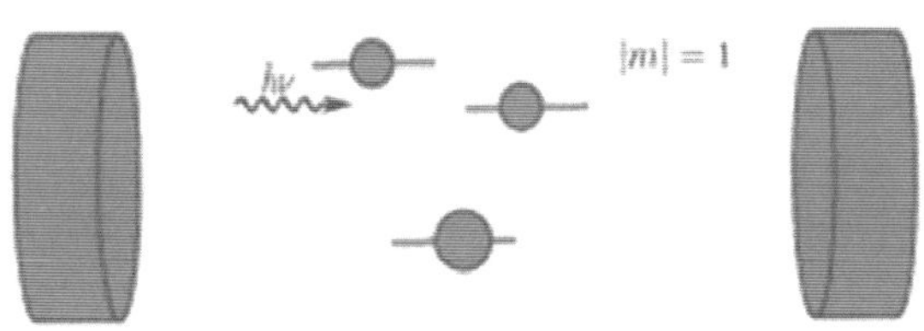

Figure 5. It is schematically shown molecular levels of interests (more molecular levels can be seen in Figure 1) with some populations and resonant microwave radiation in the microwave cavity.

that gives the field to be of the order of $\mathcal{E}_0 \simeq 10$ kV/cm. Our estimations have been made for two frequencies: $\omega_0 = (2\pi)5$ MHz, corresponding to be of the order of $\mathcal{E}_0 \simeq 1$ kV/cm and $\omega_0 = (2\pi)10$ GHz, to be of the order of $\mathcal{E}_0 \simeq 30$ kV/cm.

Applying a dc electric field modifies the rotational energy levels of molecules (see

Figure 1(b,c)). The states $|\tilde{m}\rangle$ in the dc electric field with the rotational states $|m\rangle$ (without electric field) are related as

$$|\tilde{0}\rangle \to |m=0\rangle, \quad \text{for } m>0, \;\; |\tilde{m}=2|m|-1,2|m|\rangle \to |\pm|m|\rangle$$

The energy modification can be easily understood by naively considering aligned molecules in the dc electric field [5, 6] (see Figure 2). Aligned molecules can be viewed as simple pendulums. For levels with low energy excitation, they have the energy structure of a simple harmonic oscillator (see Figure 1(a)). For higher excitation levels, the molecules retain rotational structure.

The dependence of the molecular energy $\frac{2IE}{\hbar}$ on the $\frac{2I\wp\mathcal{E}_0}{\hbar}$ is shown in Figure 1(d). It shows that for the larger fields $\wp\mathcal{E}_0 \gg \frac{\hbar}{I}$, the electric dipole under the presence of the electric field looks like a simple harmonic oscillator, with the Hamiltonian given by

$$\hat{H} = -\frac{\hbar}{2I}\frac{\partial^2}{\partial\phi^2} + \frac{I\omega_E^2}{2}\phi^2 - \wp\mathcal{E}_0, \tag{15}$$

where $\omega_E^2 = \frac{\wp\mathcal{E}_0}{I}$. As can be seen in Figure 1(b), the ground and lower states have energies

$$E_{\tilde{m}} \simeq -\wp\mathcal{E}_0 + \hbar\omega_E(\tilde{m}+1/2). \tag{16}$$

Meanwhile, for larger $\tilde{m}$ such that $\frac{\hbar}{2I}\tilde{m}^2 \gg \wp\mathcal{E}_0$, the fast molecular rotation completely averages out the effect of the dc electric field, and the molecular energy is $E_{\tilde{m}} \simeq \frac{\hbar}{2I}\tilde{m}^2$. One can see that changes in Figure 2 where the molecular level structure is shown in the different dc electric fields. (A) $\mathcal{E}_0 = 0$, no dc field, level structure of rotator; (B) $\mathcal{E}_0$ is weak ($B \gg \wp\mathcal{E}_0$), one can see small splitting of the low levels of rotator; (C) $\mathcal{E}_0$ is strong ($B \ll \wp\mathcal{E}_0$), one can see that the low levels of rotator look similar to the levels of the simple harmonic oscillator. (see Eq. (15) and Eq. (16)).

Then, the proposed maser scheme works in the following stages (see Figures 3 and 4). First, we can chose the levels that can be used for our maser transition (see Figure 4(A)), these can be levels with $|m| = 1$ (see Figure 1). Maser starts at the stage (1), at some relatively strong dc electric field $\mathcal{E}_0$ that causes the weak alignment of molecules. The maser levels a and b are split (see Figure 4(B)). The molecules have the corresponding thermal distribution between molecular levels. Then, the electric field $\mathcal{E}_0$ changes to the vicinity of the electric field $\mathcal{E}_P$ that corresponds coherent pump field (see Figure 4(C)). Time dependence of the adiabatically changing strong electric field (arbitrary units) $\mathcal{E}_0$ shown in Figure 3(A). The corresponding population inversion between corresponding levels (arbitrary units) is shown in Figure 3(B). Stage (2): the dc field adiabatically changes the electric field to tune the molecular transition through the resonance with relatively strong ac pumping field to produce adiabatic population transfer (see Figure 4(D)) that creates the population inversion between corresponding levels (as shown in Figure 1(e)). Stage (3): the dc field adiabatically changes the electric field to tune the molecular transition till it reaches the resonance with the maser cavity mode (see Figure 4(E,F)). One can see the molecular rotational level structure with, for example, $|m| = 1$ (see in Figure 5) that can be used for the maser operation. During this stage the microwave amplification occurs depleting all population inversion and then returning to the pumping stage (1).

Using the set of equations for density matrix and the equation for the microwave cavity field, We estimate the gain for some simple molecules such as [25], using $\wp \simeq 1D$ as a typical value, and for frequency $\omega_0 = (2\pi)10$ GHz, and $N \simeq 2\ 10^{16}$ cm^{-3} and the gain is of the order of $G \simeq 2\ 10^8$ cm^{-1} correspondingly.

4. Conclusion

In conclusion, in the paper, we have developed and demonstrated a new mechanism of manipulation of population in molecular rotational levels in a weakly aligned molecules. Treating molecules as simple rotators, we have described their behavior using the density matrix taking into account the relaxation processes. We have considered the interaction of the weakly aligned molecules with a microwave field in a high finesse cavity. Using the mechanism based on an adiabatically changing electric field interacting with the molecules with dipole moments, we have shown that, on one hand, the population inversion can be reached in the ensemble of the weakly aligned molecules to be used for the maser operation at room temperature. On the another hand, we have found that the enhancement of the absorption can reach the theoretical limit and be used for gas sensing with high sensitivity and selectivity.

It is interesting to note here that the same idea of manipulation of population in molecular rotational levels can be used not only to produce the maser effect, but also to detect signal from molecules and do spectroscopy or gas sensing. This technique can work as a sensitive and very selective molecular detector. The resonance depends on the dc electric field applied and the molecular parameters such as rotational constants and the magnitude of the molecular dipole moment. This opens an interesting opportunity of molecular detection at the chosen frequency [7, 8]. Different molecules have their resonances at the chosen frequency but at the different magnitudes of the electric field. Knowing electric field and frequency allows researcher to identify the molecules. We estimate the sensitivity of the technique for some simple molecules such as [25], using $\wp \simeq 1D$ as a typical value, and for frequencies $\omega_0 = (2\pi)5$ MHz and 10 GHz, and obtain $N \simeq 2\ 10^{11}$ cm^{-3} and $N \simeq 2\ 10^{8}$ cm^{-3} correspondingly. Such sensors can efficiently analyze the multi-gas mixtures and be used for a huge range of applications – stretching from technology, sciences, control of environment, biology and medicine.

Acknowledgments

We gratefully acknowledge the financial support of the Ministry of Education, Science and Technological Development of the Republic of Serbia (Contract No. 451-03-9/2021-14/200053), and from The Advanced Materials and Manufacturing Processes Institute at the University of North Texas Seed Research Project. Y.R. acknowledges the support from the Fulbright Scholarship, 2021-2022 to Serbia, and he is grateful to the hospitality of the Institute for Multidisciplinary Research, University of Belgrade. We cordially thank Vladimir Sautenkov, Goran Branković, Andrey Matsko, Robert Murawski, and Vyrgil Sanders for fruitful discussions, comments and suggestions.

References

[1] Gordon, J. P.; Zeiger, H. J.; Townes, C. H. (1955). "The Maser—New Type of Microwave Amplifier, Frequency Standard, and Spectrometer". Phys. Rev. 99 (4): 1264 (1955) doi:10.1103/PhysRev.99.1264
[2] Mario Bertolotti, The History of the Laser (CRC Press. NewYork, 2004),
[3] Oxborrow, M., Breeze, J. & Alford, N. Room-temperature solid-state maser. Nature 488, 353–356 (2012) doi:10.1038/nature11339
[4] Jonathan D. Breeze, Enrico Salvadori, Juna Sathian, Neil McN. Alford, Christopher W. M. Kay. Continuous-wave room-temperature diamond maser. Nature, 555 (7697): 493 (2018) DOI: 10.1038/nature25970
[5] M. Lemeshko, et al., Manipulation of molecules with electromagnetic fields, *MOLECULAR PHYSICS* **111**, 1648-1682 (2013).
[6] H.J. Loesch, J. Bulthuis, S. Stolte, A. Durand, J.-C. Loison, J. Vigué, Molecules Oriented by Brute Force, *Europhysics News* **27**, 12-15 (1996).
[7] Zorica Brankovic, Yuri Rostovtsev, A resonant single frequency molecular detector with high sensitivity and selectivity for gas mixtures, Scientific Reports 10, 1537 (2020).
[8] Yuri Rostovtsev, Zorica Brankovic, A resonant single-frequency molecular detector based on adiabatically changing electric field, Proceedings Volume 11700, Optical and Quantum Sensing and Precision Metrology; 1170018 (2021) https://doi.org/10.1117/12.2586498

[9] E. Arimondo, in Progress in Optics, edited by E. Wolf (Elsevier Science, Amsterdam, 1996), Vol. XXXV, p. 257.

[10] M. O. Scully and M. S. Zubairy, Quantum Optics (Cambridge University Press, Cambridge, England, 1997).

[11] S. E. Harris, Electromagnetically induced transparency, *Phys. Today* **50**, 36 (1997).

[12] M. Fleischhauer, A. Imamoglu, and J. P. Marangos, Electromagnetically induced transparency: Optics in coherent media, *Rev. Mod. Phys.* **77**, 633 (2005).

[13] V.A. Sautenkov, H. Li, Y.V. Rostovtsev, M.O. Scully, Ultradispersive adaptive prism based on a coherently prepared atomic medium, *Phys. Rev. A* **81**, 063824 (2010).

[14] L.V. Hau, S.E. Harris, Z. Dutton, and C.H. Behroozi, Light speed reduction to 17 metres per second in an ultracold atomic gas, *Nature* **397**, 594 (1999); C. Liu, Z. Dutton, C.H. Behroozi, and L.V. Hau, Observation of coherent optical information storage in an atomic medium using halted light pulses, *Nature* **409**, 490 (2001).

[15] M.M. Kash, V.A. Sautenkov, A.S. Zibrov, L. Hollberg, G.R. Welch, M.D. Lukin, Y. Rostovtsev, E.S. Fry, and M.O. Scully, Ultraslow group velocity and enhanced nonlinear optical effects in a coherently driven hot atomic gas, *Phys. Rev. Lett.* **82**, 5229 (1999). D. Budker, D.F. Kimball, S.M. Rochester, and V.V. Yashchuk, *Phys. Rev. Lett.* **83**, 1767 (1999).

[16] L.J. Wang, A. Kuzmich, and A. Dogariu, Gain-assisted superluminal light propagation, *Nature* (London) **406**, 277 (2000); A. Dogariu, A. Kuzmich, and L.J. Wang, Transparent anomalous dispersion and superluminal light-pulse propagation at a negative group velocity, *Phys. Rev. A* **63**, 053806 (2001).

[17] G. S. Agarwal, T. N. Dey, and S. Menon, Knob for changing light propagation from subluminal to superluminal, *Phys. Rev. A* **64**, 053809 (2001).

[18] E.E. Mikhailov, V.A. Sautenkov, Y.V. Rostovtsev, and G.R. Welch, Buffer-gas-induced absorption resonances in Rb vapor, *J. Opt. Soc. Am. B* **21**, 425 (2004); Q. Sun, Y.V. Rostovtsev, J.P. Dowling, M.O. Scully, and M. S. Zubairy, Optically controlled delays for broadband pulses, *Phys. Rev. A* **72**, 031802 (2005).

[19] Dmitry Pestov, Robert K Murawski, Gombojav O Ariunbold, Xi Wang, Miaochan Zhi, Alexei V Sokolov, Vladimir A Sautenkov, Yuri V Rostovtsev, Arthur Dogariu, Yu Huang, Marlan O Scully, Optimizing the laser-pulse configuration for coherent Raman spectroscopy, Science **316**, 265-268 (2007).

[20] A.R. Edmonds, Angular Momentum in Quantum Mechanics, Series:Investigations in Physics (PRINCETON UNIVERSITY PRESS, Princeton, 1996),

[21] H.-P. Breuer, F. Petruccione, *The theory of open quantum systems,* Oxford University Press Inc., New York 2006.

[22] E.G. Warburg, Magnetische Untersuchungen über einige Wirkungen der Coerzitivkraft, Annalen der Physik (Leipzig) **13**, p. 141-164, (1881).

[23] Pierre Weiss, Auguste Piccard, "Le phénomène magnétocalorique". J. Phys. (Paris). 5th Ser. (7), 103–109 (1917).

[24] Anders Smith, Who discovered the magnetocaloric effect? The European Physical Journal **38**, (4): 507–517, (2013).

[25] Parameters of some molecules are the following: CO, $\wp_{CO} = 0.122$ D, $B_e = 1.93$ cm^{-1}; and HCN, $\wp_{HCN} = 2.98$ D, $B_v = 1.47$ cm^{-1}, N_2O, $\wp_{N_2O} = 0.17$ D, $B_e = 1.93$ cm^{-1}; and NO, $\wp_{NO} = 0.16$ D, $B_v = 1.67$ cm^{-1}. NO_2, $\wp_{NO_2} = 0.316$ D, $A = 8.0$ cm^{-1}, $B = 0.43$ cm^{-1}, $C = 0.41$ cm^{-1}, H_2O, $\wp_{H_2O} = 1.85$ D, $B_v = 15.$ cm^{-1}.

Ultrabright photoluminescence spikes and step-wise photoluminescence increase from colloidal silver nanoparticles for patch nanoantennas

Svetlana G. Lukishova[1], Jerome Brone[1], Darman Khan[2] and Zihao Li[1]

[1]The Institute of Optics, University of Rochester, 275 Hutchison Road, Rochester NY 14627, Rochester NY, USA.
[2]Department of Physics and Astronomy, University of Rochester, 500 Wilson Blvd, Rochester, NY 14627, USA

The corresponding author's e-mail address: lukishov@optics.rochester.edu

Abstract. Ultrabright intensity spikes (up to ~400—900 kcounts/s) appear sporadically in time traces of photoluminescence from 100-nm silver nanocubes used in nanopatch antennas for single-photon source applications. Both 532 and 633 nm laser excitation causes them.

1. Introduction

Room-temperature single (antibunched) photon sources (SPSs) based on single-emitter fluorescence in photonic/plasmonic nanostructures are very important for long-distance secure quantum communication. From all known nanostructures plasmonic nanoantennas provide both the highest reported emission rate enhancements and the highest photon flux increase from single emitters [1-6]. From all types of plasmonic nanoantennas the highest Purcell factor with increasing emitter radiative decay rate (~1,000 times), see, e.g., [2], and 2,300-fold enhancement in the total fluorescence intensity [3] were obtained with metal plasmonic patch (gap) nanoantennas. Up to 35 million photon counts per seconds from NV center nanodiamonds coupled to silver patch nanoantennas was reported in [4]. Plasmonic nanoantenna array metasurfaces which produce remarkably narrow plasmonic resonances [7, 8], with patch nanoantennas as array elements may be the path towards indistinguishable single photons at room temperature.

Most of reported patch nanoantennas consist of emitters situated in a vertical dielectric nanogap between a metal nanoparticle of a given shape (cube, triangle, etc.) and a metal film. Silver nanoparticles, for instance, silver nanocubes [2-6] or their arrays [6] are used in patch nanoantennas for emitter fluorescence enhancement in visible spectral range [2-6] as well as for increasing stability and brightness of organic light emitting devices (OLEDs) [9]. In this paper we describe two experimentally observed effects that may influence performance of single-photon sources with modern patch plasmonic nanoantennas based on colloidal silver nanoparticles under continuous wave (cw) laser excitation (either 532 or 633 nm):

- spontaneous photoluminescence spikes up to 400—900 kcounts/s;
- under identical conditions, a step-wise spontaneous photoluminescence increase of up to several times.

2. Experimental setup and sample preparation

A confocal fluorescence microscope was built from an inverted Nikon TE2000-U conventional optical microscope in which a high numerical aperture (1.3), oil immersion microscope objective focuses a laser beam onto a sample (Figure 1, left). The excitation lasers used in this research are cw 532 nm diode pumped solid state and cw 633 nm HeNe lasers. The excitation laser intensity on the sample with this setup was approximately not more than 100 kW/cm^2, which is usual for single-emitter excitation. The sample is mounted on an XY piezo-translation sample stage (Nano Drive, Mad City Labs) with a 50 μm by 50 μm scanning range to raster scan it through the microscope objective focus. The fluorescence light is then collected by the same objective. A dichroic mirror and interference filters are used to separate the excitation light from the fluorescence light. The fluorescence light from the sample can then be directed to the Hanbury Brown and Twiss intensity interferometer consisting of a nonpolarizing beamsplitter and two avalanche photodiode modules, APDs, (SPCM AQR-14, Perkin Elmer). It can also be directed to another microscope port with an EM-CCD camera (iXon DV 887 ECS-BV, Andor Technologies) for low-light imaging of a sample area. Additionally, another microscope port leads to a grating spectrometer (Acton SpectraPro 2150i, Princeton Instruments) equipped with a second Andor EM-CCD (iXon Ultra), which was used for recording fluorescence spectra of the samples at low-light levels. A LabView program user interface (Figure 1, right) shows each detector raster scan fluorescence micrograph as well as time traces of photon counts at selected place of the micrograph. A TimeHarp 200 time-correlated single-photon counting board with "start" and "stop" input channels was used for building histograms of interphoton times.

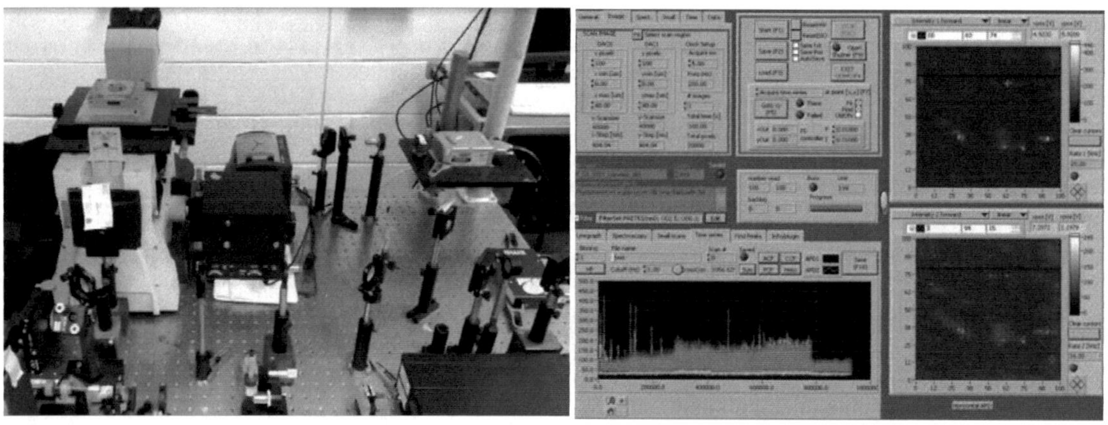

Figure 1. LEFT: Experimental setup. RIGHT: User interface showing confocal fluorescence microscope raster scan images from two APD detectors (right) and time traces of count rate (counts/pixel) in the point marked by the cross on a top image. 1 pixel corresponds to 5 ms.

100 nm silver nanocubes (NanoXast from nanoComposix) coated by a few nanometers of polyvinylpyrrolidone (PVP) for protection from environment were used in these experiments. Silver nanocubes were either spin-coated from a solution, or we used a drop of their solution on a cover glass slip. We used two concentrations of silver nanocubes in ethanol: (1) as received (1 mg/ml) and (2) 20 fold diluted in ethanol.

Results

We have records for the samples with fresh silver nanocubes of both concentrations from two different vials and for some samples prepared after two years of keeping silver nanocubes in a refrigerator at a recommended temperature (2-8°C). In any case the results are almost similar, although with the samples prepared from more than 1 year old solution the spikes intensity is much smaller than with a fresh solution. With a sample prepared with a 20 fold diluted solution we observed one, unusually bright intensity spike ~1.8 Mcounts/s although a number of photoluminescent centers were less in comparison with a higher concentration.

Figure 2 shows typical confocal fluorescence microscope raster scan micrographs of the areas from which some intensity spikes were observed. For instance, spikes in time traces of Figure 3 were obtained from the area marked with a cross in a left top micrograph of Figure 2.

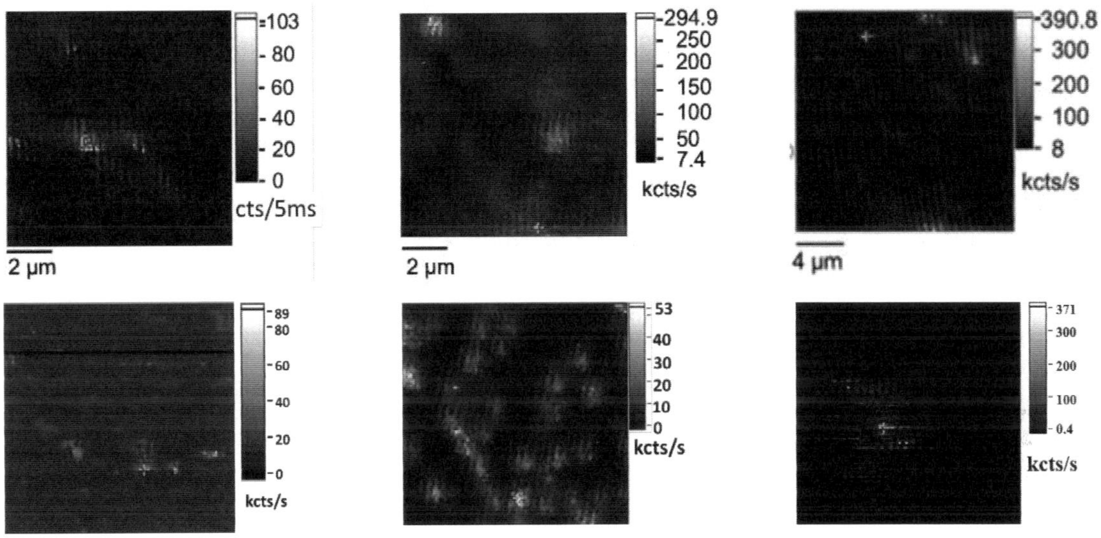

Figure 2. Raster scan micrographs of photoluminescence from silver nanocube samples on glass cover slips under cw, visible light laser irradiation. Bottom micrographs scan sizes: 40μm x 40μm, 25μm x 25 μm, 5μm x 5μm. Narrow periodic stripes on some micrographs is a result of noise in electronics.

All micrographs show single-emitter behavior of some photoluminescent features: stripes and semicircles under raster scans. In freshly prepared samples, especially at a lower concentration of silver nanocubes it is not easy to find them, but after approximately 1 hour of irradiation they appear and their number usually increases. Antibunching correlation measurements show Poissonian photon statistics after multiple samples testing.

Figure 3 shows high-count-rate spontaneously appeared photoluminescence spikes in time traces from one of the silver nanocube samples. Data are collected from the area marked by the cross (on a semicircle) of a top left micrograph of Figure 2. Intensity spikes appear sporadically, sometimes after few minutes of irradiation. Its duration varies from less than 0.1 s up to few seconds. With time accuracy of 5 ms the shortest spikes had time duration of ~ 50—100 ms.

A top part of Figure 3 shows raw data on time traces in counts/pixel from a user interface. A central part of Figure 3 shows processed data of a picture from a user interface. Processed data with a higher resolution are presented in Figure 3, bottom. 1 pixel corresponds to 5 ms, and 100,000 units on a time scale correspond to 100 s. Figure 4 shows a time trace from a different sample with spike intensity ~ 900 kcounts/s.

A second effect that was observed under irradiation of the samples with silver nanocubes by a visible laser light is a step-wise photoluminescence increase. Time traces showing a step-wise increase of photoluminescence count rate are presented in Figure 5. High intensity spikes are also seen in these time traces.

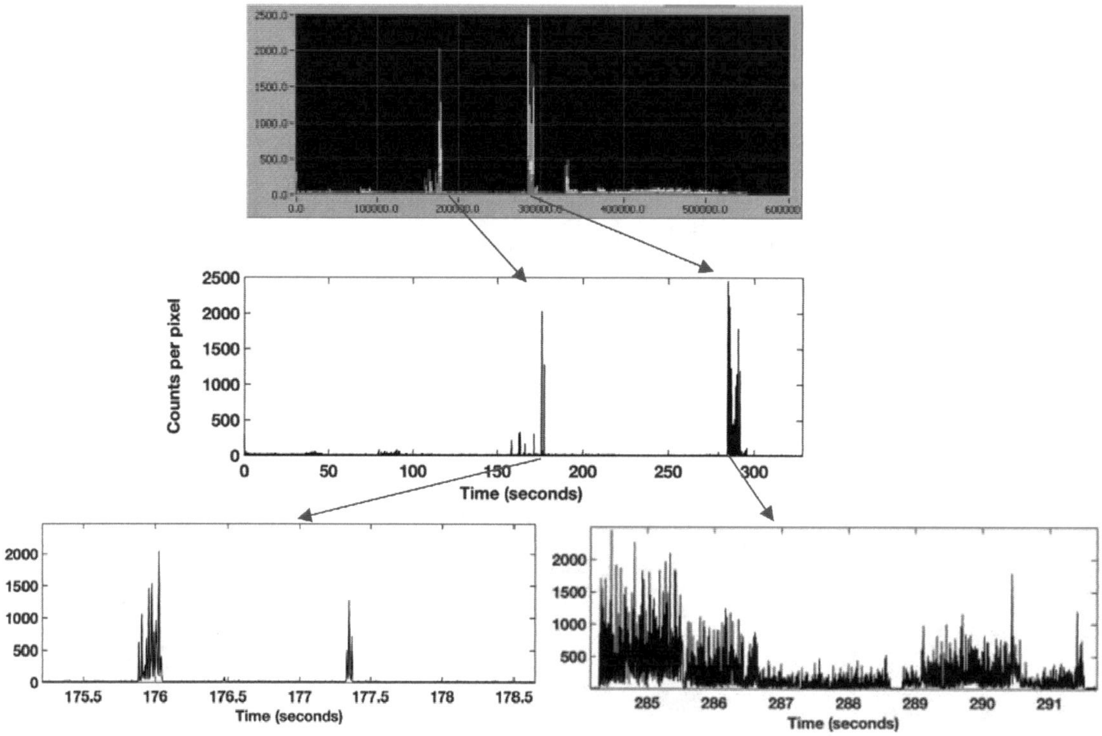

Figure 3. TOP: Raw data of a time trace of counts/pixel (1 pixel corresponds to 5 ms; 100,000 units on a time scale correspond to 100 s). CENTER: Processed data of a top picture from a user interface. BOTTOM: Processed data of a top picture with higher resolution than a plot in the center. A 633 nm laser was used for excitation. See Figure 2, top left for the area of data collection (marked by a cross).

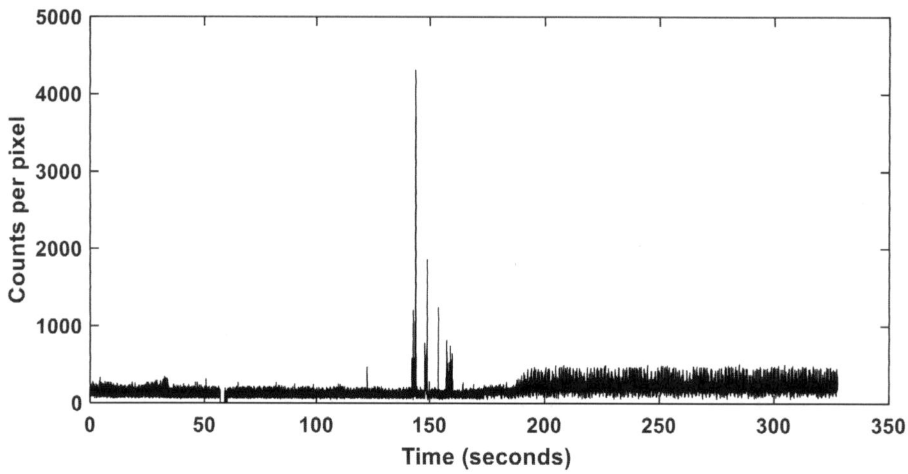

Figure 4. Processed data of time trace with a high intensity spike of ~900 kcounts/s. 1 pixel corresponds to 5 ms.

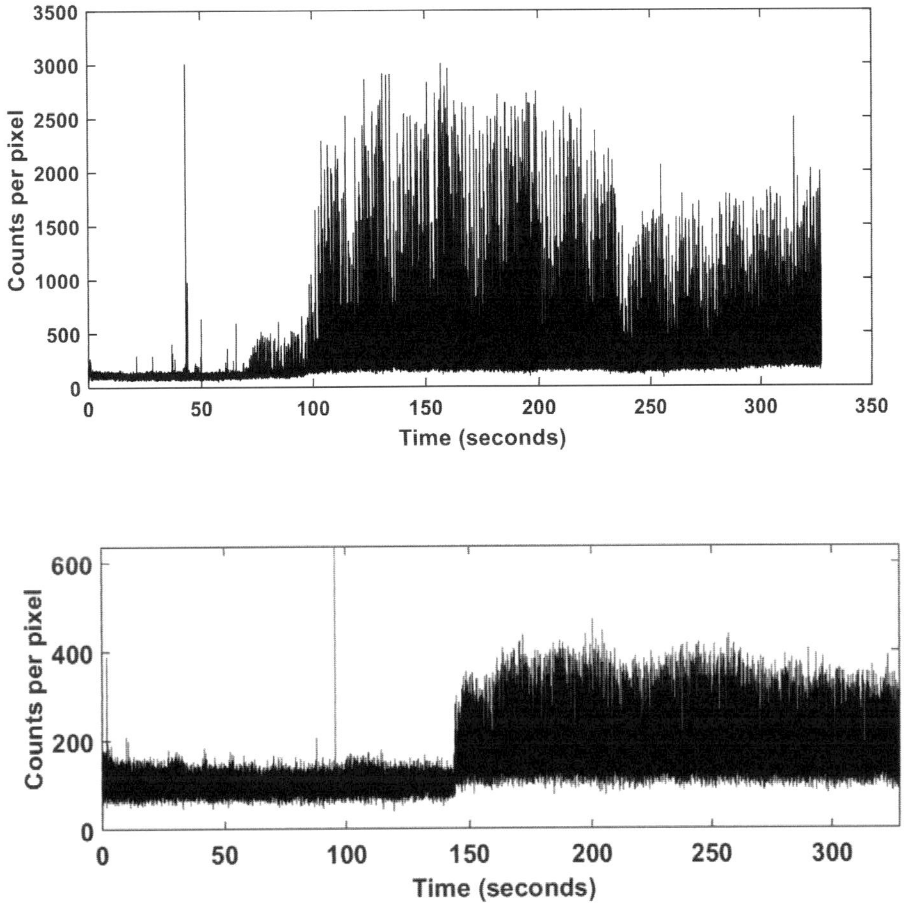

Figure 5. A step-wise photoluminescence increase during irradiation of silver nanocubes by a 633 nm laser light. Narrow spikes are also present in these time traces.

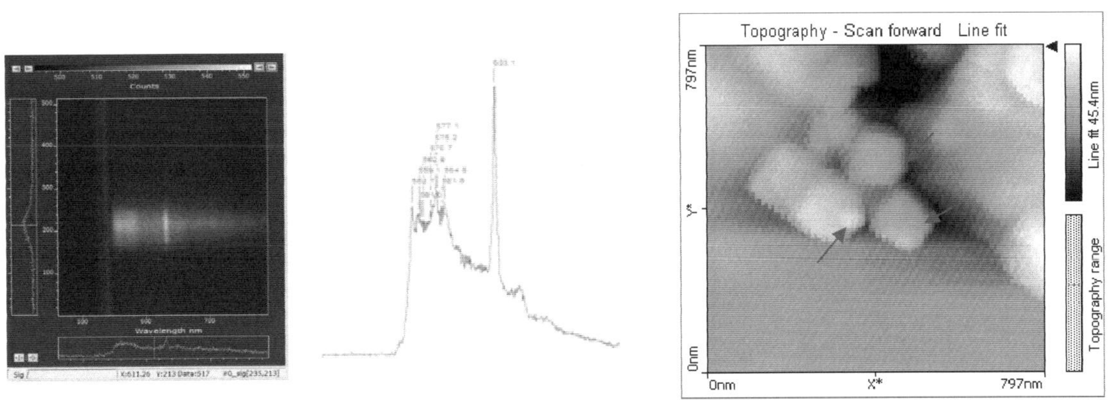

Figure 6. LEFT and CENTER: Photoluminescence spectrum of silver nanocubes under 532 nm laser excitation: on a wide spectrum of silver photoluminescence SERS lines from PVP coating are clearly seen. Exposure time was 10 s with cooling a EMCCD to -65 °C, RIGHT: AFM micrograph of topography of silver nanocubes after laser irradiation. Arrows show ~30 nm features on the surface of nanocubes.

A photoluminescence spectrum of the sample with silver nanocubes with 532 nm laser excitation is presented in Figure 6 (left and center). Surface enhanced Raman scattering (SERS) lines from PVP (1023-1490 cm^{-1}, 1685-1730 cm^{-1} , and 3017 cm^{-1}) are seen clearly on a weak background of silver photoluminescence. A similar Raman lines were observed under 633 nm excitation. Usually at least 10 s exposure time with cooling an EM-CCD to -65°C were used to record SERS lines of few nanometers thickness PVP on the surface of silver nanocubes. But it should be noted that not always these SERS lines are observed.

3. Conclusion and explanation of results

There are two possible explanation of results:

- Photoinduced creation of small silver nanoclusters on possibly oxidized silver surface reported in Refs. [10-11] (There is a whole direction of research with interesting applications of authors of Refs. [10-11] on very bright metal nanoclusters containing up to few dozens of silver atoms that behave as single emitters with observation of photon antibunching from some of these nanoclusters). Photoluminescence from fractal silver structures also shows intensity spikes in time traces [12].
- SERS instability (In Ref. [13] blinking in time was observed in SERS spectra of the PVP-coated colloidal gold triangular nanoprisms (~150-nm edge), although time scale was much longer and count rate values were significantly lower in [10] than in our experiments).

Atomic force microscopy of a topography of silver nanocubes after laser irradiation shows appearance of ~30 nm nanoclusters on the surface of silver nanocubes (see Figure 6, right). Similar size structural photochemical changes (~20 nm) were observed in Ref. [14] under 15 mins irradiation of silver iodide nanocrystals by 540 nm laser light. Although a mechanism of photochemistry in [14] with using a sensitizer to facilitate formation of a reduced metal silver was different than in our case, our bright photoluminescence features may also be from very small silver nanoclusters formed on the oxidized silver surface.

Acknowledgments
The authors thank Liam Young for his assistance.

References

[1] S.G. Lukishova and L.J. Bissell, "Nanophotonic advances for room-temperature single-photon sources", pp. 103-178, in *Quantum Photonics: Pioneering Advances and Emerging Applications*, R.W. Boyd, S.G. Lukishova, V.N. Zadkov (Eds.), Springer (2019).

[2] T.B. Hoang, G.M. Akselrod, C. Argyropoulos, J. Huang, D.R. Smith, M.H. Mikkelsen, "Ultrafast spontaneous emission source using plasmonic nanoantennas", Nat. Commun. **6**, Article number 7788 (2015).

[3] G.M. Akselrod, C. Argyropoulos, T.B. Hoang, C. Ciracì, C. Fang, J. Huang, D.R. Smith, M.H. Mikkelsen, "Probing the mechanisms of large Purcell enhancement in plasmonic nanoantennas", Nat. Photonics **8**, 835–840 (2014).

[4] S.I. Bogdanov, M.Y. Shalaginov, A.S. Lagutchev, C.-C. Chiang, D. Shah, A.S. Baburin, et al., "Ultrabright room-temperature sub-nanosecond emission from single nitrogen-vacancy centers coupled to nanopatch antennas", Nano Lett. **18**, 4837–4844 (2018).

[5] S.I. Bogdanov, O.A. Makarova, X. Xu, Z.O. Martin, A.S. Lagutchev, M. Olinde, D. Shah, S.N. Chowdhury, A.R. Gabidullin, I.A. Ryzhikov, I.A. Rodionov, A.V. Kildishev, S.I. Bozhevolnyi, A. Boltasseva, V.M. Shalaev, and J.B. Khurgin, "Ultrafast quantum photonics enabled by coupling plasmonic nanocavities to strongly radiative antennas", Optica **7** (5), 463-469 (2020).

[6] A.V. Gritsienko, N.S. Kurochkin, P.V. Lega, A.P. Orlov, A.S. Ilin, S.P. Eliseev, and A.G. Vitukhnovsky, "Optical properties of new hybrid nanoantenna in submicron cavity", Journal of Physics: Conference Series **2015**, 012052 (2021).

Journal of Physics: Conference Series **2249** (2022) 012002 doi:10.1088/1742-6596/2249/1/012002

[7] V.G. Kravets, A.V. Kabashin, W.L. Barnes, and A.N. Grigorenko, "Plasmonic surface lattice resonances: a review of properties and applications (Review), Chem. Rev. **118**, 5912-5951 (2018).

[8] S. Zou, N. Janel, G.C. Schatz, "Silver nanoparticle array structures that produce remarkably narrow plasmon lineshapes", J. Chem. Phys. **120**, 10871 (2004).

[9] M.A. Fusella, R. Saramak, R. Bushati, V.M. Menon, M.S. Weaver, N.J. Thompson & J.J. Brown, "Plasmonic enhancement of stability and brightness in organic light-emitting devices", Nature **585**, 37 (2020).

[10] L.A. Peyser, A.E. Vinson, A.P. Bartko, R.M. Dickson, "Photoactivated fluorescence from individual silver nanoclusters", Science **291**, 103, 5 January (2001).

[11] L.A. Peyser, T.-H. Lee, R.M. Dickson, "Mechanism of Ag_n nanocluster photoproduction from silver oxide films", J. Phys. Chem. B **106**, 7725-7728 (2002).

[12] C.D. Geddes, A. Parfenov, I. Gryczynski, J.R. Lakowicz, "Luminescent blinking from silver nanostructures", J. Phys. Chem. B **107**, 9589-9993 (2003).

[13] A. Mezni, T. Dammak, A. Fkiri, A. Mlayah, Y. Abid, and L.S. Smiriet, "Photochemistry at the surface of gold nanoprisms from surface-enhanced Raman scattering blinking", J. Phys. Chem. C **118**, 17956–17967 (2014).

[14] S.V. Vinogradov, M.A. Kononov, B.M. Kononov, V.V. Savranskii, "Structural changes of silver iodide nanocrystals during surface optical sensitization", Bulletin of the Lebedev Physics Institute **42** (1), 30-32 (2015).

Resonant spontaneous bremsstrahlung in the scattering of ultrarelativistic electrons on nuclei in a strong light field

S. P. Roshchupkin[1], A.V. Dubov[1], V. V. Dubov[1]

[1]Peter the Great St. Petersburg Polytechnic University, 195251, St-Petersburg, Russian Federation, Russia

Abstract. An analytical study of the process of resonant spontaneous bremsstrahlung radiation during the scattering of ultrarelativistic electrons with energies by the nuclei in strong laser fields with intensities up to $I \sim 10^{25}$ Wcm^{-2} was carried out. Under resonant conditions, an intermediate electron in the wave field enters the mass shell. As a result, the initial second-order process by the fine structure constant is effectively reduced to two first-order processes: laser-stimulated Compton effect and laser-assisted Mott process. The resonant kinematics for two reaction channels (A and B) is studied in detail. An analytical resonant differential cross-section with simultaneous registration of the frequency and the outgoing angle of a spontaneous gamma-quantum is obtained. It is shown that the resonant differential cross-section has the largest value in the region of average laser fields and can be of the order of $\sim 10^{18}$ in units $Z^2 \alpha r_e^2$. The obtained results reveal new features of spontaneous emission of ultrarelativistic electrons on nuclei in strong laser fields and can be tested at international laser installations.

1. Introduction

Theoretical study of fundamental processes of quantum electrodynamics (QED) in strong laser fields (see, for example, [1-18]) is connected not only with the development of QED, but also with the creation of powerful laser radiation sources [19-23]. It is important to emphasize that higher-order QED processes with respect to the fine structure constant in a laser field can proceed both in a resonant and non-resonant way. Here, the so-called Oleinik resonances may occur [1, 2], due to the fact that lower-order processes are allowed in the laser field by the fine structure constant. It is important to note that the resonant differential cross-section can significantly exceed the corresponding non-resonant one [8, 9, 13, 14, 17, 18]. Oleinik resonances for spontaneous bremsstrahlung during electron scattering on the nucleus in the field of a light wave has been studied for a long time (see, for example, [8, 9, 13, 14, 17, 18]). In recent articles [8, 9], the process of spontaneous emission of ultrarelativistic electrons on nuclei in the field of a weak electromagnetic wave was considered.

In this article, unlike the previous ones, we will study the resonant spontaneous emission of ultrarelativistic electrons on nuclei in medium and strong laser fields in the Born approximation ($v/c >> Z/137$, v is the velocity of the electron, c is the speed of light in a vacuum, Z is the charge of the nucleus).

We will study the resonance SB process for high-energy particles when the basic classical parameter η satisfies the relation

$$\eta << \frac{E_{i,f}}{m} >> 1, \quad \eta = \frac{eF\lambda\!\!\!^{-}}{mc^2}. \tag{1}$$

Here e and m are the charge and mass of the electron, F and $\lambda\!\!\!^{-} = c/\omega$ are the electric field strength and wavelength, ω is the frequency of the wave, E_i and E_f are the energy of the initial and final electrons. We will use the relativistic system of units: $\hbar = c = 1$.

2. Poles of the SB amplitude in a strong field

Oleinik resonances occur when an intermediate electron in the electromagnetic wave field enters the mass shell [1, 2, 14]. Because of this, for channels A and B, we get:

$$\tilde{q}_i^2 = m_*^2, \quad \tilde{q}_f^2 = m_*^2. \tag{2}$$

We will study the most interesting case of ultrarelativistic electron energies, when the spontaneous gamma-quantum and the final electron fly out in a narrow cone along the momentum of the initial electron. In this case, the direction of propagation of the wave lies far from the given narrow cone of particles (if the direction of propagation of the wave lies inside the narrow cone of particles, then the resonances disappear [8, 9, 17, 18]).

$$\theta'_{i,f} = \measuredangle\left(\mathbf{k}',\mathbf{p}_{i,f}\right) << 1, \quad \theta_{if} = \measuredangle\left(\mathbf{p}_i,\mathbf{p}_f\right) << 1, \quad \theta_{i,f} = \measuredangle\left(\mathbf{k},\mathbf{p}_{i,f}\right) \sim 1. \tag{3}$$

Expressions defining the 4-quasimomenta of the intermediate electrons $\tilde{q}_i$ and $\tilde{q}_f$ [15, 24] for channels A and B in resonance, it is convenient to write as:

$$\tilde{p}_i + rk = \tilde{q}_i + k', \tag{4}$$

and

$$\tilde{q}_f + rk = \tilde{p}_f + k'. \tag{5}$$

Since both the equalities $\tilde{p}_i^2 = \tilde{q}_i^2 = m_*^2$ and $\tilde{p}_f^2 = \tilde{q}_f^2 = m_*^2$, $k^2 = k'^2 = 0$, then equations (4) and (5) are valid only for $r \geq 1$.

We determine the resonant frequency $\left(\omega'_{\eta i(r)}\right)$ of the spontaneous gamma-quantum for channel A. After simple calculations [13, 14], we get

$$x'_{\eta i(r)} = \left[1 + \frac{\left(1+\delta'^2_{\eta i}\right)}{\kappa_{\eta(r)}}\right]^{-1}, \quad x'_{\eta i(r)} = \frac{\omega'_{\eta i(r)}}{E_i}, \quad \delta'_{\eta i} = \frac{E_i\theta'_i}{m_*} = \frac{E_i\theta'_i}{m\sqrt{1+\eta^2}}. \tag{6}$$

Here it is indicated:

$$\kappa_{\eta(r)} = \frac{r}{r_\eta}, \quad r_\eta = \frac{m^2\left(1+\eta^2\right)}{4E_i\omega\sin^2\left(\theta_i/2\right)}. \tag{7}$$

The characteristic parameter $\kappa_{\eta(r)}$ (7) is numerically equal to the ratio of the number of absorbed photons of the wave in the laser-stimulated Compton process to the parameter r_η, which is determined by the experimental conditions and the laser installation.

We obtain the the cubic equation for the resonant frequency $\left(\omega'_{\eta f(r)}\right)$ of the spontaneous gamma-quantum in the case of channel B [13, 14]. A simple analysis of the cubic equation shows that there are two intervals for the parameter $\kappa_{\eta(r)}$, in which the resonant frequencies and the outgoing angles of the spontaneous gamma-quantum qualitatively changes. So, in the range of values

$$0 < \kappa_{\eta(r)} \leq 8 \tag{8}$$

there is a single valid solution to equation for the resonant frequency of a spontaneous gamma-quantum:

$$x'_{\eta f(r)} = \frac{2}{3} + \left(\alpha_{\eta(r)+} + \alpha_{\eta(r)-}\right), \tag{9}$$

where

$$\alpha_{\eta(r)\pm} = \left[-\frac{b_{\eta(r)}}{2} \pm \sqrt{Q_{\eta(r)}}\right]^{1/3}, \quad Q_{\eta(r)} = \left(\frac{a_{\eta(r)}}{3}\right)^3 + \left(\frac{b_{\eta(r)}}{2}\right)^2, \tag{10}$$

$$a_{\eta(r)} = \frac{1}{3\delta'^2_{\eta f}}\left[3\left(1+\kappa_{\eta(r)}\right) - \delta'^2_{\eta f}\right], \quad b_r = \frac{1}{27\delta'^2_{\eta f}}\left[18 + 2\delta'^2_{\eta f} - 9\kappa_{\eta(r)}\right], \quad \delta'_{\eta f} = \frac{E_i\theta'_f}{m_*} = \frac{E_i\theta'_f}{m\sqrt{1+\eta^2}}. \tag{11}$$

At the same time, the outgoing angle of the spontaneous gamma-quantum relative to the final electron varies from values close to zero to a certain maximum value

$$0 < \delta'^2_{\eta f} \le \delta'^2_{\eta f \max}, \quad \delta'^2_{\eta f \max} = 3\left(1+\kappa_{\eta(r)}\right). \tag{12}$$

Figure 1a, 1b show the resonant frequencies of a spontaneous gamma-quantum for channels A (6) and B (8), (9) as the functions of the square of its corresponding outgoing angle and the characteristic parameter $\kappa_{\eta(r)}$ for strong laser fields $(\eta >> 1)$. From these figures, it can be seen that for strong fields, when the characteristic parameter $\kappa_{\eta(r)} << 1$, the resonant frequencies are small compared to the energy of the initial electrons. However, if $\kappa_{\eta(r)} \gtrsim 1$ then the resonant frequencies become of the same order or close to the energy of the initial electron.

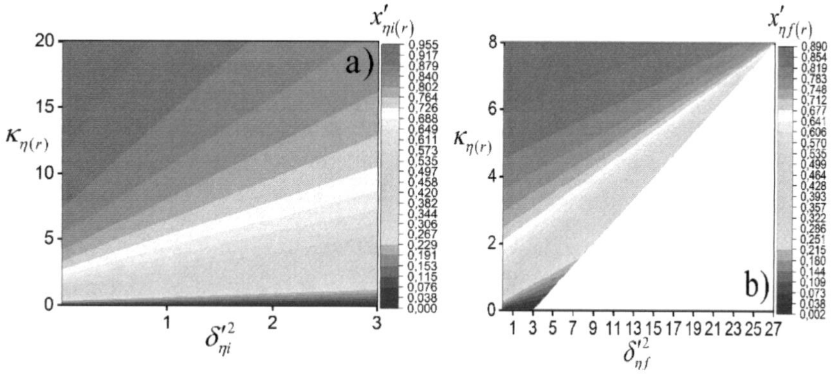

Figure 1. Resonant frequency of the spontaneous gamma-quantum (in units of energy of the initial electron) for channel A: $x'_{\eta i(r)}$ (6) (see Figure 1a) and for channel B: $x'_{\eta f(r)}$ (8), (9) (see Figure 1b) as functions of a parameter $\kappa_{\eta(r)}$ (7) and corresponding outgoing angle for strong laser fields $(\eta >> 1)$.

3. The Resonant Differential SB Cross Section in the Ultrarelativistic Energy Limit

Using the expression for the process amplitude, we obtain an expression for the resonant differential cross section in the case of unpolarized particles. After the standard calculations we obtain the maximum resonant differential cross-sections for channels A and B (see [14, 25]):

$$R^{\max}_{\eta i(r)} = \frac{d\sigma^{\max}_{\eta i(r)}}{dx'_{\eta i(r)} d\delta'^2_{\eta i}} = \left(Z^2\alpha r_e^2\right) c_{\eta i} \Psi_{\eta i(r)}, \quad R^{\max}_{\eta f(r)} = \frac{d\sigma^{\max}_{\eta f(r)}}{dx'_{\eta f(r)} d\delta'^2_{\eta f}} = \left(Z^2\alpha r_e^2\right) c_{\eta i} \Psi_{\eta f(r)}. \tag{13}$$

Here, the $\Psi_{\eta i(r)}$ and $\Psi_{\eta f(r)}$ functions determine the spectral-angular distribution of the resonant SB cross-section for channels A and B:

$$\Psi_{\eta i(r)} = \frac{x'_{\eta i(r)}}{\left(1 - x'_{\eta i(r)}\right) g_{\eta i(r)}} K\left(u_{\eta i(r)}, \kappa_{\eta(r)}\right), \quad \Psi_{\eta f(r)} = \frac{x'_{\eta f(r)}\left(1 - x'_{\eta f(r)}\right)}{g_{\eta f(r)}} K\left(u_{\eta f(r)}, \kappa_{\eta(r)}\right) \tag{14}$$

and $c_{\eta i}$ - the coefficient, which is determined by the parameters of the laser installation.

$$c_{\eta i}=\pi\left(\frac{8\pi E_i}{\alpha m_*\mathrm{K}\left(r_\eta\right)}\right)^2\approx 3.72\cdot 10^7\left[\frac{E_i}{m_*\mathrm{K}\left(r_\eta\right)}\right]^2=c_{\eta i}=5.82\times 10^{17}\frac{1}{\left(1+\eta^2\right)\mathrm{K}^2\left(r_\eta\right)}>>1. \qquad (15)$$

In expressions (14) and (15), the functions $K\left(u_{\eta i(r)},\kappa_{\eta(r)}\right)$, $K\left(u_{\eta f(r)},\kappa_{\eta(r)}\right)$ and $\mathrm{K}\left(r_\eta\right)$ are determined by the probability of the laser-stimulated Compton effect [14]. Note that the physical meaning of the parameter $c_{\eta i}$ is determined by the small transmitted momenta in the kinematic region, as well as the resonant width associated with the full probability of the laser-stimulated Compton effect.

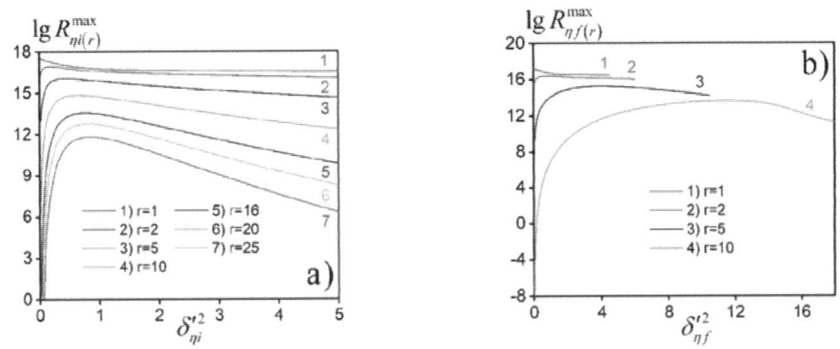

Figure 2. The maximum resonant differential cross-section (in units of $Z^2\alpha r_e^2$) as functions of the corresponding outgoing angles, plotted for the different values of absorbed wave photons. Figure 2a - $R^{\max}_{\eta i(r)}$ for channel A. Figure 2b - $R^{\max}_{\eta f(r)}$ (13) for channel B. The intensity of the laser wave $I\approx 1.861\cdot 10^{18}$ Wcm^{-2}.

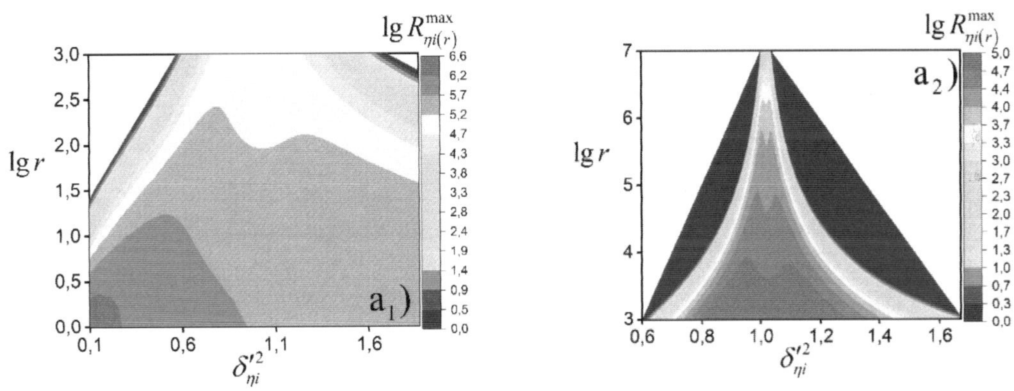

Figure 3. The maximum resonant differential cross-section (in units of $Z^2\alpha r_e^2$) as functions of the corresponding outgoing angles and the different values of absorbed wave photons. Figure 3a$_1$ and Figure 3a$_2$ - $R^{\max}_{\eta i(r)}$ (13) for channel A. The intensity of the laser wave $I\approx 1.861\cdot 10^{24}$ Wcm^{-2}.

From these figures, it can be seen that for the number of absorbed photons $r\geq 2$, each resonance cross-section curve has a maximum that shifts to the region of large outgoing angles with increasing r.
At the same time, with an increase in the number of absorbed photons, the resonant frequency increases, and the resonant differential cross-section decreases quite quickly. It is important to note that for the same number of absorbed laser photons, the maximum value of the maximum resonant differential cross-section for channel B is always less than the corresponding value for channel A. Moreover, this

ratio of values decreases with an increase in the number of absorbed photons. So, for $r = 1, 2, 5, 10, 16$ respectively we have $R^{\max}_{\eta f(r)} \big/ R^{\max}_{\eta i(r)} \approx 0.467, 0.301, 0.142, 0.065, 0.013$.

4. Conclusion

We considered Oleinik resonances for resonant spontaneous bremsstrahlung of ultrarelativistic electrons with energies less than or on the order of 100 GeV in the nucleus field and a strong light field up to the intensities $I \sim 10^{24}$ Wcm^{-2}.

- All the particles (initial and final) fly in a narrow cone. It is important to note that the resonant frequency of a spontaneous gamma-quantum significantly depends on its outgoing angle, as well as on the number of absorbed photons of the wave. In this process, there is a characteristic parameter r_η that is determined by the parameters of the laser installation and is proportional to the intensity of the wave.
- It is shown that the largest value of the resonant differential cross-section decreases with increasing wave intensity. So, for the laser intensities $I \approx 1.861 \cdot 10^{18}$ Wcm^{-2}, $1.861 \cdot 10^{20}$ Wcm^{-2}, $1.861 \cdot 10^{22}$ Wcm^{-2}, $1.861 \cdot 10^{24}$ Wcm^{-2}, respectively, we get: $R^{\max}_{\eta i(r)} \lesssim 7.13 \times 10^{17}$, 4.11×10^{13}, 1.09×10^{10}, 4.54×10^{6} (in units $Z^2 \alpha r_e^2$). With an increase in the number of absorbed photons, the spectral-angular distribution functions (14) decrease rapidly. This leads to a rapid decrease in the magnitude of the resonant differential cross sections.

This theoretical research predicts a number of new physical effects and can be tested in international research project ELI (Extreme Light Infrastructure, Czech Republic).

5. Acknowledgment

The work was supported by the State Program for the Fundamental Research (theme code FSEG-2020-0024).

6. References

[1] Oleinik V 1967 *Zh. Eksp. Teor. Fiz.* **52** 1049 [*Sov. Phys. JETP* **25** 697]
[2] Oleinik V 1967 *Zh. Eksp. Teor. Fiz.* **53** 1997 [*Sov. Phys. JETP* **26** 1132]
[3] Dondera M and Florescu V 2006 *Radiat. Phys. Chem.* **75** 1380
[4] Florescu A and Florescu V 2000 *Phys. Rev. A* **61** 033406
[5] Wang A, Ren N, Wang P, Zhu W, Hoehn X and Kais S 2013 *J. Appl. Phys.* **114** 124904
[6] Lötstedt E, Jentschura U and Keitel C 2007 *Phys. Rev. Lett.* **98** 043002
[7] Schnez S, Lötstedt E, Jentschura U and Keitel C 2007 *Phys. Rev. A.* **75** 053412
[8] Dubov A, Dubov V and Roshchupkin S 2020 *Laser Phys. Lett.* **17** 045301
[9] Roshchupkin S, Dubov A and Dubov V 2021 *Laser Phys. Lett.* **18** 045301
[10] Krachkov P, Piazza A and Milstein A 2019 *Phys. Lett. B* **797** 134814
[11] Kanya R, Morimoto Y, and Yamanouchi K 2010 *Phys. Rev. Lett.* **105** 123202
[12] Hartin A 2018 *International Journal of Modern Physics A* **33** 1830011
[13] Roshchupkin S, Dubov V, Dubov A et al. 2021 *in High Power Lasers and Applications* **11777**, ed J Hein, T Butcher, P Bakule, C Haefner, G Korn and L Silva, International Society for Optics and Photonics (SPIE, 2021) pp. 40 – 55
[14] Roshchupkin S, Dubov A, Dubov V, Starodub S 2021 Resonant Spontaneous Bremsstrahlung Effect in The Scattering of Ultrarelativistic Electrons on Nuclei in a Strong Laser Field. arXiv:2108.10186 [hep-ph]
[15] Ritus V and Nikishov A 1979 *Quantum Electrodynamics Phenomena in the Intense Field Trudy FIAN* vol 111 ed V Ginzburg (Nauka, Moscow)

[16] Ehlotzky F, Krajewska K and Kamiński J 2009 *Rep. Prog. Phys.* **72** 046401
[17] Roshchupkin S 1996 *Las. Phys.* **6** 837
[18] Roshchupkin S, Lebed' A, Padusenko E and Voroshilo A 2012 *Las. Phys.* **22** 1113
[19] Bula C, McDonald K, Prebys E, Bamber C, Boege S, Kotseroglou T, Melissinos A, Meyerhofer D, Ragg W, Burke D et al. 1996 *Phys. Rev. Lett.* **76** 3116
[20] Burke D, Field R, Horton-Smith G, Spencer J, Walz D, Berridge S, Bugg W, Shmakov K, Weidemann A, Bula C et al. 1997 *Phys. Rev. Lett.* **79** 1626
[21] Mourou G, Tajima T and Bulanov S 2006 *Rev. Mod. Phys.* **78** 309
[22] Bagnoud V, Aurand B, Blazevic A, Borneis S, Bruske C, Ecker B, Eisenbarth U, Fils J, Frank A, Gaul E et al. 2010 *Appl. Phys. B* **100** 137
[23] Piazza A, Müller C, Hatsagortsyan K and Keitel C 2012 *Rev. Mod. Phys.* **84** 1177
[24] Volkov D 1935 *Zeit. Phys.* **94** 250
[25] Berestetskii V, Lifshitz E and Pitaevskii L 1980 *Quantum Electrodynamics* (Nauka, Moscow)

High Harmonic Generation from Oriented Asymmetric Molecules in the Presence of Static Electric Field

A A Silaev[1,2], A A Romanov[1,2] and N V Vvedenskii[1,2]
[1]Institute of Applied Physics of the Russian Academy of Sciences, Nizhny Novgorod 603950, Russia
[2]Lobachevsky State University of Nizhny Novgorod, Nizhny Novgorod 603950, Russia

Abstract. We investigate the high harmonic generation during the interaction of the femtosecond laser field with oriented asymmetric diatomic molecules in the presence of an external static electric field. Based on the numerical solution of the time-dependent Schrödinger equation for the model one-electron system, we calculate the high harmonic generation spectrum for a wide range of static-field strengths. The possibilities for visualizing low-frequency (terahertz and mid-infrared) radiation using time-delay measurements of even high harmonics intensities are discussed.

1. Introduction

The interaction of an intense laser pulse with a gaseous medium leads to the generation of high harmonics of the laser pulse [1]. The mechanism of high harmonic generation (HHG) is well understood in terms of the "three-step" model [2]: the electron is first ionized, then accelerated in the laser field, and finally recombined on the parent ion with the emission of an energetic photon. This process occurs periodically in each half cycle of the driving field; hence, only odd harmonics are generated for the isotropic gas medium. The use of asymmetric molecules breaks the symmetry of the ionization process in consecutive half laser cycles, and the HHG spectrum contains both odd and even harmonic numbers [3].

This paper is devoted to investigating the HHG during the interaction of the femtosecond laser field with oriented asymmetric diatomic molecules in the presence of an external static electric field. The relevance of these studies is associated with the development of powerful sources of pulsed low-frequency terahertz (THz) and mid-infrared (mid-IR) radiation, capable to influence the HHG by infrared laser pulses [4–8]. Our numerical study is based on the solution of the three-dimensional time-dependent Schrödinger equation (3D TDSE) for the model one-electron system, whose parameters are adjusted to simulate a carbon monoxide (CO) molecule. Based on the calculated HHG spectra, we find the dependences of harmonic intensities on the static-field strength. Finally, we discuss the possibilities for using time-delay measurements of HHG yield to visualize (reconstruct the waveform) of THz or mid-IR pulses.

2. Statement of the problem

We consider the interaction of an asymmetric diatomic molecule interacting with an intense linearly polarized laser pulse in the presence of a quasi-dc electric field, which does not change on the time scale of the laser pulse. We assume that both the dc and the laser-pulse electric fields are directed along the axis of the molecule $\hat{\mathbf{z}}$:

$$\mathbf{E}(t) = \hat{\mathbf{z}}E(t) = \hat{\mathbf{z}}[E_{\rm L}(t) + E_{\rm DC}], \tag{1}$$

$$E_{\rm L}(t) = A_0\cos(\omega_0 t)f(t), \quad f(t) = \cos^2(\pi t/T), \quad t \in [-T/2, T/2], \tag{2}$$

where $T = \pi\tau_p/2\,\mathrm{acos}(2^{-1/4})$ is full pulse duration, A_0, ω_0, and τ_p are the peak amplitude, central angular frequency, and intensity full-width at half-maximum (FWHM) duration, respectively. We calculate the electron current density excited in the produced plasma by solving the 3D TDSE for the electron wave function $\psi(\mathbf{r}, t)$ of the most active electron in a strong external field $\mathbf{E}(t)$ (atomic units, a.u., are used throughout unless stated otherwise):

$$\mathrm{i}\frac{\partial\psi}{\partial t} = -\frac{1}{2}\nabla^2\psi + \Big[V(\mathbf{r}) + zE(t)\Big]\psi. \tag{3}$$

The binding potential is presented as a some of two effective potentials centered at the molecule nuclei positions: $V(\mathbf{r}) = Z_1U(\mathbf{r}-\hat{\mathbf{z}}d/2)+Z_2U(\mathbf{r}+\hat{\mathbf{z}}d/2)$, where $Z_1+Z_2 = 1$, d is the internuclear distance, $U(\mathbf{r}) = -\alpha\mathrm{sech}^2(r/a)-\tanh(r/a)/r$. We use potential parameters values $Z_2/Z_1 = 0.93$, $\alpha = 3.5$, $a = 0.6$, and $d = 2.13$, corresponding to electron binding energy in the ground state is 0.5, and permanent dipole moment of the molecule 0.06. This parameters and internuclear distance $d = 2.13$ approximately correspond to the realistic values for the CO molecule (the ionization potential 0.51 and permanent dipole moment 0.05 [9]).

The three-dimensional TDSE (3) is solved in the cylindrical coordinate system using the split-step method with fast Fourier transform along the longitudinal direction, and discrete Hankel transform along transverse direction [10, 11]. We use the numerical grid with the transverse size of 60 and the longitudinal size of 410. Near all boundaries, we use the absorbing layers to prevent the reflection and transmission of waves through the boundaries [11]. The size of the absorbing layer is 30 for the transverse direction and 40 for the longitudinal direction. The HHG spectrum $S(\omega)$ is found as the squared Fourier spectrum of the dipole acceleration $\mathbf{a}(t) = \hat{\mathbf{z}}a(t)$ given by

$$S_\omega = |a(\omega)|^2, \quad a(\omega) = \int_{-\infty}^{\infty} a(t)\mathrm{e}^{\mathrm{i}\omega t}\mathrm{d}t, \tag{4}$$

where $a(t)$ is expressed in terms of ψ using Ehrenfest theorem

$$a(t) = -E(t) - \Big\langle\psi\,\Big|\,\nabla V\,\Big|\,\psi\Big\rangle. \tag{5}$$

3. Numerical results

We first calculated the HHG spectra excited by the laser pulse in the presence of the dc field. The laser pulse has FWHM duration $\tau_p = 20$ fs, central wavelength $\lambda = 2\pi c/\omega_0 = 800$ nm, and peak laser-pulse intensity $I = A_0^2 I_a = 8 \times 10^{13}$ W/cm^2 (where A_0 is normalized to the atomic electric field, $I_a = 3.51 \times 10^{16}$ W/cm^2 is the atomic intensity, and c is the speed of light). In the numerical calculations, the dc field is turned on at the time interval of several femtoseconds before the arrival of the laser pulse.

Figure 1(a) shows the calculated HHG spectra S_ω for different dc-field strengths from -1 to 1 MV/cm. It can be seen that in the absence of a dc field (thick solid blue line), the HHG spectrum, in addition to strong odd harmonics, also contains pronounced even harmonics. When a positive (negative) dc field is introduced, the intensity of even harmonics increases (decreases), while the intensity of odd harmonics practically does not depend on the static-field strength $E_{\rm DC}$ in the considered range of $E_{\rm DC}$. Figure 1 (b) shows the dependences of the normalized (to the

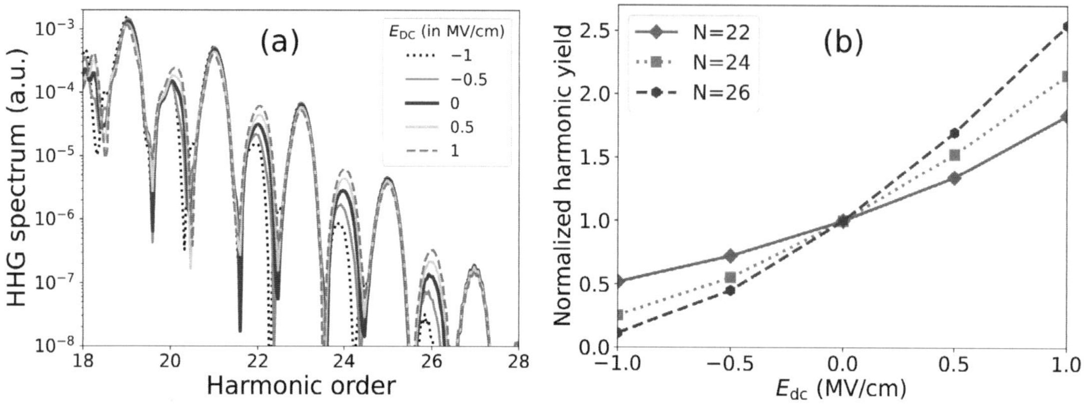

Figure 1. (a) HHG spectra for the asymmetric molecule in a linearly polarized laser pulse (2) of intensity $I = 8 \times 10^{13}$ W/cm^2, duration 20 fs and wavelength 800 nm in the presence of an external static field with five different strength values from −1 to 1 MV/cm. (b) The dependence of the spectral intensities of even 22, 24, 26 harmonics in the cutoff region on the static-field strength $E_{\rm DC}$.

corresponding value for zero dc field) spectral intensity of even 22, 24, 26 harmonics near cutoff on $E_{\rm DC}$. It is seen that in a wide range of $E_{\rm DC}$, the even harmonics intensities are proportional to $E_{\rm DC}$. The obtained results indicate that in a sufficiently wide range of dc-field values, the dipole acceleration can be represented as the sum of two terms: (a) the dipole acceleration in pure laser field and (b) an additional term linear in $E_{\rm DC}$, which gives a linear term in the spectral intensity:

$$a(\omega) \approx a_0(\omega) + b(\omega)E_{\rm DC}, \tag{6}$$

$$S_\omega = |a(\omega)|^2 \approx |a_0(\omega)|^2 + 2E_{\rm DC}{\rm Re}[a_0(\omega)b^*(\omega)]. \tag{7}$$

Quadratic terms contribute when constant and linear terms cancel each other and for $E_{\rm DC} >$ 500 kV/cm, corresponding to the dominance of the quadratic term.

The linear dependence of the intensity of even harmonics on the dc-field strength opens up the possibility of measuring the waveform of a low-frequency (THz or mid-IR) field propagating together with a laser pulse in gas from oriented asymmetric molecules using the measurement of the intensities of even harmonics as a function of the delay time of a probe laser pulse relative to the measured field. This method differs from the existing THz detection method based on measuring the THz-field induced second harmonic of the laser field generated as a result of four-wave mixing in the air or atomic gases [12] in the following points. (a) The duration of the high harmonic signal near the cutoff turns out to be significantly shorter than the duration of the second harmonic; as a result, a higher detection resolution can be achieved. (b) When using a gas consisting of oriented asymmetric molecules, there is no need to add an external static field that creates a constant signal for coherent detection (complete waveform retrieving) of low-frequency pulses.

In order to illustrate the possibility of reconstructing the waveform of a low-frequency field, we simulate the HHG by asymmetric molecule in a two-color electric field, which is a superposition of a low-frequency field $E_{\rm LF}(t)$ and a time-delayed probe laser pulse (defined by Eq.(2)):

$$E(t) = E_{\rm L}(t - \tau_d) + E_{\rm LF}(t), \tag{8}$$

$$E_{\rm LF}(t) = A_{\rm LF}\sin(\omega_{\rm LF}t). \tag{9}$$

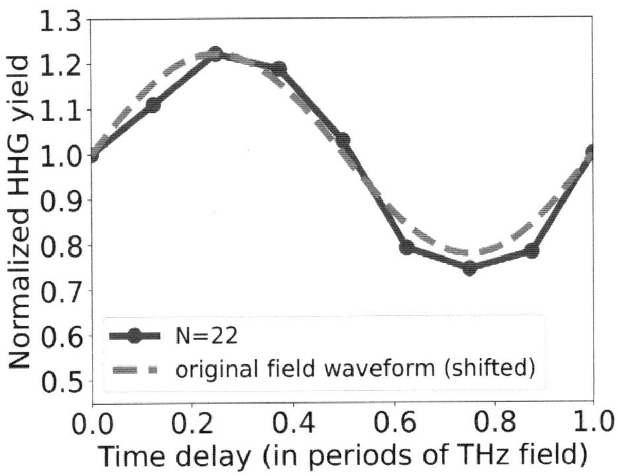

Figure 2. The dependence of the 22th harmonic intensity on the delay time τ_d between the probe ultrashort laser pulse (2) and the low-frequency field (9). The low-frequency field has the amplitude 500 kV/cm and the wavelength 10 μm. The laser pulse has the peak intensity 8×10^{13} W/cm^2, the wavelength 800 nm, and the FWHM duration 50 fs.

The low-frequency field has the amplitude $A_{\mathrm{LF}} = 500$ kV/cm and a wavelength $\lambda_{\mathrm{LF}} = 2\pi c/\omega_{\mathrm{LF}} = 10$ μm (corresponding to the long-wavelength part of the mid-IR range). The laser pulse has the peak intensity of $I = 8 \times 10^{13}$ W/cm^2, the wavelength of 800 nm, and the FWHM duration of 50 fs. The 3D TDSE solution was performed on the time interval $\tau_d - T/2 < t < \tau_d + T/2$ at which the laser field is non-zero. Figure 2 shows the dependence of the spectral intensity of the 22th harmonic as a function of the time delay τ_d (dots connected by solid lines) in the range τ_d from 0 to $2\pi/\omega_{\mathrm{LF}}$. The value of the spectral intensity is normalized to the corresponding value at $\tau_d = 0$. The red dashed line in Fig. 2 also marks the values of the low-frequency field at time moments $t = \tau_d$, $E_{\mathrm{LF}}(\tau_d)$, which, for convenience of comparison, are shifted and scaled along the vertical axis with a change in sign. As can be seen from the figure, the dependence of the intensity of the 22th harmonic on the time delay reproduces with high accuracy the waveform of the low-frequency field, which confirms the robustness of the discussed procedure of low-frequency pulses measurement.

4. Conclusions

In this paper, we have investigated the high harmonic generation (HHG) during the interaction of the femtosecond laser field with oriented asymmetric molecules in the presence of an external static electric field. Based on the numerical solution of three-dimensional time-dependent Schrödinger equation for the model one-electron system, we calculated the HHG spectrum for different static-field strengths. It is shown that the static field can enhance or suppress even harmonic intensities. For harmonics in the cutoff region, the dependences of harmonic intensities on the static-field strength are linear in a wide range of static-field strengths. We have demonstrated that this range of parameters is attractive for implementing sampling methods of measuring the waveforms of low-frequency (terahertz or mid-infrared) radiation based on the HHG of probe laser pulse with variable time delay related to the low-frequency field.

Acknowledgments

The numerical calculations were supported by Russian Science Foundation (Grant No. 20-72-00131). The development of the computer code was supported by the Foundation for the

Advancement of Theoretical Physics and Mathematics "BASIS" (19-1-2-52-1).

References

[1] Krausz F and Ivanov M 2009 *Rev. Mod. Phys.* **81** 163
[2] Corkum P B 1993 *Phys. Rev. Lett.* **71** 1994
[3] Bian X B and Bandrauk A D 2010 *Phys. Rev. Lett.* **105** 093903
[4] Sanchez D, Hemmer M, Baudisch M, Cousin S, Zawilski K, Schunemann P, Chalus O, Simon-Boisson C and Biegert J 2016 *Optica* **3** 147–150
[5] Agranat M, Chefonov O, Ovchinnikov A, Ashitkov S, Fortov V and Kondratenko P 2018 *Phys. Rev. Lett.* **120** 085704
[6] Silaev A A, Kostin V A, Laryushin I D and Vvedenskii N V 2018 *JETP Lett.* **107** 151–156
[7] Fülöp J A, Tzortzakis S and Kampfrath T 2020 *Adv. Opt. Mat.* **8** 1900681
[8] Musheghyan M, Geetha P P, Faccialà D, Pusala A, Crippa G, Campolo A, Ciriolo A G, Devetta M, Assion A, Manzoni C *et al.* 2020 *J. Phys. B: At. Mol. Opt. Phys.* **53** 185402
[9] Scuseria G E, Miller M D, Jensen F and Geertsen J 1991 *J. Chem. Phys.* **94** 6660–6663
[10] Silaev A A and Vvedenskii N V 2015 *Physics of Plasmas* **22** 053103
[11] Silaev A A, Romanov A A and Vvedenskii N V 2018 *J. Phys. B: At. Mol. Opt. Phys.* **51** 065005
[12] Lu X and Zhang X C 2014 *Front. Optoelectron.* **7** 121–155

Simulation of High Harmonic Generation in Xenon Based on Time-Dependent Density-Functional Theory

A A Romanov[1,2], A A Silaev[1,2], T S Sarantseva[2,3], M V Frolov[2,3] and N V Vvedenskii[1,2]

[1]Institute of Applied Physics of the Russian Academy of Sciences, Nizhny Novgorod 603950, Russia
[2]Lobachevsky State University of Nizhny Novgorod, Nizhny Novgorod 603950, Russia
[3]Department of Physics, Voronezh State University, Voronezh 394018, Russia

E-mail: romanoval@ipfran.ru

Abstract. The high harmonic generation (HHG) in xenon is studied by using the time-dependent density-functional theory for active 4th and 5th atomic shells. The HHG spectra are calculated for different intensities and carrier-envelope phases of a laser pulse. The HHG spectrum shows enhancement at the frequency region near 100 eV, where the HHG yield is by more than an order of magnitude higher than in the single-active electron approximation. The high accuracy of the obtained results is confirmed by comparing the calculated HHG spectra with experimental results.

1. Introduction

The high harmonic generation (HHG) is one of the well-studied phenomena of strong-field physics, whose practical applications in attoscience, biology, generation of coherent light, etc., are of great interest [1]. The physical mechanism of HHG is qualitatively explained in terms of the three-step scenario [2]. At the first step, atomic/molecular electron is transitioned in the continuum by an intense laser field. At the second step, the alternating electric field of the laser pulse accelerates the liberated electrons in the continuum. At the third step, the accelerated electron recombines with the emission of a high-energy photon. In the first and third steps, the inner multielectron dynamics can significantly affect the form of the HHG spectrum [3, 4]. An example of the large influence of multielectron dynamics at the recombination step consists in the giant enhancement of HHG yield in xenon. This enhancement is associated with the interaction of the photoelectron, originating from the external $5p_0$ orbital, with the inner electrons of the atom [4–11]. The giant enhancement in the HHG spectrum was predicted in [5] based on the analytical parametrization of the HHG spectrum in terms of the electron wave packet and the differential photorecombination cross section [12]. Experimental measurements of HHG spectra in Xe using near-IR driving wavelengths confirm the existence of a giant (more than an order of magnitude) harmonic yield enhancement near 100 eV [4, 13]. This phenomenon was also simulated numerically by using the time-dependent configuration-interaction singles approach in [7, 13] and time-dependent R-matrix theory in [9].

In this work, we discuss an implementation of the time-dependent density functional theory (TDDFT) for the simulation of HHG in Xe atoms in intense laser fields. A more detailed analysis of HHG in Xe (including analytical description and calculations of contributions of different atomic orbitals) can be found in our recent article [11]. Here, we provide additional calculations for different laser pulse intensities, carrier-envelope phases (CEPs) and compare the results with the single-active electron (SAE) approximation and experiment [4]. Atomic units (a.u.) are used throughout the paper unless specified otherwise.

2. Statement of the problem

We consider the interaction of Xe atom with an intense linearly polarized laser pulse with an electric field

$$\mathbf{E}(t) = -\frac{1}{c}\frac{\partial A}{\partial t}\mathbf{z}, \quad A(t) = -\frac{cE_0}{\omega_0} f(t) \sin\left(\omega_0 t + \varphi_{\text{CEP}}\right), \tag{1}$$

where ω_0, E_0, φ_{CEP} are the carrier frequency, the peak amplitude of electric field, and the CEP of the laser pulse, respectively, c is the speed of light. The envelope is $f(t) = \cos^2(\pi t/\tau)$ for $-\tau/2 < t < \tau/2$, where τ is the full pulse duration, and $f(t) = 0$ otherwise. Within the framework of the TDDFT, the dynamics of spin-unpolarized quantum system is described by the time-dependent Kohn-Sham (TDKS) equations

$$\mathrm{i}\frac{\partial}{\partial t}\psi_j(\mathbf{r},t) = \hat{H}\psi_j(\mathbf{r},t), \quad j = 1, 2, ..., N/2, \tag{2}$$

$$\hat{H} = -\frac{\nabla^2}{2} - \frac{N}{r} + \mathbf{r}\cdot\mathbf{E}(t) + V_{\text{ee}}[\rho(\mathbf{r},t)], \tag{3}$$

where $N = 54$ is the total number of electrons in Xe, $\psi_j(\mathbf{r},t)$ is the wave function of jth Kohn-Sham orbital, $V_{\text{ee}}[\rho] = V_{\text{H}}[\rho] + V_{\text{xc}}[\rho]$ is the electron-electron interaction potential, which is the functional of the electron density

$$\rho(\mathbf{r},t) = 2\sum_{j=1}^{N/2} |\psi_j(\mathbf{r},t)|^2 . \tag{4}$$

The electron-electron interaction potential is presented by a sum of Hartree potential

$$V_{\text{H}}[\rho](\mathbf{r},t) = \int \frac{\rho(\mathbf{r}',t)}{|\mathbf{r}-\mathbf{r}'|} d\mathbf{r}', \tag{5}$$

and the exchange-correlation potential $V_{\text{xc}}[\rho]$, which is considered in the spin-unpolarized LB94 approximation [14]. In this work, we do not consider the spin-orbit interaction and other relativistic effects and describe the interaction of Xe atom with laser pulse in the dipole approximation.

Interaction of an intense laser pulse with an atomic target induces a time-dependent dipole moment $\mathbf{D}(t)$. For the linearly polarized field, the dipole moment is directed along z axis, $\mathbf{D}(t) = \hat{\mathbf{z}}D(t)$. Properties of the radiation generated by an atom subjected into intense laser field are determined by the dipole acceleration $a(t)$:

$$a(t) = \mathrm{d}^2 D(t)/\mathrm{d}t^2 = \sum_j a_j(t), \tag{6}$$

$$a_j(t) = -2F(t) - 2\int |\psi_j|^2 \frac{\partial}{\partial z}\left(V_{\text{ee}} - \frac{N}{r}\right) \mathrm{d}\mathbf{r}, \tag{7}$$

where $a_j(t)$ is the partial dipole acceleration associated with jth Kohn-Sham orbital. The spectral density of the dipole acceleration is given by squared Fourier transform of $a(t)$,

$$R(\omega) = \left| \int_{-\infty}^{\infty} a(t) e^{i\omega t} dt \right|^2 . \tag{8}$$

The algorithm for the numerical solution of TDKS equations (2) is based on the expansion of wave functions in spherical harmonics and was well described in Refs. [3, 11, 15, 16]. To speed up numerical calculations, we use a nonuniform spatial grid, which has a higher density of nodes near the nucleus. The radial nodes of the spatial grid are specified as

$$r_k = k\Delta r + (\delta r/\Delta r - 1) r_\alpha \tanh(k\Delta r / r_\alpha), \tag{9}$$

where k is an integer, $\delta r = 10^{-4}$ is tiny step of the radial grid, which is realized near the nucleus; $\Delta r = 0.1$ is the radial step for large distances; $r_\alpha = 20$ is the scale for changing spatial step from δr to Δr. The size of radial grid was limited by 200. The maximum orbital momentum in expansion of wave functions in spherical harmonics is $l_{\max} = 256$. The orbitals on 1-3 shells are frozen. Freezing of internal orbitals makes it possible to use a relatively large time step $\Delta t = 0.02$ without losing accuracy in numerical calculations. The absorbing layer is modeled by multi-hump imaginary potentials [17] with total absorption layer width $r_{\text{abs}} = 50$, providing a wide range of effectively absorbed de Broglie wavelengths.

3. Numerical results

We have performed numerical calculation of the HHG spectrum $R(\omega)$ for various laser pulse intensities $10^{14} < I < 2 \times 10^{14}$ W/cm^2 and CEPs $\varphi_{\text{CEP}} = 0, \pi/2$. The laser pulse has the intensity full-width at half-maximum (FWHM) duration $\tau_p = \tau[2\arccos(2^{-1/4})/\pi] = 12$ fs, central wavelength $\lambda = 2\pi c/\omega_0 = 1.8$ μm. The considered parameters of the laser pulse correspond to the experiment [4]. The results of calculations based on TDDFT are shown in Fig. 1 by solid blue lines. For eye guiding in Fig. 1 we show smoothed spectra obtained by averaging HHG spectra using a smooth rectangular window function with width $\approx 2\omega_0$. We have also calculated the HHG spectra using the SAE approximation. For the SAE case, the three-dimensional single-electron time-dependent Schrödinger equation is solved for the wave function initially being in the $5p_0$ state. The atomic potential was calculated using stationary Kohn-Sham equations as in [3]. The corresponding results are shown by red lines in Fig. 1. Gray lines show the experimental results [4], in which the peak laser-pulse intensity was calibrated using the analytical formula for the rate of nonadiabatic ionization (the value of laser-pulse CEP was not specified in the experiment).

It can be seen from the figures that in the low-frequency region (slightly above the ionization potential $I_p = 12.1$ eV), the calculated HHG spectra contain a narrow near-threshold enhancement region. As shown in [11], this enhancement is associated with a sharp increase of the recombination transition matrix element from the continuum state to the $5p_0$ state with decreasing frequency of emitted photons. At harmonic frequencies below I_p, the SAE approximation significantly overestimates the spectral intensity (up to one order of magnitude) compared to TDDFT (there are no experimental data for this range). In this range, the main mechanism of harmonic generation relates with the response of free electrons (Brunel harmonics) [18, 19], which in the case of the SAE approximation turns out to be overestimated due to the absence of polarization screening of the external field that ionizes the atom [3]. As the harmonic frequency increases from 15 eV, the spectral intensity decreases as in the experiment, as well as in the results of numerical calculations based on TDDFT and SAE approximation for all the considered laser pulse intensities and CEPs.

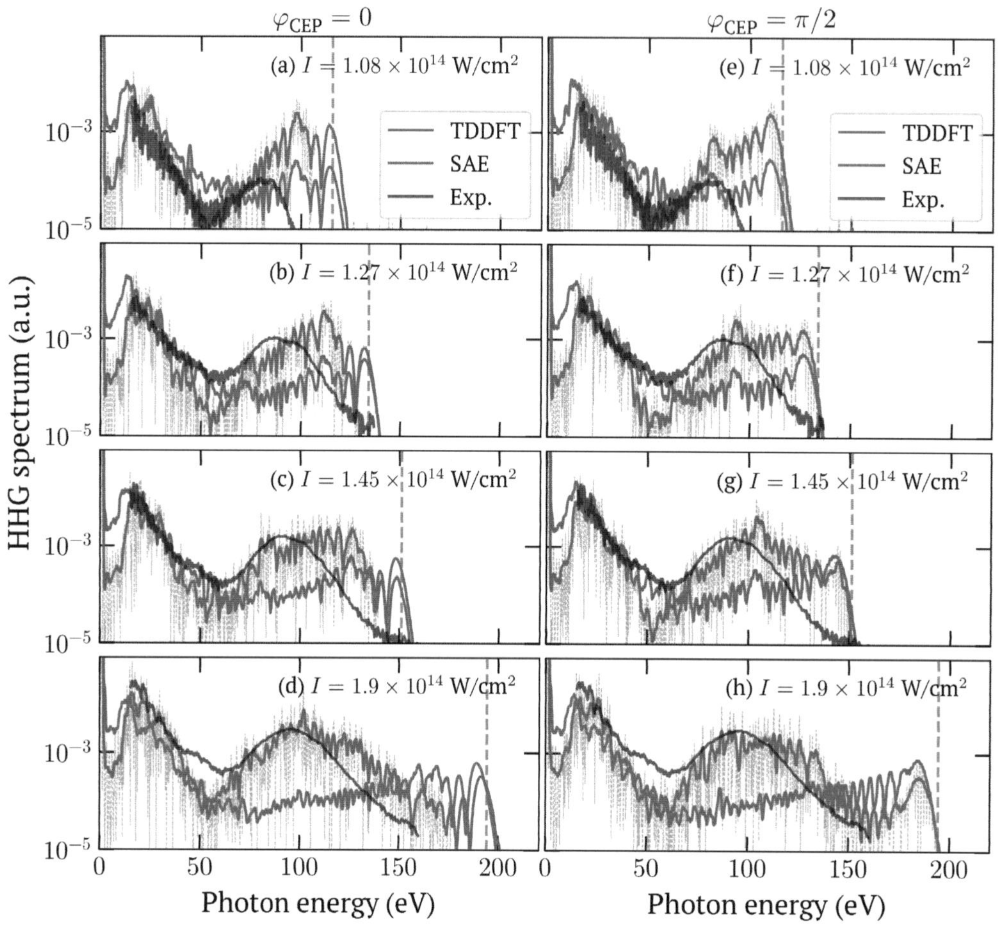

Figure 1. HHG spectra for Xe atom subjected to the two-cycle linearly polarized laser pulse with peak intensity $I = 1.08, 1.27, 1.45$ and 1.9×10^{14} W/cm^2, central wavelength $\lambda = 1.8$ μm, and with CEP $\varphi_{CEP} = 0$ (a-d) and with $\varphi_{CEP} = \pi/2$ (e-h). The dotted blue lines are the TDDFT results, solid blue lines are the smoothed TDDFT results, solid red lines are results based on the SAE approximation, solid gray lines denote the normalized experimental HHG spectrum from Ref. [4]. Dashed vertical gray lines mark the classical cutoff given by Eq. (10).

The spectral intensity in the experiment and TDDFT calculation increases at a frequency of about 60 eV due to the presence of a broad region of giant enhancement in the HHG yield. The giant enhancement is a result of the giant dipole resonance in the photoelectron recombination cross-section with a maximum near 100 eV due to the correlation of the photoelectron with inner orbitals [5, 20, 21]. In the TDKS equations, this enhancement is provided by the orbitals response from the 4th shell, caused by the perturbation of the electron-electron interaction potential induced by recolliding photoelectron wavepacket originating from the $5p_0$ orbital [11]. The giant enhancement is not observed in the results of the SAE approximation, which does not take into account multielectron dynamics. As a result, HHG yield in the harmonic frequency range of about 100 eV is greatly underestimated.

As can be seen from figures, the HHG spectra calculated using both TDDFT and SAE approximation extend up to the frequency, which is in good agreement with the classical cutoff of the HHG plateau [2]

$$\omega_{\text{cut}} \approx I_p + 3.17E_0^2/4\omega_0^2. \tag{10}$$

Due to the short (two-cycle) laser-pulse duration, the cutoff frequency at CEP $\varphi_{\text{CEP}} = \pi/2$ is slightly smaller than at $\varphi_{\text{CEP}} = 0$, in accordance with [6, 22]. In addition, for a short pulse, CEP strongly affects the extreme closed trajectories corresponding to the merging of short and long electrons trajectories. Therefore, the shape of the entire HHG spectrum is sensitive to CEP; in particular, the spectral intensity near the cutoff for $\varphi_{\text{CEP}} = \pi/2$ turns is higher than for $\varphi_{\text{CEP}} = 0$. In the experiment, the spectral intensity decreases smoothly without a pronounced cutoff. At frequencies $\sim \omega_{\text{cut}}$, where the numerical calculation predicts a high spectral intensity, the experimental spectrum gives an almost complete absence of a signal. This can be caused by the propagation macroscopic effects associated with phase mismatch, defocusing, and others in the experiment [23–26]. Also, our calculations do not consider the spatial intensity distribution of the laser pulse, which can be one of the reasons for the deviation of the HHG spectrum from the experimental results.

4. Conclusions

To conclude, we have used the TDDFT for studying the high harmonic generation by Xe atom at different intensities and CEPs of the two-cycle infrared laser pulse. We compared the results given by TDDFT for the active 4th and 5th shells of the Xe atom with the one-electron approximation based on the solution of the time-dependent Schrödinger equation and with the results of the experiment [4]. Our results show that using the TDDFT leads to a good qualitative agreement with the experiment. The SAE approximation gives qualitatively correct dependences of HHG yield on the frequency in the range 15-50 eV, while it is absolutely inapplicable at frequencies $\sim$ 100 eV, where the HHG spectrum contains a region of giant enhancement.

Acknowledgments

The research was carried out within the state assignment of Ministry of Science and Higher Education of the Russian Federation (theme No. 0030-2021-0002).

References

[1] Krausz F and Ivanov M 2009 *Rev. Mod. Phys.* **81** 163
[2] Corkum P B 1993 *Phys. Rev. Lett.* **71** 1994
[3] Romanov A A, Silaev A A, Frolov M V and Vvedenskii N V 2020 *Phys. Rev. A* **101** 013435
[4] Shiner A D, Schmidt B E, Trallero-Herrero C, Wörner H J, Patchkovskii S, Corkum P B, Kieffer J C, Légaré F and Villeneuve D M 2011 *Nat. Phys.* **7** 464
[5] Frolov M V, Manakov N L, Sarantseva T S, Emelin M Y, Ryabikin M Y and Starace A F 2009 *Phys. Rev. Lett.* **102** 243901
[6] Frolov M V, Manakov N L, Silaev A A, Vvedenskii N V and Starace A F 2011 *Phys. Rev. A* **83** 021405
[7] Pabst S and Santra R 2013 *Phys. Rev. Lett.* **111**(23) 233005
[8] Faccialà D, Pabst S, Bruner B D, Ciriolo A G, De Silvestri S, Devetta M, Negro M, Soifer H, Stagira S, Dudovich N and Vozzi C 2016 *Phys. Rev. Lett.* **117**(9) 093902
[9] Hassouneh O, Brown A C and van der Hart H W 2014 *Phys. Rev. A* **90**(4) 043418
[10] Bray A W, Freeman D, Eckart S and Kheifets A S 2019 *Phys. Rev. A* **100**(1) 013404
[11] Romanov A A, Silaev A A, Sarantseva T S, Frolov M V and Vvedenskii N V 2021 *New J. Phys.* **23** 043014
[12] Morishita T, Le A T, Chen Z and Lin C D 2008 *Phys. Rev. Lett.* **100** 013903
[13] Faccialà D, Pabst S, Bruner B D, Ciriolo A G, De Silvestri S, Devetta M, Negro M, Soifer H, Stagira S, Dudovich N and Vozzi C 2016 *Phys. Rev. Lett.* **117**(9) 093902
[14] van Leeuwen R and Baerends E J 1994 *Phys. Rev. A* **49** 2421
[15] Sarantseva T S, Silaev A A, Romanov A A, Vvedenskii N V and Frolov M V 2021 *Opt. Exp.* **29** 1428–1440
[16] Sarantseva T S, Romanov A A, Silaev A A, Vvedenskii N V and Frolov M V 2021 *Opt. Exp.* **29** 38298–38313
[17] Silaev A A, Romanov A A and Vvedenskii N V 2018 *J. Phys. B: At. Mol. Opt. Phys.* **51** 065005
[18] Brunel F 1990 *J. Opt. Soc. Am. B* **7** 521–526

Journal of Physics: Conference Series **2249** (2022) 012005 doi:10.1088/1742-6596/2249/1/012005

[19] Kostin V A and Vvedenskii N V 2018 *Phys. Rev. Lett.* **120**(6) 065002
[20] Amusia M 1990 *Atomic Photoeffect* (Plenum, NY)
[21] Starace A F 1982 *Theory of atomic photoionization* (Springer-Verlag, Berlin) pp 1–121
[22] Frolov M V, Manakov N L, Popov A M, Tikhonova O V, Volkova E A, Silaev A A, Vvedenskii N V and Starace A F 2012 *Phys. Rev. A* **85** 033416
[23] Farrell J P, Spector L S, McFarland B K, Bucksbaum P H, Gühr M, Gaarde M B and Schafer K J 2011 *Phys. Rev. A* **83**
[24] Gibson E A, Paul A, Wagner N, Tobey R, Backus S, Christov I P, Murnane M M and Kapteyn H C 2004 *Phys. Rev. Lett.* **92**(3) 033001
[25] Arpin P, Popmintchev T, Wagner N, Lytle A, Cohen O, Kapteyn H and Murnane M 2009 *Phys. Rev. Lett.* **103** 143901
[26] Sun H W, Huang P C, Tzeng Y H, Huang J T, Lin C D, Jin C and Chen M C 2017 *Optica* **4** 976–981

Journal of Physics: Conference Series **2249** (2022) 012006 doi:10.1088/1742-6596/2249/1/012006

Effects of Low-Level Laser Irradiation on Mammalian Cell Cultures: Comparative Experimental Studies with Different Types of Lasers at 1260-1270 nm

A Khokhlova[1], I Zolotovskii[1], S Sokolovski[2], Yu Saenko[3], E Rafailov[2], D Stoliarov[1,2], E Pogodina[3], A Gilmutdinova[1], V Svetukhin[4]and A Fotiadi[1,5,6]

[1]Laboratory of Nonlinear and Microwave Photonics, Ulyanovsk State University, Ulyanovsk, Russia

[2]Aston Institute of Photonic Technologies, Aston University, Birmingham, UK

[3]Laboratory of Molecular and Cell Biology, Ulyanovsk state university, Ulyanovsk, Russia

[4]Scientific-Manufacturing Complex "Technological Centre", Moscow, Russia

[5]Université de Mons, Mons, Belgium

[6]Ioffe Physical-Technical Institute, St.-Petersburg, Russia

E-mail: avhohlova@gmail.com

Abstract. The effects observed under near-infrared laser irradiation of mammalian cells have been demonstrated to depend to a large extent on the type of irradiation source. In our experiments, we have measured concentration of reactive oxygen species in cell cultures of different origin (rodent and human, cancerous and non-cancerous) exposed to the radiation of low-level lasers at 1265 nm. Surprisingly, the radiation effects of narrowband laser occur to be more pronounced compared with those provided by the lasers of broader linewidth. Also, we have found that the aggressive types of cancer require a more accurate selection of irradiation parameters and laser operation regime.

1. Introduction

In recent years, the laser sources emitting in the near infrared range have been enthusiastically exploited in medical applications. An infrared light exhibits ability essential for medical needs to penetrate tissues deeper than the visible light [1-3].

The laser sources operating 1265-1270 nm range are of particular interest. Such sources enable singlet oxygen generation in inhomogeneous media and, in particular, in biological objects [4-8].

Within these wavelengths, the effects in biological objects have been observed at very low power. Therefore, the 1265-1270 nm radiation can be used in low-level laser therapy or in the so-called photobiomodulation therapy (PBMT) allowing minimal tissue damages [9-11].

Photobiomodulation applications range from regenerative medicine to cancer treatment [12-16]. It is very important to study the radiation-induced effects provided by different laser sources. The radiation emitted by different sources may possess different physical parameters thus having different effects on the biological objects.

In our study, the 1265 nm laser effects on mammalian cell cultures are explored. The performed in vitro studies have provided simple and convenient preclinical models that allow eliminating involvement of laboratory animals.

2. Methods

2.1. Cell cultures
The research has been performed on mammalian cell cultures.

1. CHO-K1 (ATCC® CCL-61™) is the Chinese hamster ovary epithelial cell line. The cells are not cancerous but immortalized.
2. HCT116 (ATCC® CCL-247™) is the human colon cancer cell line.
3. B16-F10 (ATCC® CRL-6475™) is the murine melanoma cell line. This transplantable cell culture is the most commonly exploited to study melanoma and to develop the laboratory animal models.
4. A875 (ATCC) is the human melanoma cell line.

2.2. Laser Sources
A semiconductor narrow-band (highly coherent) laser (Yenista Optics (Osics T100)) and a laser diode (Innolume, ThorLabs) operating at the wavelength of 1265 nm are used in the experiments. To adjust the diameter of the laser beam, the two collimators with waists of 3.5 and 7 mm are used.

2.3. Experiment
Twenty-four hours before the experiment, the cells were seeded in a slide chamber for adhesion and growth to the exponential phase. During the experiment, the cells were kept in a general or benchtop incubator in a humidified atmosphere, with 5% CO_2 and at a temperature of 37°C.

Since the laser irradiation at a higher power is accompanied by the heating effect, the culture medium temperature was measured in the well of the slide chamber before the experiment in order to keep the temperature inside the incubator at 37°C.

To avoid heating under the irradiation at higher powers, the control well was placed in the general incubator at 37°C to provide the same temperature. The slide-chamber was placed on the stand; the collimator was fixed at the bottom at a distance of 1 cm from the sample. During the irradiation of the experimental well, the control well was shielded with a metal foil.

The energy density of laser radiation is calculated as follows:

$$E = \frac{\mathrm{P}\times\mathrm{t}}{\mathrm{S}}, \qquad (1)$$

where P is the average output power (W), t is the exposure time (sec), S is the laser spot area on the cell culture (cm^2) [4].

The irradiation time in all experiments was 30 minutes, unless otherwise indicated in Results and Discussion. Each experiment was performed in triplicate.

2.4. Fluorescent staining of cells
After irradiation, the cells were stained for fluorescence microscopy with dichloro-dihydro-fluorescein diacetate to evaluate intracellular concentration of the reactive oxygen species as described [17].

3. Results and Discussion
Figure 1 shows the concentration of reactive oxygen species in cell cultures immediately after irradiation with a low-power narrowband laser. The dotted line is the baseline (hereinafter) where the experimental and control values are nearly the same.

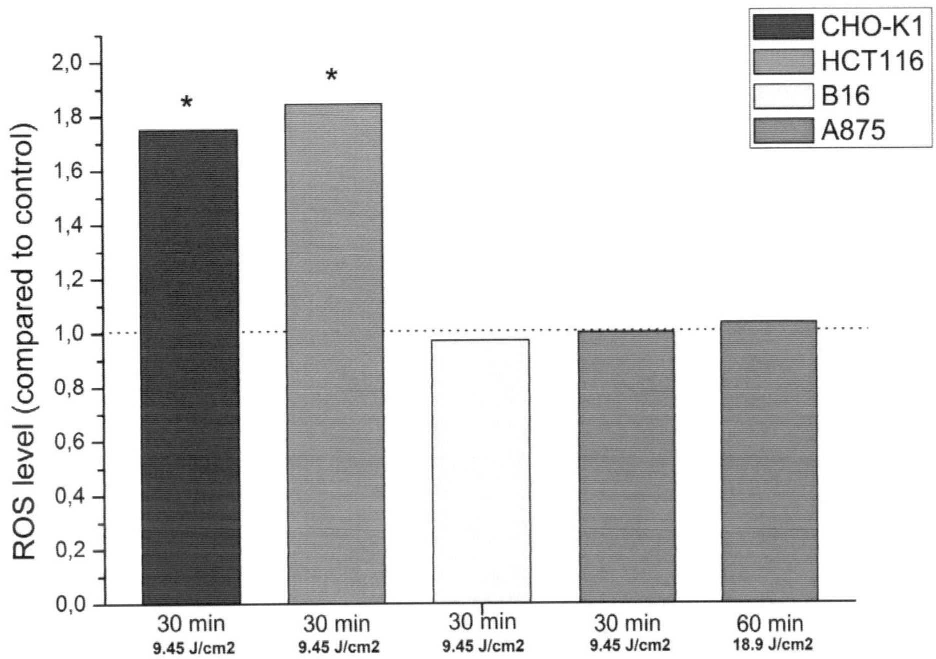

Figure 1. Level of ROS in the CHO-K1 (Chinese hamster ovary), HCT116 (human colon cancer), B16/F10 (murine melanoma), and A875 (human melanoma) cell lines exposed to 1265 nm narrowband laser irradiation (t=30min); P=4.2 mW; S=0.8 cm^2; E=9.45-18.9 J/cm^2. * - statistically significant difference between experiment and control.

The ROS level in the Chinese hamster ovary and human colon cancer cells significantly increases after the laser irradiation. However, in melanoma cells, no increase of the oxidative stress is registered even after the exposure for 60 minutes (energy density is 18.9 J/cm^2). It could be explained by the different intracellular mechanisms.

Figure 2 shows the ROS concentration in Chinese hamster ovary cells and human colon cancer cells after the exposure to laser sources of different linewidths. One can see that the effects of narrowband laser irradiation are more pronounced. Noteworthy, the power of the laser diode was more than twice as high in these experiments than the power of narrowband laser.

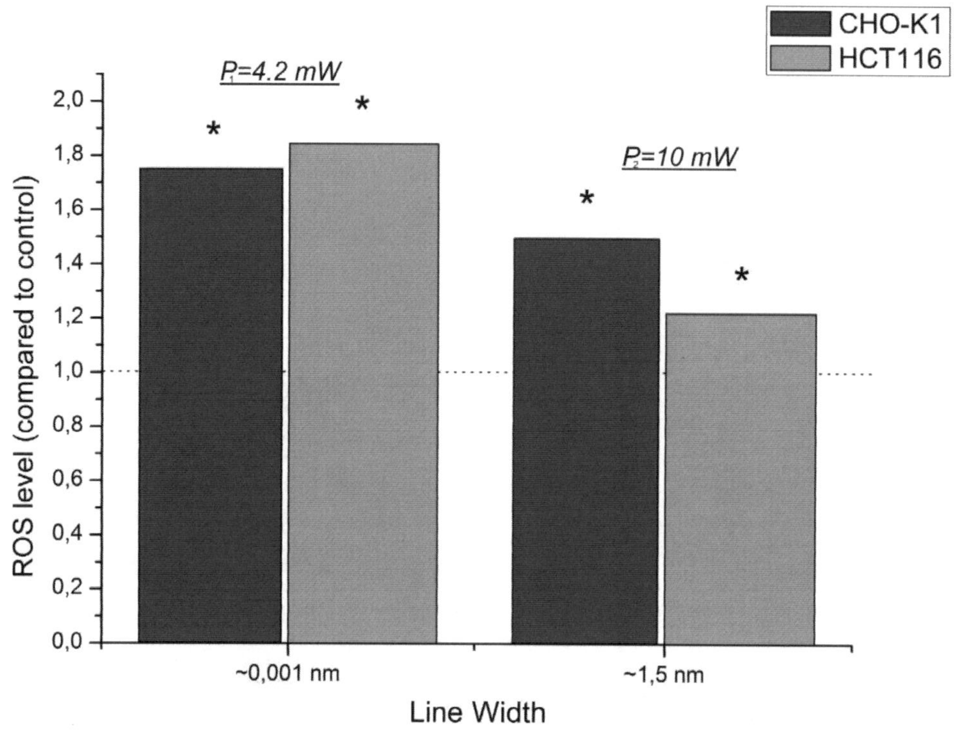

Figure 2. Level of ROS in the CHO-K1 (Chinese hamster ovary) and HCT116 (human colon cancer) cell lines exposed to 1265 nm narrowband laser and broadband laser diode irradiation (t=30min); P_1=4.2 mW, P_2=10 mW; S=0.8 cm^2; E_1=9.45 J/cm^2, E_2=18.9 J/cm^2. * - statistically significant difference between experiment and control.

It can be explained by the fact that the power spectral density of the narrowband low-power laser radiation at the point of its maximum is higher than that of the radiation provided by the high-power (broadband) laser diode: A ratio describing the power spectral densities for these laser radiation sources is as follows:

$$\frac{P_{\lambda l}}{P_{\lambda h}} \sim \frac{\Delta\lambda_h}{\Delta\lambda_l}\frac{E_l}{E_h} \sim 2\times10^3\frac{E_l}{E_h}, \qquad (2)$$

where $P_{\lambda l}$ is the power spectral density of the narrowband source (*W*), $P_{\lambda h}$ is the power spectral density of the broadbsand source (*W*), E_l is the surface energy density of PBM (*J/cm2*), E_h is the surface energy density of high-power laser irradiation (*J/cm^2*), $\Delta\lambda_l$ is the linewidth of PBMT source (*nm*), $\Delta\lambda_h$ is the linewidth of high-power laser source (*nm*).

Low-power and high-power laser linewidths are 0.001 nm and 1.5 nm, respectively. Therefore, in the present study, the power spectral density of the low-power laser radiation at the point of its maximum is higher than that of the radiation provided by the high-power laser. All this allows us to conclude existence of the narrowband resonance effect presumably on the surface of mitochondrial membranes [18-19].

Further experiments were performed with the 1265 nm laser diode to provide higher powers.

Figure 3 demonstrates resistance of the human melanoma cell line to laser irradiation. The power of 10 mW is required for normal cells or non-aggressive cancer to develop oxidative stress (Fig. 2), whereas melanoma demonstrates this effect at 250 mW.

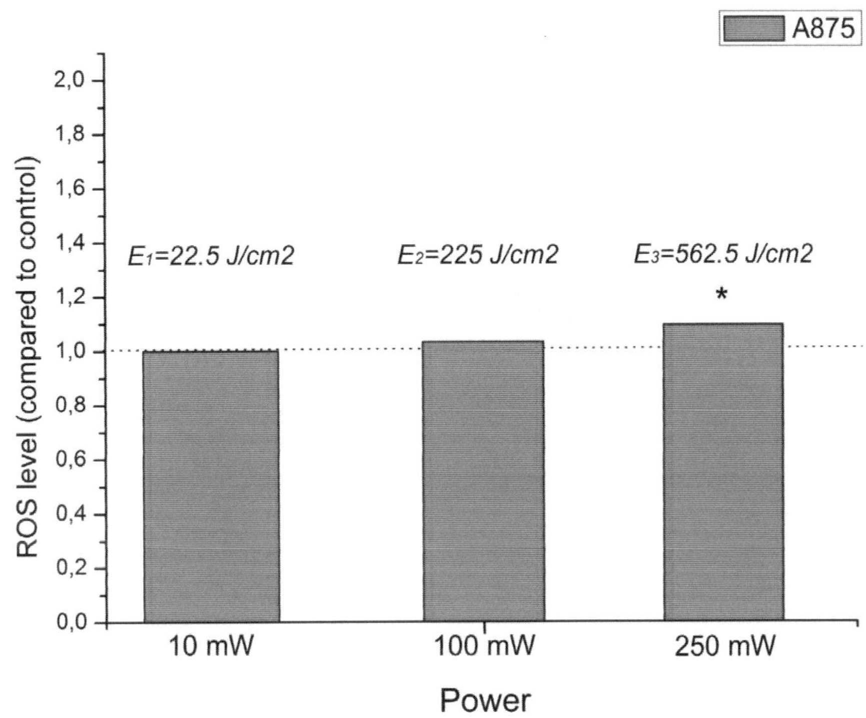

Figure 3. Level of ROS in the A875 (human melanoma) cell line exposed to 1265 nm laser diode irradiation (t=30min); P=10 mW, 100 mW, 250 mW; S=0.8 cm^2; E_1=22.5 J/cm^2, E_2=225 J/cm^2, E_3= 562.5 J/cm^2.* - statistically significant difference between experiment and control.

Figure 4 shows the ROS concentration at different intensities of laser irradiation. A significant difference is registered in the oxidative stress level of human and murine melanoma cells. This can be explained by different origin of species (human and murine) or mutation and requires further research.

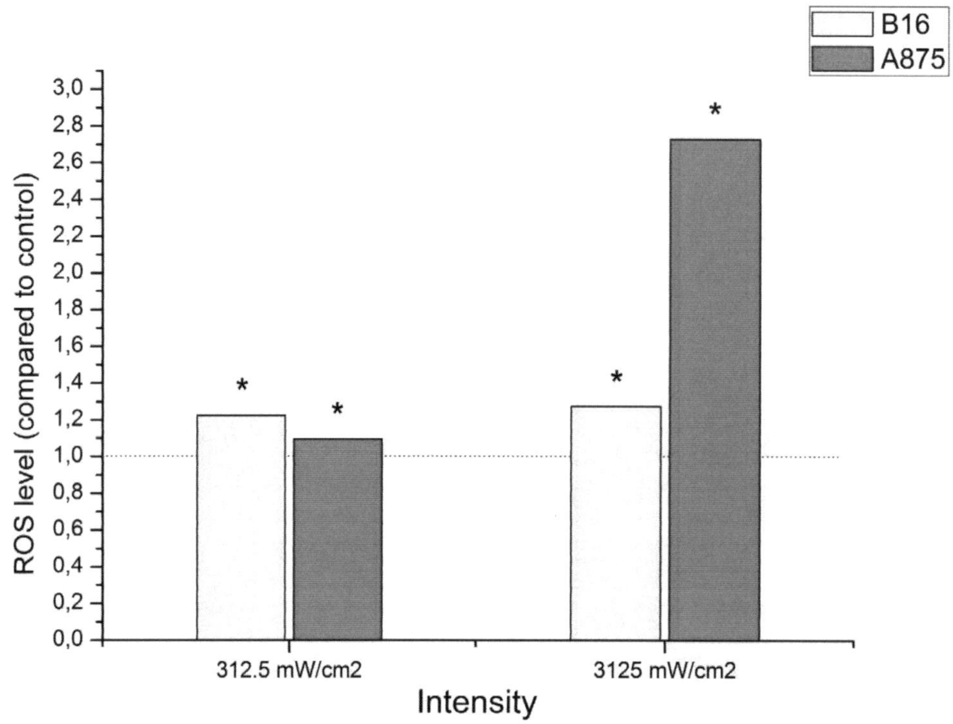

Figure 4. Level of ROS in the A875 (human melanoma) and B16/F10 (murine melanoma) cell lines exposed to 1265 nm laser diode irradiation (t=30min); P=250 mW; S=0.8 cm^2, 0.0812 cm^2; E_1=562.5 J/cm^2, E_2=5625 J/cm^2.* - statistically significant difference between experiment and control.

Figure 5 shows the concentration of reactive oxygen species 1 hour after irradiation. One can see that in the Chinese hamster ovary cells the oxidative stress is still increasing. It may indicate the antioxidant system failure leading to cell death. In melanoma cells, the level of oxidative stress decreases. Moreover, in the murine melanoma cells 1 hour after laser irradiation exposure there is no statistical significance between experiment and control groups.

Journal of Physics: Conference Series **2249** (2022) 012006 doi:10.1088/1742-6596/2249/1/012006

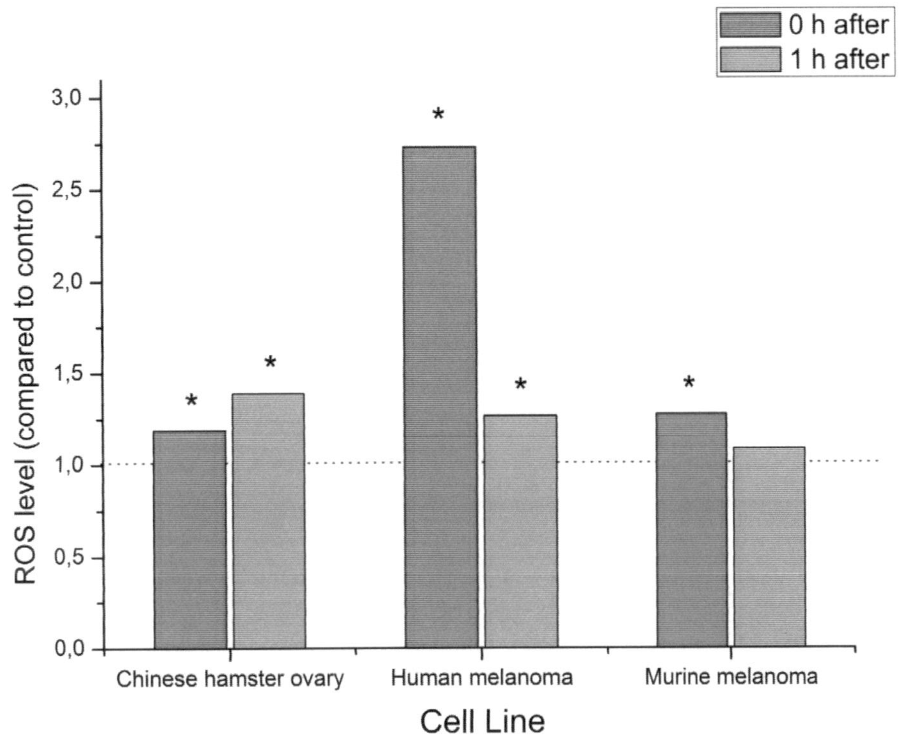

Figure 5. Level of ROS in the CHO-K1 (Chinese hamster ovary), A875 (human melanoma), and B16/F10 (murine melanoma) cell lines exposed to 1265 nm laser diode irradiation (t=30min); P=250 mW; S=0.8 cm^2; E=562.5 J/cm^2.* - statistically significant difference between experiment and control.

4. Conclusions

Our experiments have shown that the effects of the 1265 nm laser irradiation on biological cells depend not only on the irradiation dose but also on the laser linewidth, power, and intensity, as well as on the cell type.

Acknowledgments

The reported melanoma research was funded by Russian Foundation for Basic Research (RFBR) and Ulyanovsk Region, project number 19-415-730003 (to A. Khokhlova).

The reported non-melanoma research was funded by THE MINISTRY OF SCIENCE AND HIGHER EDUCATION OF THE RUSSIAN FEDERATION (MEGA-GRANT PROGRAM), grant number 075-15-2021-581 (to I. Zolovovskii, D. Stoliarov, A. Gilmutdinova, and A. Fotiadi).

Conflict of interests

The authors declare no conflict of interests.

References

[1] Holick M F. Biological Effects of Sunlight, Ultraviolet Radiation, Visible Light, Infrared Radiation and Vitamin D for Health 2016 *Anticancer Research* **36** (3) 1345-1356

[2] Martignago C C S, Tim C R, Assis L, Da Silva V R, Santos E C B D, Vieira F N, Parizotto N A, and Liebano R E. Effects of red and near-infrared LED light therapy on full-thickness skin graft in rats. 2020 *Lasers in Medical Science* **35** (1) 157-164

[3] Barolet D, Christiaens F, and Hamblin M R. Infrared and skin: Friend or foe. *Journal of Photochemistry and Photobiology* B: Biology 2016 **155** 78-85

[4] Saenko Y V, Glushchenko ES, Zolotovskii I O, Sholokhov E, and Kurkov A. Mitochondrial dependent oxidative stress in cell culture induced by laser radiation at 1265 nm 2016 *Lasers in Medical Science* **31** (3) 405-13

[5] Sokolovski S G, Zolotovskaya S A, Goltsov A, Pourreyron C, South A P, and Rafailov E U. Infrared laser pulse triggers increased singlet oxygen production in tumour cells 2013 *Scientific Reports* 12 **3** 3484

[6] Anquez F, El Yazidi-Belkoura I, Randoux S, Suret P, and Courtade E. Cancerous cell death from sensitizer free photoactivation of singlet oxygen 2012 *Photochemistry and Photobiology* **88**(1):167-74

[7] Yusupov A S, Goncharov S E, Zalevskii I D. Paramonov V M, and Kurkov A S. Raman fiber laser for the drug-free photodynamic therapy 2010 *Laser Physics* **20**: 357–359

[8] Oliveira C S, Turchiello R, Kowaltowski A J, Indig G L, and Baptista M S. Major determinants of photoinduced cell death: Subcellular localization versus photosensitization efficiency 2011 *Free Radical Biology and Medicine* **51**(4) 824-33

[9] Hennessy M and Hamblin M R. Photobiomodulation and the brain: a new paradigm 2017 *Journal of Optics* **19**(1) 013003

[10] Hamblin M R, Nelson S T, and Strahan J R. Photobiomodulation and Cancer: What Is the Truth? 2018 *Photomedicine and Laser Surgery* **36**(5) 241-245

[11] Glass G E. Photobiomodulation: A review of the molecular evidence for low level light therapy 2021 *Journal of Plastic, Reconstructive and Aesthetic Surgery* **74**(5) 1050-1060

[12] Awotidebe A W, Inglis-Jassiem G, and Young T. Low-level laser therapy and exercise for patients with shoulder disorders in physiotherapy practice (a systematic review protocol) 2015 *Systematic Reviews* **4** 60

[13] Ramos G A, Arliani G G, Astur D C, Pochini A C, Ejnisman B, and Cohen M. Rehabilitation of hamstring muscle injuries: a literature review 2016 *Revista Brasileira de Ortopedia* **52**(1) 11-16

[14] Ash C, Town G, Whittall R, Tooze L, and Phillips J. Lasers and intense pulsed light (IPL) association with cancerous lesions 2017 *Lasers in Medical Science* **32**(8) 1927-1933

[15] Błochowiak K, Andrysiak P, Sidorowicz K, Witmanowski H, Hędzelek W, and Sokalski J. Selected applications of Er:YAG and CO_2 lasers for treatment of benign neoplasms and tumorous lesions in the mouth 2015 *Postepy Dermatologii I Alergologii* **32**(5) 337-43

[16] Schomaker M, Heinemann D, Kalies S, Willenbrock S, Wagner S, Nolte I, Ripken T, Murua Escobar H, Meyer H, and Heisterkamp A. Characterization of nanoparticle mediated laser transfection by femtosecond laser pulses for applications in molecular medicine 2015 *Journal of Nanobiotechnology* **13** 10

[17] Khokhlova A, Zolotovskii I, Stoliarov D, Vorsina S, Liamina D, Pogodina E, Fotiadi A, Sokolovski S, Saenko Y, and Rafailov E. The Photobiomodulation of Vital Parameters of the Cancer Cell Culture by Low Dose of Near-IR Laser Irradiation 2019 *IEEE Journal of Selected Topics in Quantum Electronics* **1**(25) 1-10

[18] Khokhlova A, Zolotovskii I, Sokolovski S, Saenko Y, Rafailov E, Stoliarov D, Pogodina E, Svetukhin V, Sibirny V, and Fotiadi A. The light-oxygen effect in biological cells enhanced by highly localized surface plasmon-polaritons 2019 *Scientific Reports* **9**(1) 18435

[19] Khokhlova A, Zolotovskii I, Pogodina E, Saenko Yu, Stoliarov D, Vorsina S, Fotiadi A, Liamina D, Sokolovski S, and Rafailov E. Effects of high and low level 1265 nm laser irradiation on HCT116 cancer cells 2019 *Proceedings SPIE* **10861** Mechanisms of Photobiomodulation Therapy XIV 108610L

Journal of Physics: Conference Series **2249** (2022) 012007 doi:10.1088/1742-6596/2249/1/012007

Research of Surface Plasmonic Resonance at The Silicon Nano-Silver Interface in the Information Recording Unit of the Diagnostic Complex Intest

T Y Mogilnaya[1], A M Vasiliev[2], L L Pagava[2], AND A G Botikov[3]

[1]Institute No. 12 "Aerospace Science-intensive Technologies and Production", , Moscow Aviation Institute,Moscow, Russia
[2]Institute No. 12 "Aerospace Science-intensive Technologies and Production", Moscow Aviation Institute(National Research University) , 4, Volokolamskoe shosse, 125993, Moscow, Russia.Contact Phone: +79035010554
[3]The State Collection of Viruses department, 2National Research Center for Epidemiology and Microbiology named after the honorary acad-emician N. F. Gamaleya , Moscow, Russia
Contact Email: mogilnay@mail.ru

Abstract. The research presents a study of the fundamental possibility of realizing the contactless recording of information a complex for express diagnostics of an organism for the presence of infectious agents, which was called diagnostic complex "Intest". The mathematical model of the interaction of the laser radiation with the surface of the crystal was built. Calculations were performed in the COMSOL 5.5 software package. The results of the optimization calculations showed at what crystal parameters it is possible to implement contactless recording

Introduction

The research presents a study of the fundamental possibility of realizing the contactless recording of information a complex for express diagnostics of an organism for the presence of infectious agents, which was called "Diagnostic complex Intest". The mathematical model of the interaction of the laser radiation with the surface of the crystal was built. Calculations were performed in the COMSOL 5.5 software package. The results of the optimization calculations showed at what crystal parameters it is possible to implement contactless recording.

New and emerging-reemerging infections pose a particular threat to global security. They can arise as a result of natural processes of evolution of microorganisms and deliberate actions. These events are happening more and more often and are becoming more formidable. The potential for pathogenicity of these pathogens is enormous.

During the global coronavirus infection SARSCoV-2, which has now swept the entire planet, as well as the influenza pandemic announced by WHO in June 2009, billions of people fall ill and millions die. The global economy has suffered catastrophic damage.

As we have already seen, in a pandemic, the most likely development of events is the rapid spread of the virus, since most of the population is susceptible to infection. Timely and effective diagnosis is one of the main factors and is critical in limiting the potential negative impact on the population and the operation of critical services.

Access to effective diagnostic tools for mass screening of the population in the early stages of a pandemic will significantly control the spread of infection.

The most promising direction is non-invasive methods of mass screening based on quantum optical effects, the main advantages over traditional diagnostic methods such as ELISA and PCR provide the absence of reagents and consumables, obtaining a result in real time, the possibility of continuous operation, heuristic analysis and prediction based on BigData, integration into security systems and transport infrastructure.

Research and development work was carried out in the development of new methods of medical diagnosis of infectious diseases, as well as equipment associated with early and rapid detection of infectious diseases in humans and determining the qualitative and quantitative characteristics of infections in the human body. A working sample of a compact medical diagnostic complex, called "Diagnostic complex Intest", that has no analogues in the world for express diagnostics of the human body for the presence of pathogens of infectious diseases, was created.

The use of the "Diagnostic complex Intest" in medical practice will allow you to immediately (under 1 minute instead of 1-3 days) receive information about the presence or absence of various pathogens in the patient's body without "sampling material" (blood, sputum, tissue, etc.) and without ruining the skin (cuts, punctures) and/or introducing any diagnostic device into the human body.

The main fundamental diagnostic mechanism is the registration of emissions of electromagnetic fields of the DNA structure. It is known that the characteristics of radiation are individual. Currently, the team of authors is faced with the task of switching to a contactless method of recording and retrieving information using the phenomenon of surface plasmon resonance (SPR). Currently, SPR is widely used in the development of chemical and biological sensors. When in contact with biological objects (DNA, viruses, antibodies), plasmonic effects make it possible to increase the intensity of fluorescent signals by more than an order of magnitude, that is, significantly expand the possibilities of detecting, identifying and diagnosing biological objects. The phenomena of plasmon resonance are also used when recording information in the "Diagnostic complex Intest" [1] which was developed by us.

At the same time, the occurrence of plasmon resonance depends on the large number of factors, such as the shape of the particles, the material and the frequency of the exciting radiation. Moreover, the properties of localized plasmons critically depend on the shape of nanoparticles, which allows tuning the system of their resonances for effective interaction with the light or elementary quantum systems. Therefore, it is of interest to build a mathematical model and carry out calculations to optimize the above parameters and to develop new technologies. This approach made it possible to reduce the time spent on research and optimization of such technological processes as the creation of colloidal solutions for the manufacture of energy storage devices, which increased the efficiency [2], and for modification surface of silicon photo-electric converters (PEC)

To better understand the principles of application of the plasmon resonance phenomena when recording information in the "Diagnostic complex Intest", it is proposed to consider the structural diagram of recording the characteristics of pathogens in the instrument database and the diagnostic process. Device creation can be divided into two stages: entering the information into the database and the diagnostic process itself. It is of interest to investigate the fundamental possibility of investigating the appearance of plasmon resonance, which is necessary to implement contactless recording.

1. The principle of operation of the device

A device for detecting microorganisms and other organic substances includes a power source, active and reference electrodes, a unit for analyzing the measured current, a control unit, a database of cells to record information that was previously obtained as follows: on the base medium (water, alcohol, saline solutions, etc.) containing a biological agent (virus, bacterium) direct radiation from a monochromatic coherent source in the visible or near-IR radiation range. After passing through the environment with a biological agent, the radiation enters the recording cell, which consists of a semiconductor, magneto-optical material with the ability of storing information in real time, included in a direct current circuit.

Each of these cells contains information corresponding to the spectral characteristics of the electromagnetic radiation of a particular chemical or microorganism. The device also contains a switch for connecting the cells. The power source is configured to generate alternating or direct current in the circuit. The unit for analyzing the measured current and the control unit are a single unit that is connected to the active electrode, the power supply and to the database switch. The reference electrode is connected to the power source and the output of the switch that connects the cells.

A database of cells for recording and a switch for connecting cells can also be integrated into a single analysis and a control unit. In this case, a single unit can be made in the form of a programmable microcontroller, microprocessor, or logic integrated circuit, which combines the functions of analyzing the measured current, control, a database of cells for recording information and a switch for connecting cells.

The device contains electrodes, a single unit for analyzing the measured current parameter and control, a database consisting of cells for recording, each of which corresponds with the spectral characteristic of electromagnetic radiation of a certain type of biological agent or chemical substance, equipped with a switch and a power source. The electrodes are in contact with the test object. When a low-power electromagnetic radiation (electric current) is applied to an object, a wave packet of spectral composition is excited in it, corresponding with the particular spectral characteristics of all objects. In this case, the chain repeats the features of the spectral characteristics of the object. If the power source generates a constant current in the circuit, one of the electrodes serves as an active (measuring) electrode, and the other serves as a reference electrode. When operating on alternating current, each of the electrodes performs these functions alternately in accordance with the instantaneous direction of the current. In the device, the first electrode, a single analysis and control unit and the second electrode are connected in series, forming a current loop together with the object under study. The database is connected via a switch to the current loop (preferably directly to a single unit). The power supply is also connected to the loop through a single block. The addition of two spectral characteristics, from the source and the cell for recording, gives a beat as a result of the interference of two close signals, which causes an increase in the current in the circuit.

Each cell for recording is formed by preliminary registration of the corresponding spectral characteristics of electromagnetic radiation of a certain type of chemical substance or biological agent and subsequent certain technological processing of the elements constituting the cell. The cell for recording contains at least two elements, one of which is a dielectric, and the other is a semiconductor or conductor. The cell for recording can be made of silver or aluminum, fixed on a base of silicon monoxide.

The device works as follows: the electrodes are placed on the object under study at arbitrary points. A voltage is applied to the electrodes from the reference voltage source, and an alternating or direct current of 2-15 μA is passed through the object under study, selected from the condition of ensuring the best information response. A single unit, after analyzing the input signal generates a control command to the electronic switch of the database, which connects to the current loop a cell for recording, corresponding with a certain type of biological agent or chemical substance, the presence of which must be determined in the object under study. From the database cell, a response enters the current loop, the form of which repeats the nature of the spectral characteristic of the connected cell. In the presence of a biological agent or chemical substance corresponding with the connected cell in the investigated object, the dynamic characteristic will cause an increase in current as a result of signal interference, the value of which is digitized and processed according to the block operation algorithm. Based on the correspondence between the parameters of the dynamic characteristics of the structure of the connected cell and a change in the measured current, the presence of a chemical substance or biological agent corresponding with the connected cell is diagnosed in the test object.

2. Information recording concept

During the recording process, radiation from a laser source passes through the sample, which consists of a neutral solution and a component to be diagnosed. (See Figure 1.) In this case, the cuvette is placed in a conductive metal capsule and connected to the cell for recording information by a conductor made of the same metal. The cell for recording information consists of two layers of a crystalline composition with different lattice parameters and different types of conductivity (semiconductor, metal, dielectric metal, etc.). This cell is placed in the information processing unit, which allows you to preserve the features of the distribution of the electromagnetic field of our object (in this case, the pathogen). If this information is compared with information about the electromagnetic field of an object in a person or an animal, then by identifying general characteristics, one can judge the presence or absence of a pathogen in a given object.

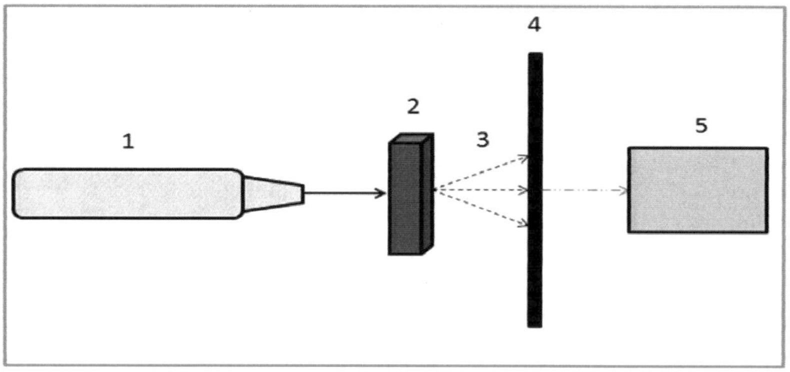

Figure 1. *Block diagram of information recording of the Intest-1 device: 1. Laser emitter. 2. Cuvette with a pathogen. 3. Electromagnetic field with the characteristics of the exciter. 4. Cell of a microcircuit for recording information. 5. Information processing unit.*

A laser beam, falling into a sample with an infectious disease agent, excites additional optical radiation in it, as a rule, this is a two-photon excitation, which in the literature is usually called superluminescence, i.e. we have a modulated electromagnetic field with modulation parameters individual for each infection, as mentioned above, i.e. radiation parameters (excitation level, frequency, polarization, coherence) have individual characteristics for each pathogen.

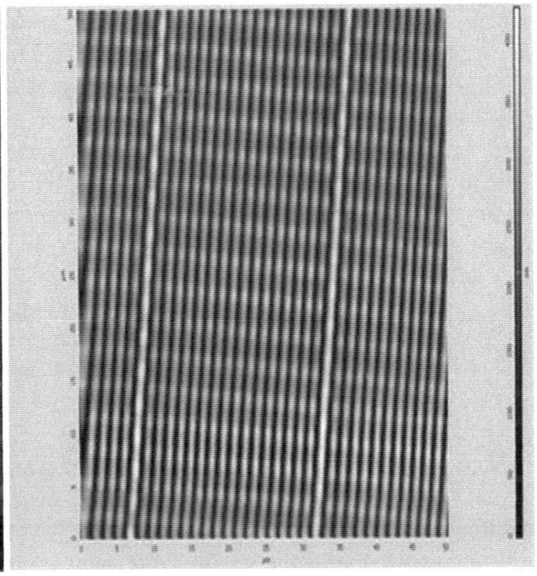

Figure 2. Circuit photo Figure 3. Crystal photo

Thus, optical radiation modulated by information about the parameters of the electromagnetic field of the object induces a current in the conductor, which enters a microcircuit (Figure 2.) containing a silicon crystal with silver or aluminum filaments deposited on it. (Figure 3.) Measurements and photographs were carried out with an atomic force microscope. The substrate made with the silicon-on-insulator technology is a three-layer package, which consists of a monolithic silicon wafer, a dielectric and a thin surface layer of silicon placed on it.

To switch to contactless recording, you can use the direct incidence of optical radiation on the surface of the crystal, similar to how it is done in photoelectronic converters. For the study, it was necessary to investigate the processes occurring when electromagnetic radiation falls on the surface of the crystal. The oscillation frequency of the surface plasmon is determined by the values of the dielectric constant of the particle and matrix. In the case of metallic inclusions, the real part of the dielectric constant is negative, which can lead to resonant absorption. With total internal reflection, an electromagnetic wave propagates along the surface reflecting light, the speed of which depends on the angle of incidence. If the velocity of this wave at a certain angle of incidence coincides with the velocity of the surface plasmon on the metal surface, the conditions for total internal reflection will be violated, and the reflection will cease to be total, and surface plasmon resonance will arise [2].

3. The results of measurements of the topography of the chip crystal.

The crystal was removed from the microcircuit case, washed from extraneous contamination in a neutral environment. After these operations, the topography of the surface of two rectangular areas was measured.

The carried out measurements revealed the identity of the topography of both sites. In this regard, the results shown in the figures apply equally to both areas. Figure 3 shows a view of the surface topography measured over an area of 50×50 μm^2.A linear periodic structure is visible. The measured period (see Figure 3.) is 1.46 μm. It is noteworthy that every 17th period is marked with a line with a higher relief. In addition, you will notice that the lines have periodic thickenings. The height of the linear relief determined from the histogram was 220 nm. The height of the prominent lines at more 50 nm. Obviously, it is on these protruding lines that the phenomenon of plasmon resonance can occur.

4. Model of radiation spread over the crystal surface.

To solve the problem of the fundamental possibility of realizing contactless information recording, a model was developed and carried out in the COMSOL 5.5 software package. Figure 4 shows a diagram of the incidence of laser radiation on a plate with a nanowire on a dielectric material with a refractive index n_β. The lattice constant or the distance between the wires is d. A plane electromagnetic wave falls on a wire grating on a dielectric substrate and propagates in a medium with a refractive index nα, falls on the grating at an angle α in a plane perpendicular to the grating. In this case, $n_\alpha = 1$ for air and n_β for a dielectric substrate. The angles of incidence, grating constants d, and the wavelength of the incident radiation λ_0 are arbitrary, and it is these parameters that were used for optimization. For the wire, the complex dielectric constant was chosen equal to the dielectric constant of the silver or aluminum material. Both transverse electric (TE) and transverse magnetic (TM) fields were calculated. For the TE wave, the component of the electric field in the z direction, outside the XY plane, was calculated. In addition to the parameters for estimating the realization of plasmon resonance, the model calculates the transmission and reflection coefficients for refraction, specular reflection, and first-order diffraction. The carried out measurements revealed the identity of the topography of both sites. In this regard, the results shown in the figures apply equally to both areas. Figure 3. shows a view of the surface topography measured over an area of 50 × 50 μm^2.A linear periodic structure is visible. The measured period (see Figure 3.) is 1.46 μm. It is noteworthy that every 17th period is marked with a line with a higher relief. In addition, you will notice that the lines have periodic thickenings. The height of the linear relief determined from the histogram was 220 nm. The height of the prominent lines at more 50 nm. Obviously, it is on these protruding lines that the phenomenon of plasmon resonance can occur.

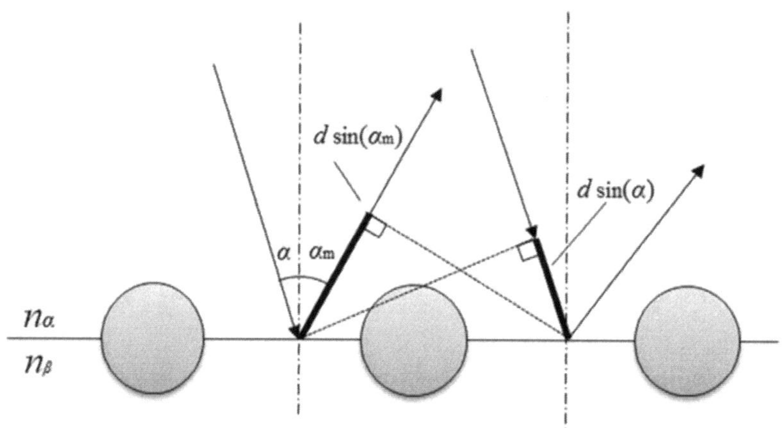

Figure 4. *Diagram of radiation incidence on a crystal.*

The model of the electric field in the Drude - Lorentz approximation for the field dispersion was chosen as the basis for calculating the parameters of the scattered and incident radiation, under the assumption that all processes of formation of plasmon resonance occur in the skin layer, where the coefficients of interdiffusion depend on the degree of ionization of atoms (1-3):

$$\varepsilon_r = (n - ik)^2 \tag{1}$$

$$\varepsilon_r = \varepsilon_\infty + \sum_{J=1}^{M} \frac{f_i \omega_P^2}{\omega_0^2 j - \omega^2 + i\Gamma_{j\omega}} \tag{2}$$

$$B = \mu_0 \mu_r \mathrm{H} \tag{3}$$

where n is the real refractive index, k is the imaginary part of the refractive index, and is the complex refractive index (4):

Journal of Physics: Conference Series **2249** (2022) 012007 doi:10.1088/1742-6596/2249/1/012007

The equation for the incident two-dimensional Gaussian beam (4) (paraxial approximation) has the

$$E_G(x,y) = E_0\sqrt{\frac{\omega_0}{\omega(y)}}\cdot e^{-(x/\omega(y))^2}\cdot \exp\left(-i\left(ky-\eta(y)+\frac{kx^2}{2R(y)}\right)\right) \tag{4}$$

and is approximated by the expansion of a plane w [6]

$$E_{pw} = \sum_{j=-Mk=0}^{M}\cdot\sum_{j=-Mk=0}^{1}\alpha_{jk}\hat{u}_k(k_j)\cdot\exp\left(-i(k_j\cdot r)\right) \tag{5}$$

Where is each wave vector k_j points in different directions for each index value j, α_{jk} - amplitude, which has a different value for each wave vector, as well as for each of the two possible directions of polarization per wave vector, $\hat{u}_k(k_j)$. The dispersion for metal particles, for the relative permittivity, can be written as (6):

$$\varepsilon(\omega) = 1 - \frac{\omega_p^2}{\omega^2} \tag{6}$$

where the angular frequency of the plasma is determined by ω_p - the plasma frequency.

$$\omega = \frac{2\pi c}{\lambda} \tag{7}$$

The equation for the absorption coefficient has the form:

$$p = D^3\frac{\varepsilon-\varepsilon_h}{\varepsilon+2\varepsilon_h}E_{loc} \tag{8}$$

The formation of plasmon resonance in the wavelength range of interest occurs in the skin layer, the depth of which can be calculated as follows:

$$\delta = \frac{1}{\mathrm{Re}\sqrt{-k_0^2\varepsilon_r}}\varepsilon \tag{9}$$

where k_0 is the wave number in free space, and r is the complex relative permittivity of the skin layer.

ε is the effective dielectric constant of a two-component medium containing spherical nanoparticles, is determined by the equation [3]:

$$\varepsilon = \varepsilon_h\left[1-\frac{3f(\varepsilon_h-\varepsilon_p)}{2\varepsilon_h+\varepsilon_p+(\varepsilon_h-\varepsilon_p)}\right] \tag{10}$$

where $(\varepsilon_{h,}\varepsilon_p)$ -the complex dielectric constants of the nanomaterial and the environment, f is a relative concentration of nano objects.

To spark the SPR which drops on the electromagnetic wave, it is necessary that the laws of conservation of energy and the projection of the quasimomentum of the bulk and surface plasmon waves are observed: Where ω-circular frequency of light; ω_{sp} -frequencies of plasmon-polaritons; $\vec{k}_c$ - wave vector of light; k_0 - wave vector of plasmon-polaritrons.

$$k_{sp} = k_x \pm iG_x \pm jG_y \tag{11}$$

$G_{x,y}$ - the wave vectors of the reciprocal lattice of the structure surface;

i, j - whole numbers, k_{sp} - the wave vector of the surface plasmon;

k_x - Is the wave vector of the exciting photon.

5. Analysis of the results

In this work, a series of calculations was carried out in order to solve the following questions: firstly, is it possible to use strips made of aluminum, and secondly, how is the field distributed around the strips in intensity and angle, depending on the angle of incidence and the distance between the threads, in order to reduce diffraction loss. The model calculates transmittances and reflections for refraction, specular, reflection and first-order diffraction Below are the calculations in the COMSOL 5.5 software package.

Of all metals, silver has the highest SPR intensity. Silver has the highest extinction coefficient at the maximum of the SPR, not only among metals, but also among all other known materials that absorb light in the visible and near infrared region. However, in devices based on the excitation of resonance using an electric current, crystals with aluminum filaments gave good characteristics, which greatly simplified the technological process of the device manufacturing. To test the possibility to use the aluminum, calculations of the field distribution around nanowires were made when electromagnetic radiation is incident on them in the visible and near- infrared spectral regions.

Earlier, we calculated the distribution of the x-component of the electric field of Gaussian beam in aluminum nanoclusters of the electric field for silver and aluminum nanowires under the conditions of plasmon resonance [6]. Calculations have shown that a stable field arises around the silver bands in a narrow space, which is often called the SPR band.

The appearance of the SPR is the result of the interaction of light incident on the surface of nano objects with the conduction electrons of the metal. According to the theory of electron gas, the electrons of the inner shells are localized near the atomic nucleus, and the electrons of the outer valence shells can freely move inside the metal nano-wire, in particular, the high electrical conductivity of metals. Under the action of an alternating electric field of a light beam, mobile conduction electrons are displaced. If the particle size is much less than the wavelength of the incident light, then the movement of electrons leads to the appearance of a dipole that vibrates with the frequency of the exciting electric field. It is known that the areas of positive and negative charge formed near the surface of a nano-wire as a result of displacement of conduction electrons cause polarization of the environment in which the metal particle is located. This polarization leads to a decrease in the magnitude and frequency of oscillations of the induced dipole and, as a consequence, to a shift of the SPR to longer wavelengths [3]. These effects are more pronounced, the greater the ability of the environment to polarize, the greater the dielectric constant of the environment. Therefore, the absence of a pronounced peak is most likely determined by the lower dielectric constant of aluminum compared to silver. In addition, some features are observed in the interaction of light with highly elongated metal nano-rods ($a \gg b$) - nano-wires, the length of which is comparable to or exceeding the wavelength of the incident light.

In this case, the values of the electric field strength of the incident light at the ends of the nanowire differ significantly; therefore, the dipole induced at one of the ends causes polarization of the adjacent sections of the nanowire, and instead of an oscillating dipole localized in space, a traveling wave appears on the metal surface, which propagates from one end of the nano-wire to the other [4].

If the light falls on only one end of the nano-wire when irradiated with a laser, then the traveling wave, reaching the other end, induces the formation of an oscillating dipole, which emits light of the same wavelength as the incident one [4]. Thus, silver nano-wires can serve as effective waveguides for a narrow band of resonant frequencies, while aluminum ones will lead to "spreading" of the field over the surface of the substrate, and, as a consequence, a loss in the accuracy of determination of the device. For particles of a non-spherical shape or particles in a non-uniform environment, for example, fixed on a substrate, different orientations of the particle with respect to the incident light wave are unequal.

In addition to what was said above, we are interested in the reflection coefficients depending on the angle of incidence on the plate surface

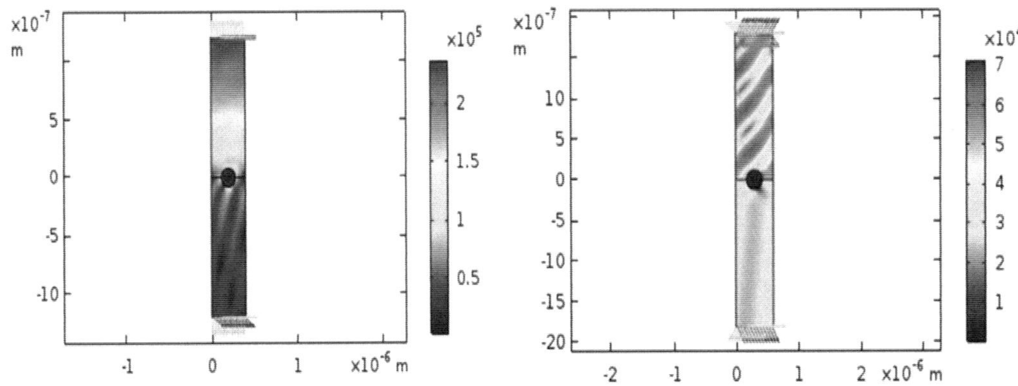

Figure 5. *Distribution of the electric field around the thread for a - silver, b - for aluminum.*

We estimated the distribution of the field around the wire, the intensity of the reflection coefficient from the angle of incidence on the wire as a function of the wavelength, and the value of diffraction loss as a function of the distance between the wires. Calculations have shown that for the selected wavelength range at a distance between nanowires greater than 1.2 μm, the diffraction losses can be neglected. Figure 5. a, b shows the field distribution around the filament a - for silver, b - for aluminum. It can be seen that for the aluminum, in addition to the fact that a part of the field energy goes into the substrate, the field around the filament is not constant, which can lead to an increase in the error margins in signal diagnostics.

Figure 6 shows the distribution of the reflection coefficient from the angle of incidence on the crystal for wavelengths of 0.441 μm, 0.553 μm, 0.670 μm, 0.830 μm and 1.200 μm. It can be seen that the reflection coefficient is a rapidly changing function of the angle of incidence and is maximum for wavelengths in the green region (0.441 μm); however, in the near-IR region, it is more stable (0.830 μm) in the range from 20° to 40° . Thus, taking into account the divergence of laser radiation this wavelength is more promising for practical use.

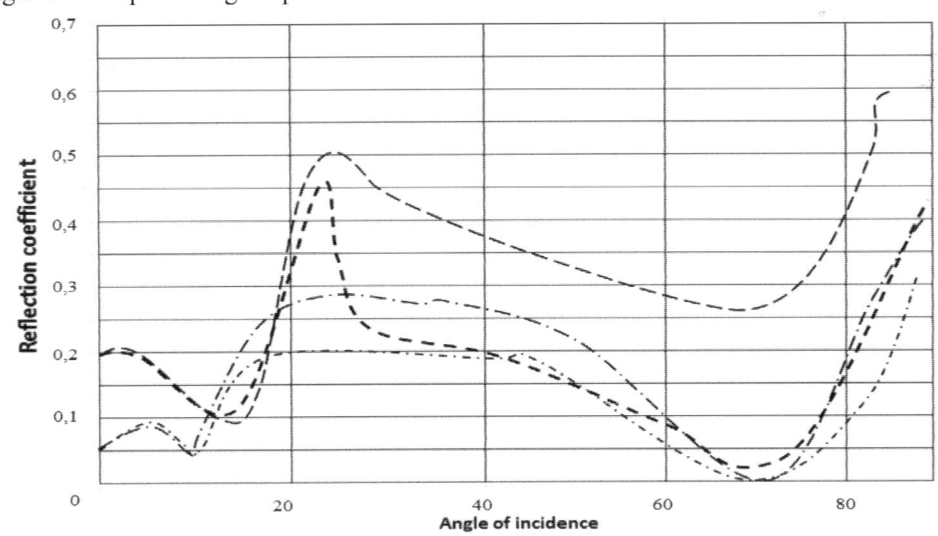

Figure 6

1) λ_0=0.441μm — — — — 2) λ_0=0.670μm - - - - - - -
3) λ_0=0.830 μm -·-·-·-·- 4) λ_0=1.200 μm —·—·—·

The calculated reflection coefficients have a maximum value in the range of angles 180°-23°. Experiment The maximum reflectance for a series of crystals, without deposition and with deposition of nano silver, depending on the angle between the incident and reflected radiation, was measured on a setup, the diagram of which is shown in Figure 7. The angle at which Table 1 shows the ratio of the intensity of the reflected radiation from a cat plate without deposited silver and a plate with deposited silver for three types of lasers λ_0 = 0.441, 0.350, 0.830 μm

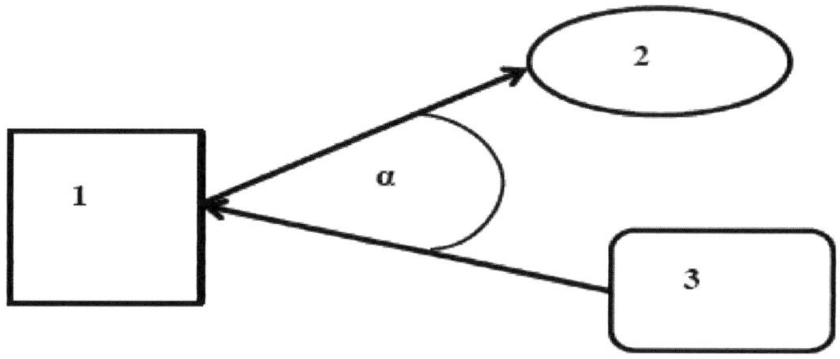

Figure 7. Installation diagram for measuring the reflection coefficient:
1- plate, 2- Laser power meter, 3- laser with turntable.

Figure 7. Installation diagram for measuring the reflection coefficient: 1- plate: 2- Laser power meter (Newport 840); 3- laser with a rotary device. The laser radiation (1) was directed onto the plate at different angles (a). The reflected power was measured at these angles. In the course of the experiment, the background power, the power of the laser radiation at the place where the plate was fixed, and the reflected power at different distances of the laser to the plate and from the plate to the laser power meter were measured. (2) In this case, the value of the angle between the incident and reflected radiation was also varied. The maximum of the reflected radiation was observed at approximately α = 17°. This value depended on the type of laser and varied in the range 16°-18° and was mainly determined by the divergence of the laser radiation source.

Our previous estimates [6] that the optimal wavelength for obtaining plasmon resonance is 0.350 μm. The largest change in intensity was observed for a 0.830 μm laser in this experiment. This is most likely due to the fact that the blue laser has a larger divergence angle than the green and IR. The power of the reflected radiation at a distance of 11 cm from the nanosilver-coated plate increased 11-14 times compared to the power of the uncoated plate at an angle of incidence α = 17^0 and this cannot be explained by the properties of the reflecting surface. In this case, the laser power at a distance of a distance of 11 cm at an angle of incidence α = 17° -25 dbm, and when falling on the plate, -32 dbm. The reflected power from the plate is 21.3 dbm almost 11-12 times more than that incident on the plate and it is impossible to explain the increase in power by the properties of the reflecting surface.

6. Conclusion

The crystal was removed from the microcircuit case, washed from extraneous contamination in a neutral environment. After these operations, the topography of the surface of two rectangular areas was measured.

The power of the reflected radiation from the anosilver-coated plate increased 11-14 times compared to the power of the uncoated plate at an angle of incidence α = 17° and this cannot be explained by the properties of the reflecting surface.

The analysis has shown that for the implementation of contactless information recording in the Intest complex, it is possible to use a nano-coating of silver. In this case, the distance between them should not be less than 1.2 μm. For information recording, lasers with a wavelength of 0.830μm, 0.553 μm are promising, lasers with a wavelength of 0.830 μm are the best option so far for recording.

Journal of Physics: Conference Series **2249** (2022) 012007 doi:10.1088/1742-6596/2249/1/012007

Reference:

[1] T. Mogilnaya, A. Botikov, E. Sagitova Proceeding of Spie 31 .07 -4.07, USA Volume 5908-41 (2005)

[2] T. Yu. Mogilnaya, A. M. Vasiliev, D. Yu. Kukushkin, A. G. Botikov/ Using Coherent Spectroscopy to Determine the Presence of Nanoparticles in Colloidal Solutions. // Bulletin of the Russian Academy of Sciences: Physics, , Vol. 83, No 10, pp 1302–1306 (2019)

[3] I . Demichev, A. Sidorov /Numerical simulation of the optical properties of metallic nanoparticles *// Textbook. manual for the implementation of a laboratory workshop. - St. Petersburg: ITMO University, 2016 52 p.*

[4] L.N. Grigoriev, G.A. Knyazev, A.L. Tolstik / Propagation of surface plasmon–polaritons at the boundary of a metal–dielectric with thermal optical nonlinearity// Bulletin of the Russian Academy of Sciences: Physics V. 79, no. 12, p. 1622-1626 (2015)

[5] T. Yu. Mogilnaya A. M. Vasiliev , D Yu Kukushkin Journal of Physics: Conference Series. 1396. 012024. 10.1088/1742-6596/1396/1/012024

[6] T. Y. Mogilnaya, B. L. Krit, V. V. Sleptsov, A. M. Vasiliev, A. O. Diteleva, M . V. Fedotikova, L. L. Pagava, M. Y. Gorozheev, N. V. Morozova, V. V. Kuvshinov / The effect of nonmetallic impurities on the occurring of the surface plasmon resonance at the deposition of nanocluster coatings onto the surface of photo-electric converters // Optics Communications, 2021 Volume 494, article id. 127065

Parametric gap solitons in PT-symmetric optical structures

M V Komissarova[1], T M Lysak[1], I G Zakharova[1], A A Kalinovich[1]

[1]Lomonosov Moscow State University, Moscow 119992, Russia

E-mail komissarova@physics.msu.ru

Abstract. It is well known that quadratic nonlinearity and feedback through Bragg periodicity are the basis for parametric gap solitons. The major part of the relevant investigations refers to passive systems. At the same time, optical systems supplemented with active elements can demonstrate unusual properties. Asymmetry intrinsic to structures with parity-time (PT) symmetry is a bright confirmation of this statement. The interplay of nonlinearity, Bragg reflection and gain/loss profile can lead to the complicated pattern of wave interactions and novel results. In this study we address the properties of two-color solitons in complex PT symmetric periodic structures with quadratic nonlinearity. We focus on the case of single Bragg resonance. We reveal the region of parameters where stable parametric solitons may exist. We demonstrate that characteristics of forming solitons depend on the order of alteration of amplifying and absorbing layers.

1. Introduction

Gap (Bragg) solitons have been studied for a long time. They are known to be a result of the balance between dispersion in a photonic bandgap and medium nonlinearity. Since gap solitons are generated in structures, where counterpropagating waves interact, they suggested practical applications, for example, optical delay lines and memories. One of the most intriguing feature of gap solitons is a possibility of their slow propagation [1,2] which allows us to consider them as optical bits trapped within the nonlinear Bragg grating.

Slowly propagating gap solitons were investigated in parametric media where two forward waves at fundamental and doubled frequencies and two backward waves at both frequencies interact between each other through quadratic nonlinearity [3-7]. Scenarios of the process were described mathematically in [7], and the stability and excitation of two-color gap solitons were discussed. At that, a relationship was demonstrated between the features of grating response and the existence of stationary parametric gap solitons when these solitons can be described analytically.

Nowadays light propagation in active artificial periodical media is intensively investigated. By active media we mean such structures where conservatism is violated. They can be fabricated by adding active elements (for example, rare-earth metal ions) to a medium. Referred to those media, parity-time (PT) structures are discussed. In these structures the alternation of the layers of amplifying and absorbing materials results in an additional gain (absorption) profile and asymmetric properties. Using PT - symmetric periodic media allows us to enhance the control, amplification and generation of optical radiation.

Processes in optical active periodic media are in the focus of numerous studies [8-11]. Speaking about nonlinear effects, they are usually investigated at Kerr nonlinearity [12,13]. Optical bistability [14,15] and solitons [16] are discussed.

In the present paper we study the formation and propagation of Bragg solitons in quadratically nonlinear PT-symmetric structures. Media with quadratic nonlinearity are known to possess a lower intensity threshold of nonlinear effect manifestation in comparison with Kerr nonlinear media. At the same time, when we theoretically investigate intensive light interaction with a quadratic nonlinear layered structure, it is natural to use a model of coupled counterpropagating waves at the fundamental and doubled frequencies [4,5]. More complicated model corresponds to the interplay of Bragg and parametric coupling in an active medium [17]. We numerically study diffracting light beams in such medium on the basis of a model generalizing those presented in [4,5,17]. To this aim we have developed an original finite difference scheme keeping motion integrals of the equation system. The purpose of our simulation is to find out a range of beam and medium parameters where forming two-color gap solitons stably propagate.

2. Problem statement

2.1. Basic equations

Using the method of coupled waves one can derive from the wave equation the following dimensionless system describing the evolution of forward $E^{+}_{1,2}$ and backward $E^{-}_{1,2}$ waves of both harmonics.

$$\begin{gathered}
i\left(\frac{\partial E_{1+}}{\partial z}+\frac{\partial E_{1+}}{\partial \tau}\right)+D_{x,1}\frac{\partial^2 E_{1+}}{\partial x^2}+\delta E_{1+}+(\kappa+g)E_{1-}+E_{1+}^{*}E_{2+}=0,\\
i\left(-\frac{\partial E_{1-}}{\partial z}+\frac{\partial E_{1-}}{\partial \tau}\right)+D_{x,1}\frac{\partial^2 E_{1-}}{\partial x^2}+\delta E_{1-}+(\kappa-g)E_{1+}+E_{1-}^{*}E_{2-}=0,\\
i\left(\frac{\partial E_{2+}}{\partial z}+v\frac{\partial E_{2+}}{\partial \tau}\right)+D_{x,2}\frac{\partial^2 E_{2+}}{\partial x^2}+\delta_2 E_{2+}+(\kappa_2+g_2)E_{2-}+\sigma(E_{1+})^2=0,\\
i\left(-\frac{\partial E_{2-}}{\partial z}+v\frac{\partial E_{2-}}{\partial \tau}\right)+D_{x,2}\frac{\partial^2 E_{2-}}{\partial x^2}+\delta_2 E_{2-}+(\kappa_2-g_2)E_{2+}+\sigma(E_{1-})^2=0,
\end{gathered} \tag{1}$$

A similar system appeared in [4,5] in a passive medium. At that, diffraction was not taken into account. Recently in [17] a system relevant to an active medium was studied but SHG and diffraction were not considered. In (1) $\delta=(k_1-\alpha)\frac{1}{l_{nl}}$, $\delta_2=(k_2-2\alpha)\frac{1}{l_{nl}}$ are detunings from Bragg resonance at the fundamental and doubled frequencies, $\Delta k=k_2-2k_1$ is the phase mismatch, $k_{1,2}$ are the wave numbers of both harmonics, $\alpha=\frac{\pi}{d}$, d is the period of Bragg structure. z is the coordinate of radiation propagation normalized to the nonlinear length $l_{nl}=\left(\gamma_1\sqrt{I_1}\right)^{-1}(z\to z l_{nl})$, the direction transversal to propagation x is measured in units of the beam width a_x $(x\to x/a_x)$, $\tau=\frac{t}{\eta}l_{nl}$ is the dimensionless time (t is the real time). Here $\eta=\frac{\varepsilon_0\omega_0}{c^2k_1}=\frac{\sqrt{\varepsilon_0}}{c}$, at that , $k_1{}^2=\frac{\varepsilon_0}{c^2}\omega_0{}^2$, ω_0 is the Bragg frequency, ε_0 is a mean value of permittivity in the structure. Coefficients $\kappa=\frac{\omega_0{}^2}{c^2k_1}\frac{\Delta\varepsilon_R}{4}\frac{1}{l_{nl}}=\frac{\omega_0}{c\sqrt{\varepsilon_0(\omega_0)}}\frac{\Delta\varepsilon_R}{4}\frac{1}{l_{nl}}$ and $\kappa_2=\frac{4\omega_0{}^2}{c^2k_2}\frac{\Delta\varepsilon_{R2}}{4}\frac{1}{l_{nl}}=\frac{2\omega_0}{c\sqrt{\varepsilon_0(2\omega_0)}}\frac{\Delta\varepsilon_{R2}}{4}\frac{1}{l_{nl}}$ are the parameters of Bragg coupling, $g=\frac{\omega_0{}^2}{c^2k_1}\frac{\Delta\varepsilon_I}{4}\frac{1}{l_{nl}}=\frac{\omega_0}{c\sqrt{\varepsilon_0}}\frac{\Delta\varepsilon_I}{4}\frac{1}{l_{nl}}$, $g_2=\frac{4\omega_0{}^2}{c^2k_2}\frac{\Delta\varepsilon_{I2}}{4}\frac{1}{l_{nl}}=\frac{2\omega_0}{c\sqrt{\varepsilon_0(2\omega_0)}}\frac{\Delta\varepsilon_{I2}}{4}\frac{1}{l_{nl}}$ are the parameters describing the asymmetric coupling of counter-propagating waves. We use the following expression for permittivity [17]

$\varepsilon(z) = \varepsilon_0 + \Delta\varepsilon_R \cos\left(\frac{2\pi}{d}z\right) + i\Delta\varepsilon_I \sin\left(\frac{2\pi}{d}z\right) + \Delta\varepsilon_{R2}\cos\left(\frac{4\pi}{d}z\right) + i\Delta\varepsilon_{I2}\sin\left(\frac{4\pi}{d}z\right).$

$\Delta\varepsilon_R, \Delta\varepsilon_{R_2}, \Delta\varepsilon_I, \Delta\varepsilon_{I_2}$ are the small amplitudes of the changes in real and imaginary parts of permittivity. Coefficients $\gamma_1 = \frac{4\pi}{c^2 k_1}\omega_0{}^2\chi^{(2)}(-\omega_0, 2\omega_0)$, $\gamma_2 = \frac{8\pi}{c^2 k_2}\omega_0{}^2\chi^{(2)}(\omega_0, \omega_0)$ describe quadratic nonlinearity, $\chi^{(2)}$ is the medium susceptibility, $\sigma = \gamma_2/\gamma_1$. $D_{x,1} = (2k_1 l_{nl} a_x^2)^{-1}$, $D_{x,2} = (2k_2 l_{nl} a_x^2)^{-1}$ refer to diffraction at each of harmonics.

2.2. Initial and boundary conditions

Assuming that the periodic medium is surrounded by a homogeneous linear one with permittivity ε_0 we neglect diffraction spreading in this domain (figure 1). Then the propagation of laser radiation in the linear medium (for $0 < x < L_x$ and $0 < z < L_{left}$, $L_{right} < z < L_z$) can be described by equations (1), where only the first two terms with parenthesis are left.

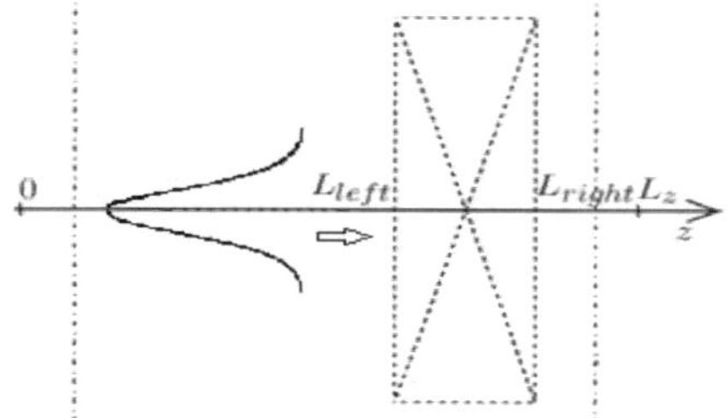

Figure 1. Scheme of laser radiation propagation in a periodic medium at SHG.

Boundary conditions are as follows.

$$E_{1+}(z=0,x,\tau) = E_{2+}(z=0,x,\tau) = E_{1-}(z=L_z,x,\tau) = E_{1-}(z=L_z,x,\tau) = 0,$$
$$E_{1\pm}(z,x=0,\tau) = E_{2\pm}(z,x=0,\tau) = E_{1\pm}(z,x=L_x,\tau) = E_{2\pm}(z,x=L_x,\tau) = 0. \quad (2)$$

At $\tau = 0$ the initial beam is launched.

$$E_{1+}(z,x,\tau=0) = E_{10}(z,x), E_{2+}(z,x,\tau=0) = E_{20}(z,x),$$
$$E_{10}(z=0) = E_{10}(z=L_{left}) = 0. \quad (3)$$

We consider the initial beam at the fundamental frequency in the form of a beam with a soliton profile in the absence of radiation at the frequency of the second harmonic.

$$E_{10}(z,x) = A_{10} ch^{-1}\left(\left(z - \frac{L_{left}}{2}\right)/h_z\right) ch^{-1}\left(\left(x - \frac{L_x}{2}\right)/h_x\right), E_{20}(z,x) = 0. \quad (4)$$

In (4) h_z, h_x characterize the longitude and transverse widths of the beam.

We have performed numerical modelling on the basis of the equations (1) supplied with the initial and boundary conditions (2) – (4). To this end we use an original nonlinear finite difference scheme realized by means of iterative process.

3. Results of numerical modelling and discussion

For numerical simulation we used the following parameter set:

$$\kappa = 1, \kappa_2 = g_2 = 0, \delta = -0.9, \delta_2 = 5, \sigma = 1/2, A_{10} = \sqrt{10}, h_z = 5. \quad (5)$$

$$h_x = 1, D_{x,1} = D_{x,2} = 0.1, g = 0,0.5,1,-1, L_{left} = 60, L_{right} = 90, L_z = 120, L_x = 10. \quad (6)$$

Parameters (5), (6) describe an active diffracting medium in a vicinity of the lower bandgap edge surrounded by a linear medium (Figure 1). In paper [4], the formation of Bragg solitons at the fundamental and double frequencies in semi-infinite passive ($g = 0$) periodic medium was demonstrated for parameters (5) without taking into account diffraction along the x-coordinate. Below we discuss the influence of diffraction on Bragg soliton formation in a passive medium (Subsection 3.1). In Subsection 3.2 – we discuss peculiarities of Bragg soliton formation in an active diffracting media for various values of asymmetric coupling of counter-propagating waves g. Initial distribution (4) for numerical simulation is shown in Figure 2.

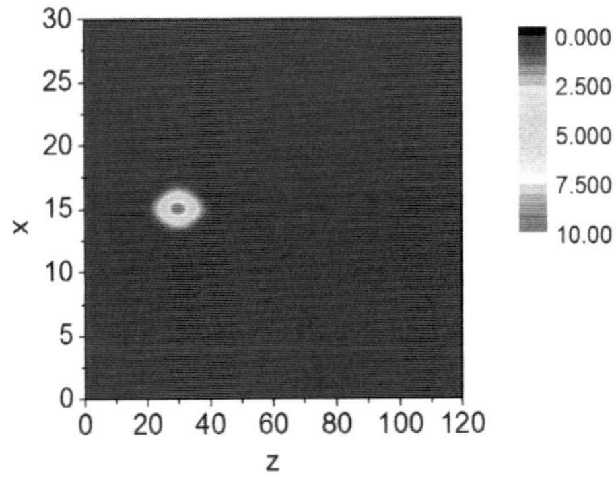

Figure 2. Incident beam at the fundamental frequency.

3.1. Bragg solitons in a passive diffractive medium

We started our numerical modelling from the results reported in [4] where the formation of Bragg solitons at the fundamental and doubled frequencies was demonstrated in a passive 1D semi-infinite periodic medium with quadratic nonlinearity. At that, the solitons were observed for forward waves and for backward ones as well.

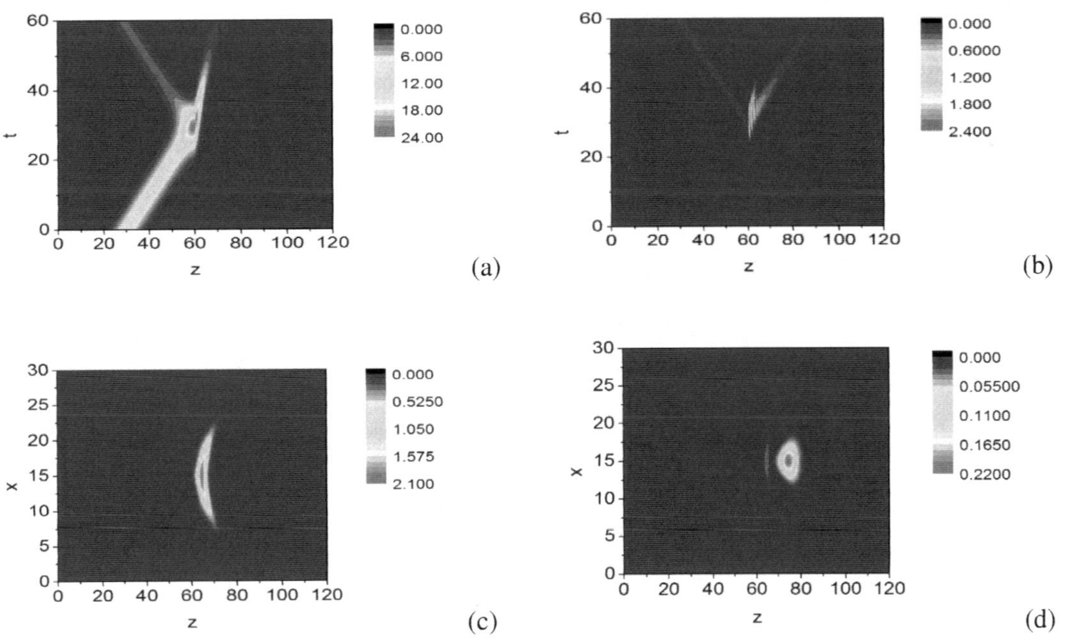

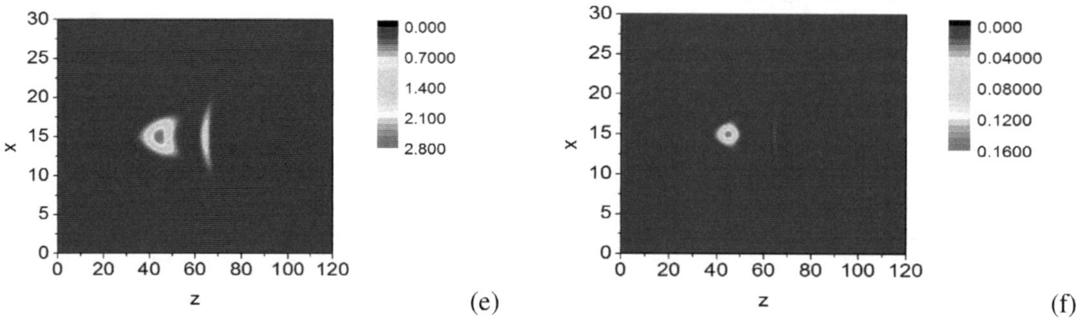

Figure 3. Bragg soliton forming in the passive medium ($g = 0$) at the cross-section $x = 15$. (a) Fundamental and (b) second harmonics. Beam profiles of the forward/backward waves at the (c, e) fundamental and (d, f) doubled frequencies at $\tau = 45$. Parameters are as in (5) – (6).

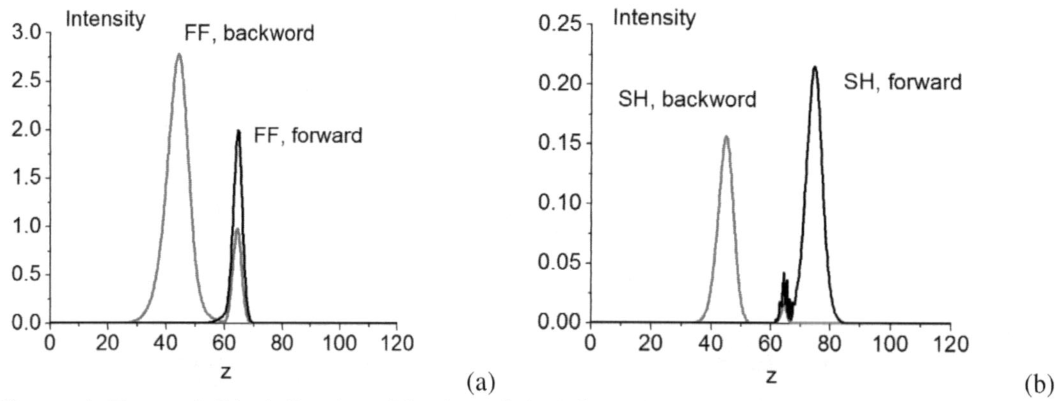

Figure 4. Forward (black lines) and backward (red lines) beam profiles in the passive medium ($g = 0$) at the fundamental frequency (a) and doubled frequency (b) at the beam center ($x = 15$) at the time moment τ = 45. Parameters are as in (5) – (6).

Figures 3,4 show Bragg solitons of the finite beam width propagating in the passive medium ($g = 0$) for the parameter set (5), (6). The contours of the total intensity at the fundamental and double frequencies shown in Figures 3a and 3b for the beam center (cross-section $x = 15$) are similar to the results reported in [4]: the incident beam is partially reflected from the periodic structure face. The rest part of the incident beam propagates through the structure generating a forward wave and a low intensity backward wave at the double frequency.

The effect of the finite beam width on the formation and propagation of Bragg solitons is shown in Figures 3c-f for the time moment τ = 45. We observe the transformation of the original soliton beam to a crescent profile. This profile develops in the forward wave at the fundamental frequency and reflects in the form of the other wave beams: the backward wave at the fundamental frequency and both waves at the doubled frequency.

3.2. Bragg solitons in active diffractive media

We proceed our series of simulation by taking into account asymmetric coupling of counter-propagating waves. Active components presenting in the periodic structure result in a significant change of the pattern of forming solitons. In Figure 5, the influence of active medium on the soliton propagation is demonstrated when the value of the parameter g is gradually increased from -1 to 1. For $g = -1$, the reflection at both faces of the structure is prominent for the wave at the fundamental frequency. The backward wave at the doubled frequency also carries a significant part of energy.

Zero (Figure 5 a,b) and positive (Figure 5 c,d) values of the parameter g result in an essential decrease of the amount of reflected energy at both frequencies. For $g = 1$ (Figure 5c), backward waves are not formed. Indeed, in this case $\kappa - g = \kappa_2 - g_2$=0, therefore, there are no generating terms in the second and fourth equations of (1).

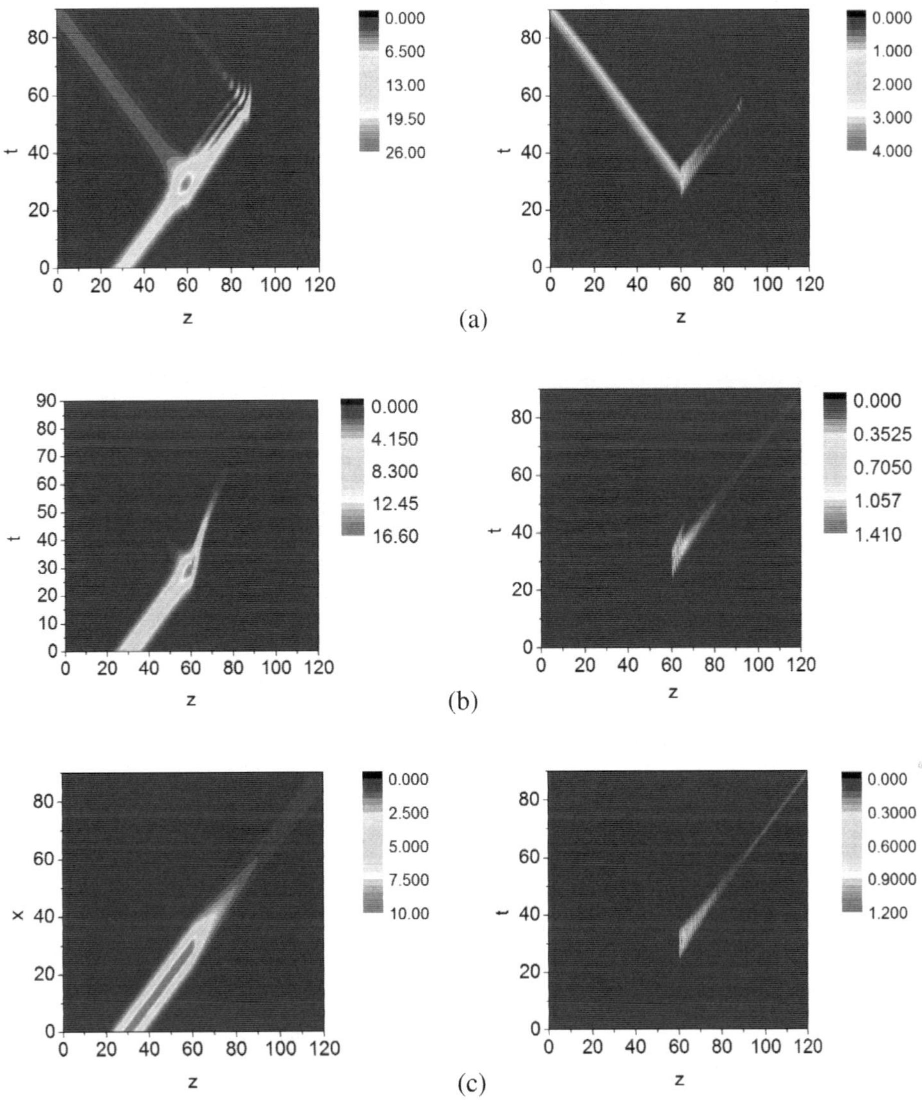

Figure 5. Bragg solitons at the fundamental frequency (left column) and doubled frequency (right column) at the beam center (cross-section $x = 15$) for $g = -1$ (a), 0.5 (b), 1 (c)).

As it is well seen in Figure 5, the scenario of propagation crucially depends on the order of alternation of the absorbing and gaining layers. In the case $g = -1$, a backward fundamental frequency wave consists of several sub-beams (Figure 6c) due to the reflection of the fundamental frequency wave from both faces of periodic structure. Slight reflection is also observed for the doubled frequency wave (Figure 6d), while forward waves propagate in a soliton-like mode (Figures 6a,b).

Journal of Physics: Conference Series **2249** (2022) 012008 doi:10.1088/1742-6596/2249/1/012008

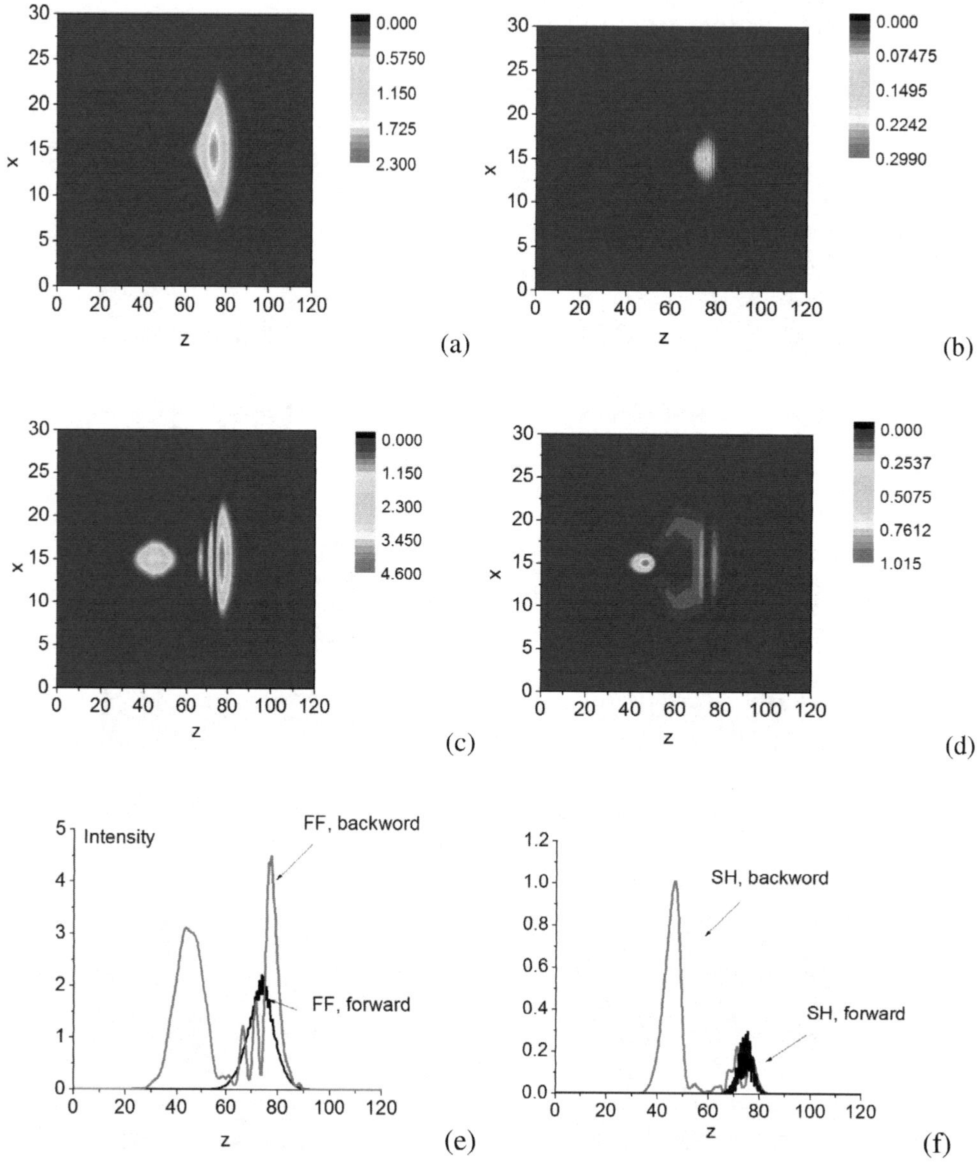

Figure 6. Bragg solitons in the active medium ($g = -1$): beam profiles of the forward/backward waves at the (a, c) fundamental and (b, d) doubled frequencies at $\tau = 45$. Forward (black lines) and backward (red lines) beam profiles at the fundamental frequency (e) and doubled frequency (f) at the beam center ($x = 15$) at the time moment $\tau = 45$ (d). Parameters are as in (5) – (6).

Two sub-beams for both waves at the doubled frequency and backward wave the fundamental frequency are well formed at the time moment $\tau = 45$ in the active medium with $g = 0.5$ (Figure 7). Lower asymmetric coupling results in less reflection from the structure faces, so only two sub-beams are formed for the forward fundamental wave. The crescent profile of the fundamental wave due to the diffraction reflects the form of other waves' sub-beams.

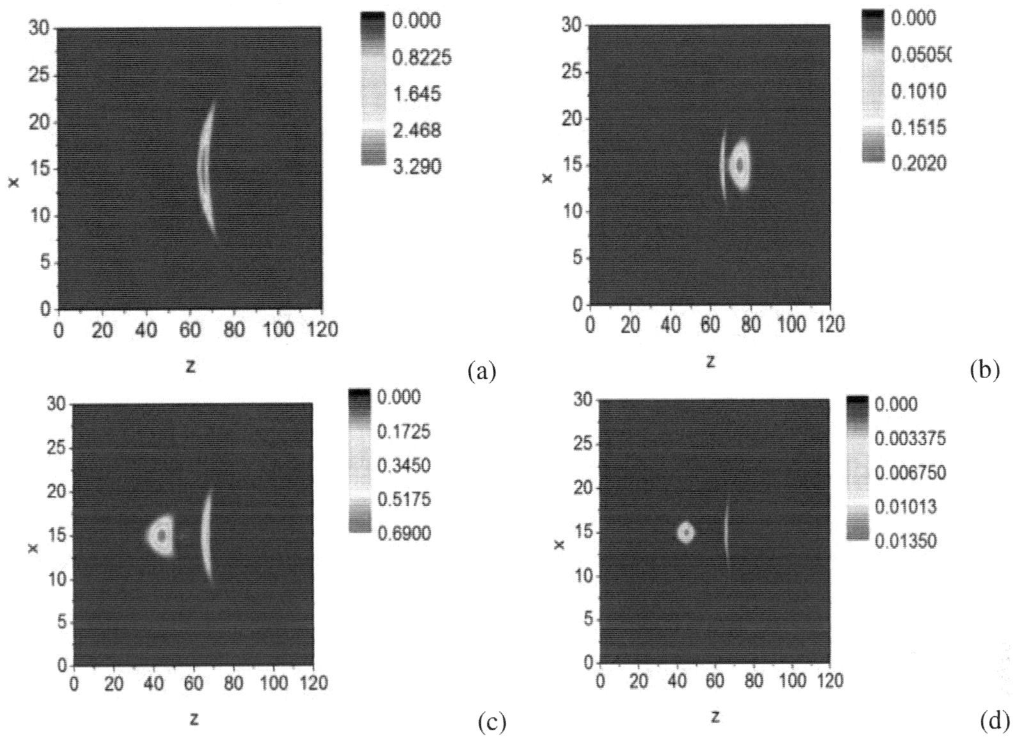

Figure 7. Bragg solitons in the active medium (g = 0.5): beam profiles of the forward/backward waves at the (a, c) fundamental and (b, d) doubled frequencies at $\tau = 45$. Parameters are as in (5) – (6).

As it is mentioned above, there is no reflection from the periodic structure faces in the case g = 1, so no backward waves are formed. Both fundamental and second harmonic waves propagate in a soliton-like mode (Figure 8).

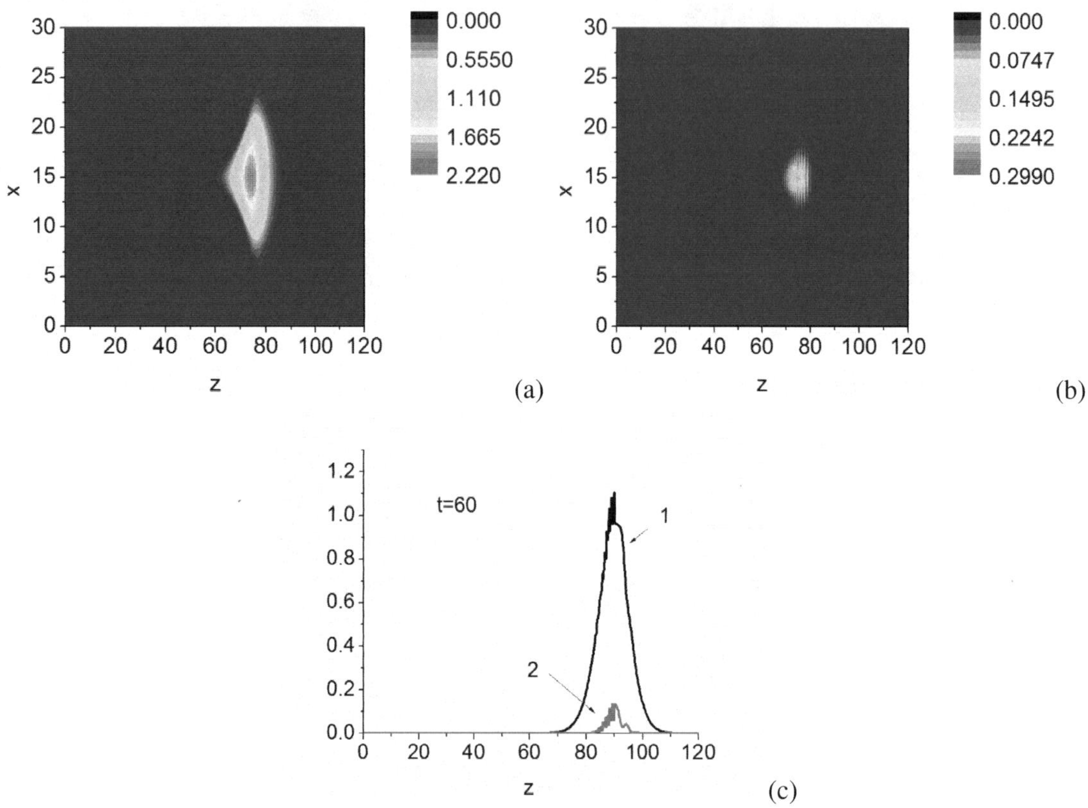

Figure 8. Bragg solitons in the active medium ($g = 1$): beam profiles of the forward waves at the (a) fundamental and (b) doubled frequencies at $\tau = 45$. Forward beam profile at the fundamental frequency (curve 1) and doubled frequency (curve 2) at the beam center ($x = 15$) at the time moment $\tau = 60$. Parameters are as in (5) – (6).

4. Conclusions

Studying the process of second harmonic generation in both passive and active layered media, we have demonstrated a possibility of the formation of Bragg solitons. We have shown the influence of the finite width of the beam in the transverse direction on the soliton capture.

We have proven the importance of the order of alternation of amplifying and absorbing layers in an active periodic structure when forming forward and backward wave solitons at the fundamental and doubled frequencies.

We have shown that changing the alternation of the amplifying and absorbing layers, it is possible to enhance radiation transmission through the periodic structure. At that, the enhancement can be achieved of radiation reflection from the boundaries of the periodic structure.

5. References

[1] de Sterke C M and Sipe J E 1994 Gap solitons in Progress in Optics (Vol XXXIII) ed E Wolf (Amsterdam, The Netherlands: Elsevier), chapter 3, pp 203
[2] Eggleton B J, Slusher R E, De Sterke C M, Krug P A and Sipe J E 1996 *Phys. Rev. Lett.* **76** 1627
[3] He H and Drummond P D 1997 *Phys. Rev. Lett.* **78** 4311
[4] Conti C, Assanto G, Trillo S 1997 Optics Letters **22 (17)** 1350
[5] Conti C, Assanto G, Trillo S 1997 Phys.Rev. Lett. **78** 2341
[6] Peschel T, Peschel U, Lederer F and Malomed B 1997 Phys. Rev. E **55** 4730
[7] Trillo S, Conti C, Assanto G and Buryak A 2000 Chaos **10** 590
[8] Zyablovsky A A, Vinogradov A P, Pukhov A A, Dorofeenko A V, Lisyansky A A 2014 *Physics – Uspekhi* **57 (11)** 1063 – 1082

[9] Suchkov S V, Sukhorukov A A, Huang J, Dmitriev S V., Lee C, Kivshar Y S 2016 *Laser Photonics Rev.* **10**. 1-37
[10] Konotop V V, Yang J, Zezyulin D A 2016 *Rev. Modern Phys.*. **88** **(3)** 035002
[11] El-Ganainy R, Makris K G, Khajavikhan M, Musslimani Z H, Rotter S and Christodoulides D N 2018 *Nature Phys.* **14** 11–19
[12] Li K, Zezyulin D A, Kevrekidis P G, Konotop V V, and Abdullaev F Kh 2013 *Phys. Rev. A* **88** 053820
[13] Ögren M, Abdullaev F Kh and Konotop V V 2017 *Optics Letters* **42** **(20)** 4079-4082
[14] Liu J, Xie X T, Shan C J., Liu T K, Lee R K. and Wu Y 2015 *Laser Phys.* **25** 015102
[15] Phang S, Vukovic A, Susanto H, Benson T M and Sewell P 2014 *Optics Lett.* **39** **(9)** 2603
[16] Miri M A, Aceves A B, Kottos T, Kovanis V and Christodoulides D N 2012 *Phys. Rev. A.* **86** **(3)** 033801
[17] Komissarova M V, Marchenko V F, Shestakov P Yu 2019 Phys. Rev. E **99** **(4)** 042205-1

Acknowledgments
This research has been supported by the Interdisciplinary Scientific and Educational School of Moscow University «Photonic and Quantum technologies. Digital medicine»

Journal of Physics: Conference Series 2249 (2022) 012009 doi:10.1088/1742-6596/2249/1/012009

The Raman and hyper-Raman scatterings of light by LO-phonons in a CdS crystal under excitation near resonance with the $A_{n=2}$ and $B_{n=1}$ exciton levels

L E Semenova

Prokhorov General Physics Institute of the Russian Academy of Sciences, 38, Vavilov St., 119991, Moscow, Russia

E-mail: sl@kapella.gpi.ru

Abstract. The theoretical treatment of the scattering of light by LO-phonons under one-photon and two-photon excitation near resonance with the $A_{n=2}$ and $B_{n=1}$ exciton levels in a CdS crystal is given. The influence of the complex structure of the top valence band on these levels is taken into account by the use of the perturbation theory. Assumption of the nonzero matrix elements of the dipole transitions between the A and B sub-bands leads to the fact that the linear combinations of the 1s and 2p exciton wave functions conform to the two "perturbed" energy levels. The possible manifestations of this mixing of states in the resonance Raman and hyper-Raman scatterings are considered.

1. Introduction

The Raman scattering (RS) of light is known to be one of methods of investigation of solids. The hyper-Raman scattering (HRS) is the three-photon process and hence complements possibilities of RS. RS as well as HRS under excitation near resonance with electronic transitions permit one to obtain information on electronic states and scattering mechanisms. Extensive material on the study of resonant RS in semiconductors has been accumulated [1], and interest in it persists up till now [2-6]. Experimental investigations of resonant HRS in semiconductors were reported earlier [7-10]. In particular, an increase in the HRS intensity was observed in a CdS crystal when the doubled frequency of the exciting radiation approached the exciton resonance region [10]. Later, this phenomenon was explained by the scattering mechanism caused by two-photon transitions into excitons of the *s*-type [11]. In recent years, the HRS spectroscopy was applied to study complex molecules [12, 13], quantum dots [14], and nanoparticulates [15].

The given paper is devoted to theoretical investigation of RS and HRS of light by longitudinal optical (LO) phonons in a CdS crystal with the wurtzite structure under excitation near resonance with the exciton levels $A_{n=2}$ and $B_{n=1}$, taking into account the Fröhlich exciton-lattice interaction. The scattering processes are considered within the framework of the model assuming the hydrogen-like Wannier excitons as the virtual intermediate states of an electronic system. But in going so by the use of the perturbation theory the effect of the complex structure of the top valence band on the exciton levels $A_{n=2}$ and $B_{n=1}$ is taken into account. In the wurtzite semiconductors A^2B^6 the valence band v is formed by *p*-states and is split into three closely-lying sub-bands of symmetry Γ_9, Γ_7 and Γ_7 [16, 17]. Excitons composed of electrons from the lowest conduction band *c* and holes from these sub-bands originate the hydrogen-like series A, B and C [16, 17]. In some semiconductors the splitting between the sub-bands A and B is less the exciton Rydberg. In CdS this results in the fact that the hydrogen-like energy levels $A_{n=2}$ and $B_{n=1}$ are closely spaced [17]. Theoretical study of HRS of light by LO-phonons in hexagonal semiconductors A^2B^6 under two-photon excitation near resonance with the $A_{n=2}$ exciton level with allowance made for the complicated top valence band was reported in Ref. [18]. But in the work presented here the matrix elements of the dipole transitions between the A and B sub-bands are assumed to be nonzero, and this leads to partial splitting and mixing of the levels $A_{n=2}$ and $B_{n=1}$.

2. Theoretical model

In consequence of the small magnitudes of the photon wave vectors the excitons with the wave vectors $K=0$ are assumed to participate in the scattering processes considered here. The exciton states in semiconductors were in detail considered in Ref. [19]. The exciton energy E_ς and the pertinent wave functions of the relative electron-hole motion can be found from the set of the equations of the effective mass approximation [19].

In the wurtzite semiconductors for the exciton formed by an electron from the lowest conduction band c and a hole from the top valence band v these equations [19] can be reduced to the form [20]

$$\left(\frac{1}{2\mu^{\perp}_{j_e j_h}}\left(\hat{p}_x^2+\hat{p}_y^2\right)+\frac{1}{2\mu^{\parallel}_{j_e j_h}}\hat{p}_z^2+U_{\text{eh}}(\boldsymbol{r})\right)\Phi^{(\varsigma)}_{j_e j_h}(\boldsymbol{r})$$

$$+\sum_{j'_h}\left(\delta_{j'_h j_h}-1\right)\left(\frac{1}{m}\boldsymbol{\pi}_{j'_h j_h}\cdot\hat{\mathbf{p}}+D^{\beta\alpha}_{j'_h j_h}\hat{p}_\alpha\hat{p}_\beta\right)\Phi^{(\varsigma)}_{j_e j'_h}(\mathbf{r})=\left(E_\varsigma-E_{j_e j_h}\right)\Phi^{(\varsigma)}_{j_e j_h}(\mathbf{r}), \quad (1)$$

where $\hat{p}_\alpha=-i\hbar\,\mathrm{d}/\mathrm{d}r_\alpha$, $U_{\text{eh}}(\boldsymbol{r})$ is the Coulomb energy of the electron-hole interaction, $\mu_{j_e j_h}$ is the effective mass of an exciton, $\boldsymbol{\pi}_{j'_h j_h}$ is the matrix element of the momentum operator, $D^{\beta\alpha}_{j'_h j_h}$ is the effective mass parameter [19, 21], $\Phi^{(\varsigma)}_{j_e j_h}(\boldsymbol{r})$ is the wave function of the relative motion of an electron and a hole, $E_{j_e j_h}=E_{j_e}-E_{j_h}$, E_j is the energy of the j band in the Brillouin center, j_e and j_h are denoted the states in the bands c and v, respectively.

If the terms connecting the equations for the different sub-bands are neglected and the influence of crystal anisotropy on the exciton effective mass is ignored, then the set of the equations (1) is reduced to the individual hydrogen-like equations [22].

For the wurtzite semiconductors it was shown in Ref. [23] that the wave functions of the valence band u_1, u_3, u_5 (u_5^*), and u_6 (u_6^*) transform over the irreducible representations Γ_1, Γ_3, Γ_5 and Γ_6, but the spin-orbit interaction leads to the splitting and the mixing of these states. In particular, the sub-band A (Γ_9) is mainly determined by u_5 (u_5^*) with the small addition of u_6 (u_6^*), and in the sub-bands B and C the mixing of u_1 and u_5 (u_5^*) takes place. The wave functions for the sub-bands A, B and C are given in Ref. [24]. The selection rules allow the dipole transitions between the sub-bands A and B. The corresponding matrix elements may turn out to be nonzero, in particular, due to presence of the u_6 (u_6^*) functions in the A sub-band. Based on the selection rules and the known band structure of the semiconductor it is not complicated to find the effective mass parameters $D^{\beta\alpha}_{j'_h j_h}$.

In this work the perturbation theory methods were applied to the closely spaced exciton levels $A_{n=2}$ and $B_{n=1}$. In doing so anisotropy of the exciton effective mass was not taken into account and the terms connected the equations for the different sub-bands were considered as perturbation. As a result, it was found that the energy levels partially split and mix. In other words, the three energy levels were obtained, and the one from them is coincident with the unperturbed $A_{n=2}$ level. Some 2p exciton wave functions and the 2s exciton wave functions of the A series conform to it. The two other energy levels are defined as

$$E_{\pm}=\frac{1}{2}\left(\left(E_{\mathrm{A}_{\mathrm{n=2}}}+E_{\mathrm{B}_{\mathrm{n=1}}}\right)\pm\left(E_{\mathrm{A}_{\mathrm{n=2}}}-E_{\mathrm{B}_{\mathrm{n=1}}}\right)\sqrt{1+|N|^2}\right), \quad (2)$$

where

$$|N|^2=16\times\left(\frac{\hbar}{ma}\right)^2\left(\frac{2}{3}\right)^8\frac{|C|^2}{\left(E_{\mathrm{A}_{\mathrm{n=2}}}-E_{\mathrm{B}_{\mathrm{n=1}}}\right)^2}. \quad (3)$$

Here $E_{\mathrm{A}_{\mathrm{n=2}}}$ and $E_{\mathrm{B}_{\mathrm{n=1}}}$ are the energies of the unperturbed exciton levels $\mathrm{A}_{\mathrm{n=2}}$ and $\mathrm{B}_{\mathrm{n=1}}$, m is the mass of an electron, a is the Bohr radius of an exciton, C is defined by the matrix elements of the dipole transitions between the A and B sub-bands

$$C=\left(q_9\sqrt{1-q_7^2}\langle u_5|\hat{p}_x|u_6\rangle-q_7\sqrt{1-q_9^2}\langle u_1|\hat{p}_x|u_5\rangle\right), \quad (4)$$

where q_7^2 and q_9^2 define the fraction of the wave functions u_1 and u_6 $\left(u_6^*\right)$ in the B (Γ_7) and A (Γ_9) sub-bands, respectively. The linear combinations of the 1s exciton wave functions of the B series (1sB) and the remaining 2p exciton wave functions of the A series (2pA) correspond to these levels. Their weight factors are determined by $|T|^2$ and $\left(1-|T|^2\right)$ where T is given by

$$T=\frac{1}{\sqrt{2}}N\left(1+|N|^2+\sqrt{1+|N|^2}\right)^{-1/2}. \quad (5)$$

As may be seen from the expressions obtained the splitting and mixing of the levels is due to the dipole transitions between sub-bands.

3. Discussion

At the first sight the mixing of 1sB and 2pA exciton wave functions in the states with the energies $E_{\pm}$ can be manifested in resonant RS of light by LO-phonons. If the intraband Fröhlich mechanism of the exciton-lattice interaction is considered in the approximation of the zero phonon wave vector [25], then due to this mixing the RS process including only the allowed dipole transitions can occur. Let us consider RS of light by LO-phonons for the scattering geometry *x(zx)z*. The frequency of incident radiation is assumed to be near resonance with the $\mathrm{A}_{\mathrm{n=2}}$ and $\mathrm{B}_{\mathrm{n=1}}$ exciton levels. Taking into account only the resonance contributions to RS and the allowed dipole transitions, one can obtain the frequency dependence of the RS cross section $\mathrm{d}\sigma_{\mathrm{RS}}/\mathrm{d}\Omega$

$$\frac{\mathrm{d}\sigma_{\mathrm{RS}}}{\mathrm{d}\Omega}\propto\frac{\omega_{\mathrm{S}}^2}{\omega_{\mathrm{L}}^2}|T|^2\left(1-|T|^2\right)\left|\kappa_S^4\left(E_{cv}^{\mathrm{A}}\right)\left(\frac{R_{\mathrm{A}}}{\left(E_{-}-\hbar\omega_{\mathrm{L}}-i\Gamma_{-}\right)}-\frac{R_{\mathrm{A}}}{\left(E_{+}-\hbar\omega_{\mathrm{L}}-i\Gamma_{+}\right)}\right)J_5\left(2,\kappa_{\mathrm{S}}\left(E_{cv}^{\mathrm{A}}\right)\right)\right|^2, \quad (6)$$

where $\kappa_{\mathrm{S}}\left(E_{cv}^{\mathrm{A}}\right)=\sqrt{R_{\mathrm{A}}/\left(E_{cv}^{\mathrm{A}}-\hbar\omega_{\mathrm{S}}\right)}$, $E_{cv}^{\mathrm{A}}=E_{cv}$, E_{cv} is an energy bandgap, $R_{\mathrm{A(B)}}$ is the exciton Rydberg for the A (B) series, $\omega_{\mathrm{L(S)}}$ is a frequency of incident (scattered) radiation, Γ is a lifetime broadening of the exciton level. Here the Green's function method [26] was used under summation over the intermediate virtual exciton states and $J_5(n,\kappa_{\mathrm{S}})$ is given in Ref. [27]. One can see from the equation (6) that the contribution of the scattering mechanism considered to RS takes place only for $T\neq 0$. The figure 1 demonstrates the RS cross section computed for $\Gamma_{-}=\Gamma_{+}=0.001\mathrm{eV}$ and the different values of $|N|^2$ as a function of the incident photon energy $\hbar\omega_{\mathrm{L}}$. Calculations were performed for the known

Journal of Physics: Conference Series **2249** (2022) 012009 doi:10.1088/1742-6596/2249/1/012009

parameters of CdS: $\hbar\omega_{\mathrm{LO}} = 0.038$ eV [28], $R_{\mathrm{A}} = R_{\mathrm{B}} = 0.028$eV, $E_{cv}^{\mathrm{A}} = 2.582$ eV, $E_{cv}^{\mathrm{B}} = E_{cv} + \Delta_{\mathrm{AB}} = 2.598$ eV [16], where ω_{LO} is the LO phonon frequency and Δ_{AB} is the splitting between A and B sub-bands.

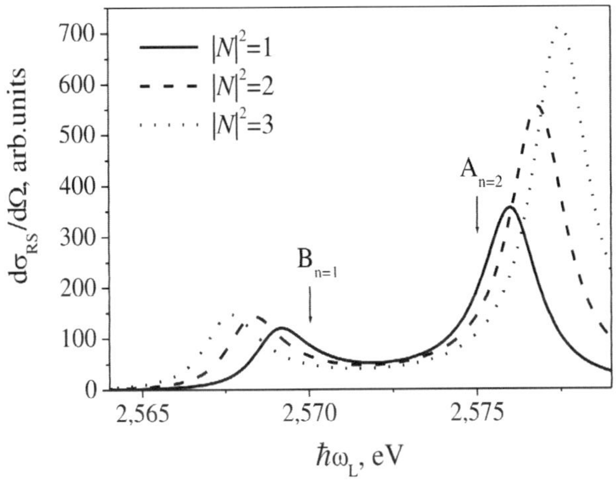

Figure 1. The resonant RS cross section as a function of the incident photon energy $\hbar\omega_{\mathrm{L}}$. The arrows show the positions of the unperturbed energy levels $\mathrm{B}_{n=1}$ and $\mathrm{A}_{n=2}$.

But the weakly-forbidden dipole transitions to *p*-exciton states were not taken into account under these calculations. Allowance made for them leads to the RS processes, which include the transitions to the states of the exciton levels $\mathrm{A}_{n=2}$ and $\mathrm{B}_{n=1}$ in the case $T = 0$. Because of this, the features on the frequency dependence of $d\sigma_{\mathrm{RS}}/d\Omega$ can be due to the mixing of the states and the scattering mechanisms induced by the weakly-forbidden transitions. It is difficult to separate the contributions of these processes.

It makes sense to consider another scattering mechanism, which is conditioned by the forbidden *q*-dependent Fröhlich interaction [25]. It is described by the following sequence of processes: the allowed dipole transition to an exciton of the *s*-type under absorption of an incident photon, then the transition to another *s*-exciton state because of the *q*-dependent Fröhlich interaction, and the dipole transition to the ground state under emission of scattered light photon. Investigations of forbidden RS in CdS were reported in a number of publications [29-31].

For the scattering geometry $y(zz)\bar{y}$ resonant RS can be induced only by the forbidden Fröhlich mechanism. Besides in this case the *s*-excitons of the A series do not give a contribution to the cross section. Consequently, the RS intensity increases as the incident photon energy is approached to resonance with the 1s exciton state of the B series. If the mixing of the states is taken into account, the frequency dependence of the resonant forbidden RS cross section $d\sigma_{\mathrm{FRS}}/d\Omega$ can be written as

$$\frac{d\sigma_{\mathrm{FRS}}}{d\Omega} \propto \left(\frac{\omega_{\mathrm{S}}}{\omega_{\mathrm{L}}}\right)^2 \left| \kappa_{\mathrm{S}}^4 \left(E_{cv}^{\mathrm{B}}\right) J_{\mathrm{q}}^{\mathrm{s}}\left(qa, \kappa_{\mathrm{S}}\left(E_{cv}^{\mathrm{B}}\right), 1\right) \left(\frac{\left(1 - |T|^2\right)}{E_{-} - \hbar\omega_{\mathrm{L}} - i\Gamma_{-}} + \frac{|T|^2}{E_{+} - \hbar\omega_{\mathrm{L}} - i\Gamma_{+}} \right) \right|^2, \qquad (7)$$

where $\boldsymbol{q}$ is the LO-phonon wave vector, $J_{\mathrm{q}}^{\mathrm{s}}(qa, \kappa_{\mathrm{S}}, 1)$ is determined in Ref. [32]. In the equation (7) allowance was also made only for the resonance contributions. The cross section $d\sigma_{\mathrm{FRS}}/d\Omega$ as a function of the incident photon energy was calculated for some values of $|N|^2$ and is shown on the figure 2. Computations were carried out for the effective masses of an electron and a hole which are given in Ref. [33]. The figure shows the peak corresponding to resonance with the E_- level and a slight increase in the cross section near E_+ for $|N|^2 \neq 0$.

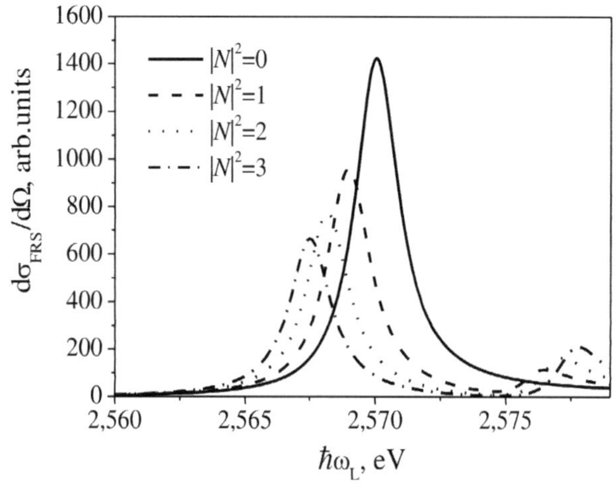

Figure 2. The cross section of resonant forbidden RS as a function of the incident photon energy. Calculations were performed for $\Gamma_- = \Gamma_+ = 0.001\text{eV}$.

Of interest is also HRS of light by LO-phonons under two-photon excitation near resonance with the levels $A_{n=2}$ and $B_{n=1}$, taking into account the mixing of the 1sB and 2pA exciton states. The two-photon transition to the *p*-exciton is described within the framework of the two-band model and includes the allowed dipole transitions [34]. Whereas the two-photon excitation of the 1s exciton can occur via the intermediate virtual *s*-exciton states formed by an electron from the lowest conduction band and a hole from the deeper valence band or by an electron from the higher-lying conduction band and a hole from the top valence band [35, 36]. Within the framework of the two-band model it is considered as the following sequence of processes: a weakly forbidden dipole transition to the *p*-exciton state and the allowed dipole transition to the 1s exciton [34, 36]. Moreover, under some conditions the contributions of the two-band and three-band models to the two-photon transition to the 1s exciton can be opposite in the sign [27, 36]. Thus, the different sequences of the intermediate exciton states contribute to resonant HRS considered here.

An analysis of HRS makes sense for scattering geometry *y(xxz)x* for which incident radiation polarized along the axis *x* travels parallel to the axis *y* and the scattered light with the polarization $\varepsilon_S || z$ is directed in the *x* axis. In this scattering configuration resonant HRS is induced by the forbidden Fröhlich interaction. In resonant conditions considered one can write the frequency dependence of the HRS cross section $d\sigma_{HR}/d\Omega$ as

$$\frac{d\sigma_{HR}}{d\Omega} \propto \frac{\omega_S^2}{\omega_L^3}\left|\kappa_S^4\left(E_{cv}^B\right)\left[\left(T^*\sqrt{1-|T|^2}A_P + A_S|T|^2\right)\frac{1}{\left(E_+ - 2\hbar\omega_L - i\Gamma_+\right)}\right.\right.$$
$$\left.\left.+\left(A_S\left(1-|T|^2\right) - T^*\sqrt{1-|T|^2}A_P\right)\frac{1}{\left(E_- - 2\hbar\omega_L - i\Gamma_-\right)}\right]J_q^s\left(qa, \kappa_S\left(E_{cv}^B\right), 1\right)\right|^2, \quad (8)$$

where A_S and A_P are related to the two-photon dipole transitions to the 1s exciton of the B series and the 2p exciton of the A series, respectively. In the case of $|N|^2 = 0$ (i.e. $T = 0$) the resonant enhancement of the cross section takes a place for $2\hbar\omega_L \sim E_{B_{n=1}}$. However, calculations of the cross section as a function of $2\hbar\omega_L$ are complicated by the fact that $d\sigma_{HR}/d\Omega$ depends not only on T but

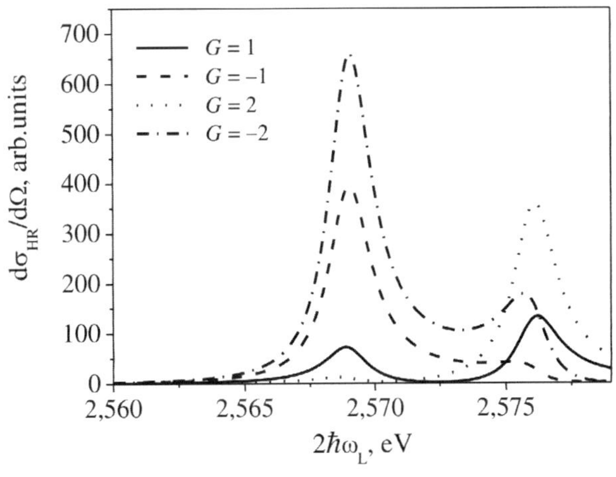

Figure 3. The resonant HRS cross section computed for $|N|^2 = 1$ and some values of G as a function of the doubled energy of incident photons $2\hbar\omega_L$. Computations were performed for $\Gamma_- = \Gamma_+ = 0.001\,\text{eV}$.

also on A_S and A_P. And the estimation of the contribution of the two-photon transition to the 1sB exciton state presents difficulties, since this requires knowing, in particular, the matrix elements of the dipole transitions in the higher-lying conduction bands and deeper valence bands [27, 36]. But under excitation conditions considered here A_S and A_P weakly depend on ω_L and hence they can assume to be constants. In order to determine an influence of A_S and A_P on the frequency dependence of the resonant HRS cross section the parameter $G = (T^* A_P)/(|T| A_S)$ was inserted, and calculations $d\sigma_{HR}/d\Omega$ as a function of $2\hbar\omega_L$ were performed for some values of G and $|N|^2 = 1$. The results are presented in the figure 3 which demonstrates the strong dependence of the resonant cross section on

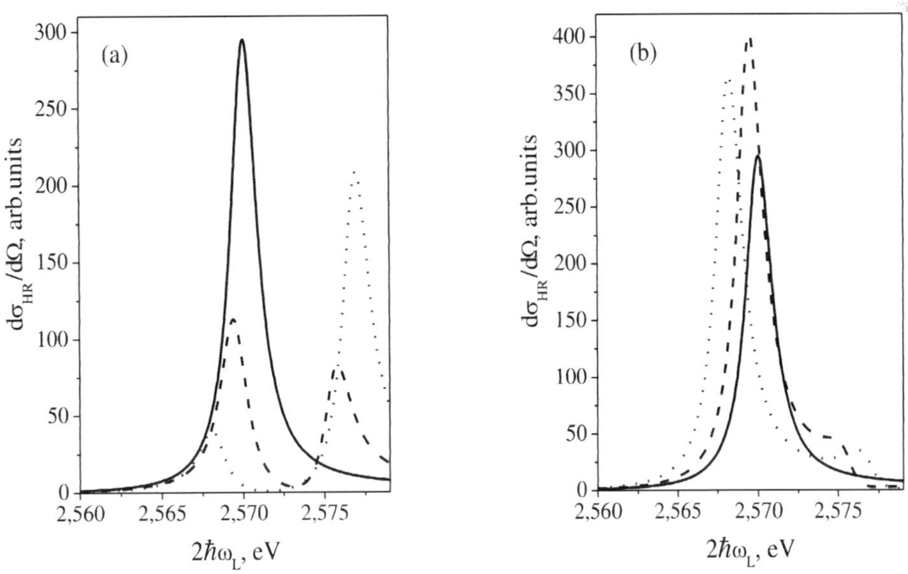

Figure 4. The resonant HRS cross section calculated for $G = 1$ (a) and $G = -1$ (b) as a function of the doubled energy of incident photons. The solid, dashed and dotted curves were calculated for $|N|^2 = 0$, 0.5 and 2.

the sign of G. As an illustrative example, the figure 4 shows the scattering cross section calculated for $G=1$ (a) and $G=-1$ (b) as a function of the doubled energy of incident photons. For comparison, the curves on the figure were computed for $|N|^2=0$, 0.5 and 2. As can be seen from the figure, in the absence of the mixing of the states ($N=0$), the cross section shows a sharp increase for the two-photon resonance with the $B_{n=1}$ exciton level. At $G=1$, the mixing can lead to the appearance of a second peak in the frequency dependence of $d\sigma_{HR}/d\Omega$. Whereas in the case of $G=-1$ it results in an additional weak increase in the cross section.

4. Conclusions

RS and HRS of light by LO phonons under excitation near resonance with $A_{n=2}$ and $B_{n=1}$ exciton levels in a CdS crystal with the wurtzite structure were theoretically investigated. The influence of the complex top valence band on these levels was taken into account by the use of the perturbation theory. It has been shown that assumption of the nonzero matrix elements of the dipole transitions between the sub-bands A and B leads to mixing of the 1sB and 2pA exciton states. This results in additional scattering processes. It has been shown that at certain conditions this mixing of the states could be manifested in the resonant RS or HRS of light.

References

[1] 1975 *Light Scattering in Solids* (*Topics in Applied Physics* vol 8) ed M Cardona (Berlin, Heidelberg, New York: Springer-Verlag)
[2] Yu C, Chen Z and Shum K 2013 *J. Raman Spectrosc.* **44** 262-5
[3] Jin S, Zhang Y, Fan F, Feng Z and Li C 2016 *J. Raman Spectrosc.* **47** 884-7
[4] Guc M, Litvinchuk A P, Levcenko S, Izquierdo-Roca V, Fontané X, Valakh M Ya, Arushanov E and Pérez-Rodríguez A 2014 *Phys. Rev.* B **89** 205205
[5] Kutrowska-Girzycka J, Jadczak J and Bryja L 2018 *Solid State Commun.* **275** 25-8
[6] Medel-Ruiz C I, Pérez Ladrón de Guevara H, Molina-Contreras J R and Frausto-Reyes C 2020 *Solid State Commun.* **312** 113895
[7] Inoue K, Minami F, Kato Y, Yoshida K and Era K 1992 *J. Cryst. Growth* **117** 738-41
[8] Inoue K, Yoshida K, Minami F and Kato Y 1992 *Phys. Rev.* B **45** 8807-10
[9] Watanabe K, Inoue K and Minami F 1992 *Phys. Rev.* B **46** 2024-33
[10] Maslov V A, Ondriash K K, Polivanov Yu N, Prokhorov K A, Semenova L E and Chuzavkov Yu L 1996 *Laser Phys.* **6** 132-43
[11] Semenova L E and Prokhorov K A 2004 *Laser Phys. Lett.* **1** 253-8
[12] Wen C-I and Hiramatsu H 2020 *J. Raman Spectrosc.* **51** 274-8
[13] Asakura M and Okuno M 2021 *J. Phys. Chem. Lett.* **12** 4780-5
[14] Tan R, Kelley D F and Kelley A M 2019 *J. Phys. Chem. C* **123** 16400-5
[15] Villa-Aleman E, Houk A L, Dick D D and Hunyadi Murph S E 2020 *J. Raman Spectrosc.* **51** 1260-3
[16] Reynolds D C 1969 Excitons in II-VI compounds *Optical Properties of Solids* ed S Nudelman and S S Mitra (New York: Plenum Press) chapter 10 pp 239-86
[17] Sobolev V V 1980 *Bands and Excitons of the $A^{II}B^{VI}$ Group Compounds* (Chisinau: Shtiintsa)
[18] Semenova L E 2018 *J. Phys.: Conf. Ser.* **999** 012012
[19] Denisov M M and Makarov V P 1973 *Phys. Stat. Sol.* (b) **56** 9-59
[20] Semenova L E 2021 *J. Russ. Laser Res.* **42** 721-9
[21] Luttinger J M and Kohn W 1955 *Phys. Rev.* **97** 869-83
[22] Landau L D and Lifshitz E M 1989 *Course of Theoretical Physics* vol 3 (Moscow: Nauka)
[23] Gutsche E and Jahne E 1967 *Phys. Stat. Sol.* **19** 823-32
[24] Lew Yan Voon L C, Willatzen M, Cardona M and Christensen N E 1996 *Phys. Rev.* B **53** 10703-14
[25] Martin R M 1971 *Phys. Rev.* B **4** 3676-85

Journal of Physics: Conference Series **2249** (2022) 012009 doi:10.1088/1742-6596/2249/1/012009

[26] Hostler L 1964 *J. Math. Phys*. **5** 591-611
[27] Semenova L E and Prokhorov K A 2003 *JETP* **96** 922-31
[28] 1991 *Physical Quantities: a Handbook* ed I S Grigor'ev and E Z Meilikhov (Moscow: Energoatomizdat)
[29] Martin R M and Damen T C 1971 *Phys. Rev. Lett*. **26** 86-8
[30] Callender R H, Sussman S S, Selders M and Chang R K 1973 *Phys. Rev*. B **7** 3788-98
[31] Klochikhin A A, Permogorov S A and Reznizky A N 1976 *Sov. Phys. Solid State* **18** 1304
[32] Semenova L E and Prokhorov K A 2001 *J. Raman Spectrosc*. **32** 942-6
[33] Nguyen B A, Nguyen V H, Nguyen T T and Nguyen A V 1982 *Phys. Rev*. B **25** 4075-80
[34] Mahan G D 1968 *Phys. Rev*. **170** 825-38
[35] Loudon R 1962 *Proc. Phys. Soc*. **80** 952-61
[36] Rustagi K C, Pradere F and Mysyrowicz A 1973 *Phys. Rev*. B **8** 2721-32

Development of a Laser Frequency Stabilization and an Optical Transmission System for the Francium Electric Dipole Moment Search

K Nakamura[1], S Nagase[2], T Nakashita[3], T Hayamizu[4], T Aoki[3], H Nagahama[1], N Ozawa[2], M Sato[3], K Yamane[3], M Fukase[2], D Uehara[2], A Takamine[4]and Y Sakemi[1]

[1]Center for Nuclear Study, The University of Tokyo, Wako, Japan
[2]Department of Physics, the University of Tokyo, Tokyo, Japan
[3]Graduate School of Arts and Sciences, the University of Tokyo, Tokyo, Japan
[4]Nishina Center for Accelerator-Based Science, RIKEN, Wako, Japan

E-mail: k-nakamura@cns.s.u-tokyo.ac.jp

Abstract. We developed a laser frequency stabilization and an optical fiber transmission system for the the francium electric dipole moment search. The absolute accuracy of a laser frequency stabilization scheme using a state-of-the-art commercial wavelength meter was 0.48 MHz at ±2 nm and -1.33 MHz at ±200 nm from calibration wavelength, respectively, and the frequency instability is below 10^{-9} with a standard deviation of 0.56 MHz over 60 hours. We also demonstrated that a 400 m long fiber laid between laboratories can transmit 30 mW of trapping laser light, which is sufficient for a magneto-optical trapping of francium. The polarization crosstalk in the fiber was stable at -25 dB over 12 hours of measurement.

1. Introduction

The search for electric dipole moment (EDM) of elementary particle is primarily sensitive to new physics beyond the Standard Model. While the Standard Model predicts very small values of the EDM, several theoretical models beyond the Standard Model, such as the supersymmetric model, predict much larger values, which are expected to be observable with modern experimental techniques. We are proceeding with the electron EDM (e-EDM) search using laser-cooled Fr atoms [1-7]. The heaviest alkali atom, Fr, has a large enhancement factor for the e-EDM of about 799 [8]. In addition, the laser cooling technique can be applied to Fr atoms, which increases the interaction time between the atoms and the applied electric field. Therefore, the e-EDM experiment with laser-cooled Fr atoms is expected to update the measurement sensitivity of the ThO molecular beam experiment [9] that reported the current upper limit.

In this paper, we report on the laser frequency stabilization and optical fiber transmission system developed for the magneto-optical trapping (MOT) of Fr. To stabilize the frequency of the laser, a stable and reproducible frequency reference is needed. Since Fr atoms have no stable isotopes, the Fr atom vapor cell cannot be used as a frequency reference. Therefore, an absorption line of I_2 [1, 10], a Fabry-Perot cavity [11-12], and an optical frequency comb [13] have been employed as frequency references for such a wavelength range where the use of atomic reference cells is difficult. Recently, wavelength meters (WLM)-based on solid-state Fizeau interferometers have been attracting attention as frequency references [14-16]. The absolute accuracy of the commercial WLM, HighFinesse WSU-2, was reported to be about 5 MHz over the wavelength range 760 nm to 940 nm when calibrated at 632 nm with an iodine-stabilized He-Ne laser [16]. In addition, when combined with the optional fiber switch unit, up to eight lasers can be measured simultaneously, and the frequency of all these lasers can be simultaneously feedback controlled via the analog voltage output of the proportional-integral-derivative (PID) regulator. For a laser stabilization using the commercial WLM (WSU-10) a frequency instability below 10^{-9} with a standard deviation of 1 MHz over10 hours was reported [15]. This indicates that it is possible to stabilize the laser frequency in a region narrower than the natural width of the MOT cooling transition (~MHz level). Therefore, this scheme can be a useful and powerful tool as a method of frequency stabilization of lasers for MOT. We installed the most advanced wavelength meter (WS8-2). We describe the absolute accuracy and stability of the frequency stabilization using the wavelength meter. We also present the transmittance and polarization stability of 400 m polarization maintaining (PM) fibers laid between laboratories.

2. Optical setup for the Fr and Rb MOTs

We are constructing an experimental system for the e-EDM search using laser-cooled Fr atoms at the RIKEN Nishina Center for Accelerator-Based Science (RNC) [3-7]. A partial energy diagram of ^{210}Fr and ^{87}Rb are shown in Figs. 1(a) and 1(b), respectively. Figure 1(c) shows a schematic of the optical setup for the Fr and Rb MOTs. There are two rooms for the experiment. One is a laser laboratory (lab A). The other is a radioisotope experimental laboratory (lab B), where we have constructed a MOT system with the neutralizer for the Fr and Rb ions. We generate Fr ions by fusion reaction using a beam supplied from RIKEN AVF cyclotron. A ^{210}Fr ion beam of 5×10^6/s was successfully extracted [5]. Since ^{210}Fr has a lifetime of about 3 minutes, it needs to be experimented online. Since the production of Fr requires the operation of a cyclotron, stable Rb atoms, which are alkali atoms and have similar chemical properties, are used to optimize the operating parameters of the entire device.
We use a Ti:sapphire (Ti:S) laser (M squared lasers, SolsTiS) and a homemade interference-filter-type external-cavity diode laser (ECDL) as a trapping laser (718 nm) and a repumping laser (817 nm) for ^{210}Fr, respectively. On the other hand, a homemade Littrow-type external-cavity diode laser (ECDL)

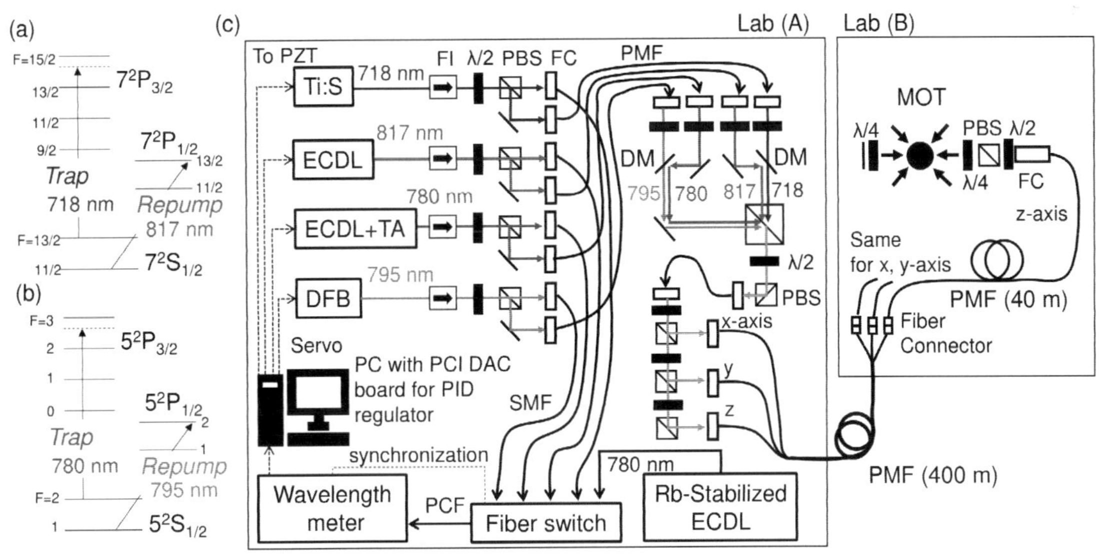

Fig. 1. (a) Partial energy diagram of ^{210}Fr and (b) ^{87}Rb. (c) Schematic of the optical setup for the Fr and Rb MOTs. PMF: polarization maintaining fiber, SMF: single mode fiber, PCF: photonic crystal fiber, FI: faraday isolator, FC: fiber collimator, DM: dichroic mirror.

and a DFB laser(Toptica photonics) are used as a trapping laser (780 nm) and a repumping laser (795nm) for ^{87}Rb, respectively. The frequencies of each of the four lasers described above are simultaneously stabilized using the WLM (HighFinesse, WS8-2 standard, 330-1180 nm available). The laser light entering the device through a photonic crystal fiber is coupled to Fizeau interferometers, and the induced interference patterns are imaged on CCD photodiode arrays. The smallest possible exposure time is 1 ms, which is reached by entering ~mW of optical power. The recorded pattern is transferred to a computer via a high-speed USB connection, and the wavelength is determined by comparing it with a stored calibration pattern.

A interference-filter-type ECDL (Moglabs) with frequency stabilized using the modulation transfer method to the D2 line (780 nm) of ^{87}Rb in the Rb vapor cell was used as the laser light source for calibrating the WLM. The WLM was automatically calibrated at regular intervals by entering the calibration laser light into one of the eight ports of the fiber switch unit through a single-mode fiber.

The trapping and repumping laser beams were combined with a dichroic mirror. The laser beams for Rb and Fr were orthogonalized by a polarizing beam splitter (PBS). Switching of the laser beams for Fr and Rb was done by rotating the half-wavelength plate. To form a 3D MOT, the laser beam was divided into three parts: x, y, and z axes. A 400m long flexible stainless tube (Nippon steel welding & engineering, Picoflec) housing seven PM fibers was laid between labs A and B. Four fibers among them were Fujikura, PM63, two were PM85, and one was PM98. The laser beam, which combines trapping and repumping lasers, was transmitted by PM63. In Lab (B), 400 m-long PM fibers were connected to 40 m-long ones via fiber optic connectors (loss due to the connector was ~25%). Linearly polarized light was aligned to the slow axis of the PM fiber. The power of the unwanted polarization mode produced by polarization crosstalk was removed with PBS after the output from the fiber. We have already succeeded in Rb MOT and are currently working towards Fr MOT.

3. Laser frequency stabilization scheme using wavelength meter

3.1. Evaluation of absolute accuracy

First of all, we evaluated the absolute accuracy of the WLM in Lab (A). The manufacturers specifications for the absolute accuracy were 2, 10, 30 MHz, for ±2 nm, ±200 nm, elsewhere around calibration wavelength, respectively. We evaluated the absolute accuracy of ±2 nm and ±200 nm with the configurations shown in Figure 2(a) and (b), respectively. ECDL, a trapping laser source for Rb,

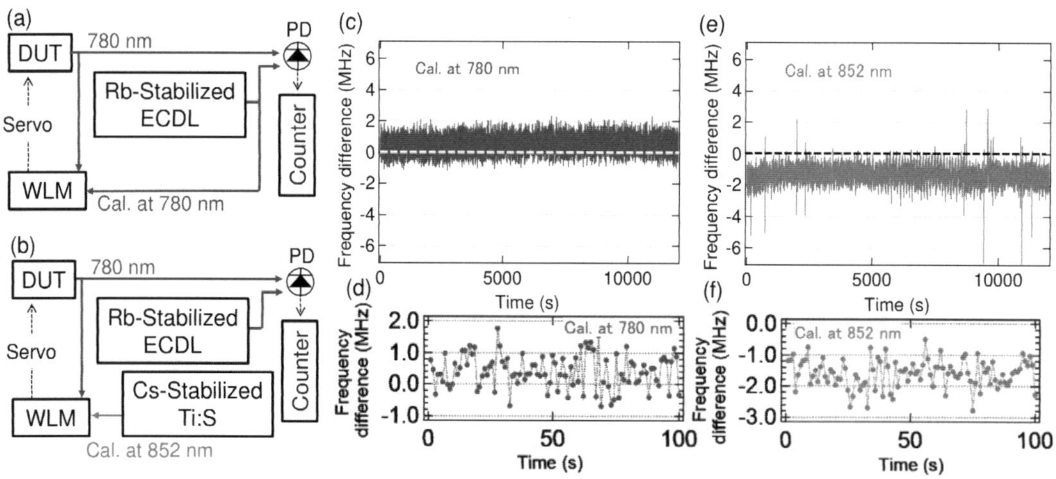

Fig.2. (a) Schematic of the experimental setup for evaluating absolute accuracy of wavelength meters (WLM) calibrated at ±2 nm and (b) at ±200 nm from wavelength of device under test (DUT). (c-f) Beat signals measured in the configuration (a-b), respectively.

was used as the device under test (DUT). The beat signal between the DUT and the Rb stabilized laser was measured every 1 s with a frequency counter (Pendulum, CNT-90). For the absolute accuracy evaluation at ±2 nm from calibration wavelength, the WLM was calibrated with the 780 nm Rb-stabilized laser (Fig. 2(a)). On the other hand, for the ±200 nm, the WLM was calibrated with a Ti:S laser stabilized to the ^{133}Cs D2 line (~852 nm) by the modulation transfer method (Fig. 2(b)). Figures 2(c) and 2(e) show the time series data of the beat signals obtained in each experiment. The vertical axis is the difference frequency, calculated by numerically subtracting the offset from the obtained beat frequency, which shows the difference between the frequency of the DUT recognized by the WLM and the absolute frequency of the DUT. Figures 2(d-f) are zoomed graphs of the portion from 1 to 100 s. When calibrated at 780 nm and 852 nm, the absolute accuracy of the WLM for 780 nm was 0.48 MHz and -1.33 MHz, respectively. As a result, we were able to confirm absolute accuracy that was better than the manufacturing specifications under our experimental environment.

3.2. Evaluation of Stability

Next, we evaluated the long-term stability of the WLM-based laser frequency stabilization scheme in the experimental setup shown in Fig. 2(a). For stable MOT, the laser source needs to be frequency stabilized below the natural width of the trapping transition of the atoms. The natural width of the trapping transition of ^{210}Fr (^{87}Rb) is 7.6 MHz (6.1 MHz). Both Figs. 3(a) and 3(b) show the time series data of the beat signal between the DUT, which was frequency stabilized with the WLM, and the Rb-stabilized ECDL. Figure 3(a) shows the measurement results when the WLM was automatically calibrated every minute with the Rb-stabilized ECDL. On the other hand, Fig. 3(b) shows the one when the WLM was calibrated once before the beat measurement and not calibrated during the measurement, i.e., free-running of the WLM. The input of the fiber switch unit was adjusted so that the exposure time was ~10 ms (input optical power ~100 μs). The output voltage range of PCI DAC board for PID regulator is -10 V to 10 V with a step resolution of ~0.3 mV. We amplified the output voltage from the PCI board about 6 times with a piezo driver and applied it to the piezo actuator of the DUT. The "amount of frequency shift vs. driving voltage to the piezo" of the DUT was ~270 MHz/V, so the dynamic range for the frequency control of the DUT was ±17 GHz and the minimum resolution was ~500 kHz. Figure 3(c) shows room temperature and atmospheric pressure in lab (A) measured at the same time as Figure 3(b). Figure 3(d) shows the optical frequency fluctuations of the free-running DUT measured by a WLM without calibration. Figure 3(e) shows the normalized Allan deviation of each time series data. While lab (A) was temperature controlled and stabilized below ±0.2 °C, it was

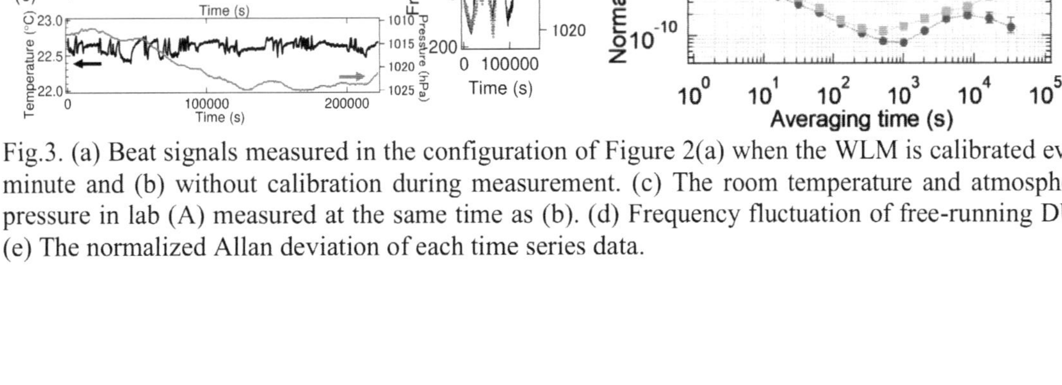

Fig.3. (a) Beat signals measured in the configuration of Figure 2(a) when the WLM is calibrated every minute and (b) without calibration during measurement. (c) The room temperature and atmospheric pressure in lab (A) measured at the same time as (b). (d) Frequency fluctuation of free-running DUT. (e) The normalized Allan deviation of each time series data.

Journal of Physics: Conference Series **2249** (2022) 012010 doi:10.1088/1742-6596/2249/1/012010

not atmospheric pressure controlled and showed the same fluctuations as outside. The free-running DUT drifted significantly due to atmospheric pressure fluctuations (Fig. 3(d)), while the free-running wavemeter was robust against ones (Fig. 3(b-c)). The spike-like frequency fluctuations in Fig. 3(a-b) were caused by the sudden changes in atmospheric pressure that occurred when the door of lab (A) was opened and closed. As a result, a frequency instability of the WLM-based laser frequency stabilization scheme, which was calibrated every minute, was below 10^{-9} with a standard deviation of 0.56 MHz over 60 h (Fig. 3(a)). The stability is sufficient for the Fr and Rb MOT. It was also shown to be stable over a long period of time, even without calibration (Fig. 3(b)).

4. Optical fiber transmission system

In order to trap a large number of atoms by MOT, the intensity of the irradiated laser beam must be sufficiently larger than the saturation intensity of the trapping transition. The saturation intensity of the trapping transition of Fr (Rb) is 2.86 mW/cm^2 (1.64 mW/cm^2). Considering our beam diameter, this corresponds to an optical power of ~6.5 mW (~3.7 mW). We investigated the transmittance of a 400 m long optical fiber laid between lab (A) and (B). In addition, we studied the wavelength dependence and fiber length one for it, respectively. Figure 4(a) shows the measured transmittance of the 400 m long fiber at input wavelengths of 718 nm, 740 nm, and 780 nm. We also measured transmittance of the 800 m long fiber at input wavelength of 718 nm. The 800 m long fiber was constructed by connecting 400 m fibers to each other with a connector. The horizontal axis in Fig. 4(a) shows the optical power before coupling to the optical fiber, and the vertical axis shows the optical power output from the optical fiber into free space. A Ti:sapphire laser was used for these investigations. The output initially increased in proportion to the input, but then saturated. The slope of ~0.5 in the proportional part was mainly due to coupling losses. On the other hand, the saturation was considered to be due to nonlinear optical effects in the optical fiber since fiber length dependence was shown. Although there was a limit to the transmission power to lab (B) due to the saturation phenomenon, we were able to confirm that 30 mW (45 mW) could be transmitted at 718 nm (780 nm).

The temporal fluctuation of the polarization crosstalk in the fiber directly leads to the fluctuation of the MOT input intensity. Therefore, it is important to investigate the temporal fluctuation of the polarization crosstalk in the fiber. We split a output 718 nm trapping laser beam from 400 m long fiber into p-wave and s-wave by PBS, and measured the optical power every 1 s, respectively. The measurement results are shown in Figure 4(b). Polarization crosstalk was stable at -25 dB over 12 hours of measurement.

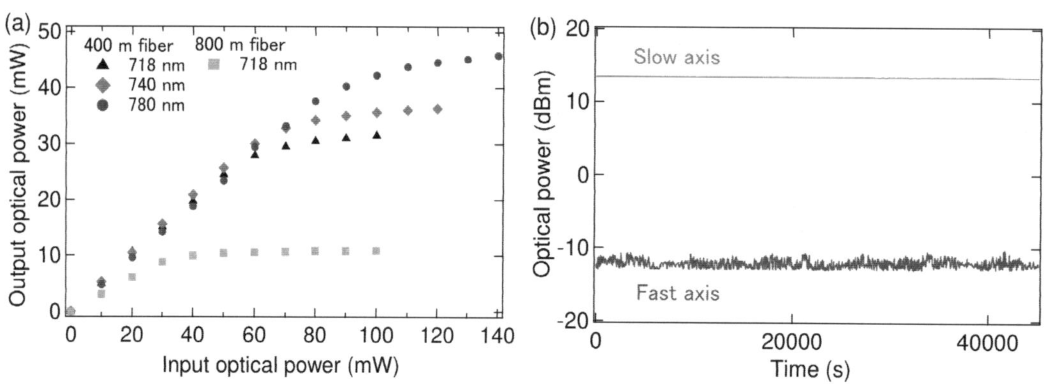

Fig.4. (a) Fiber transmittance of optical fiber laid between lab (A)-(B). (b) Polarization crosstalk in a 400 m long fiber.

5. Conclusion

We developed a laser frequency stabilization and an optical fiber transmission system for the Fr e-EDM search. The WLM-based laser frequency stabilization scheme achieved an absolute accuracy of 0.48 at ±2 nm (1.33 at ±200 nm), and a frequency instability of below 10^{-9} with a standard deviation of 0.56 MHz over 60 hours. We also demonstrated that a 400 m long fiber laid between laboratories can transmit 30 mW of trapping laser light, which is sufficient for Fr MOT. The polarization crosstalk in the fiber was stable at -25 dB over 12 hours of measurement.

Acknowledgments

This research was supported by a Grant-in-Aid for Scientific Research (Nos. 19H05601, 20K14482) and the Murata Science Foundation. This experiment was performed at RI Beam Factory operated by RIKEN Nishina Center and CNS, The University of Tokyo.

References

[1] K. Harada, T. Aoki, S. Ezure, K. Kato, T. Hayamizu, H. Kawamura, T. Inoue, H. Arikawa, T. Ishikawa, T. Aoki, A. Uchiyama, K. Sakamoto, S. Ito, M. Itoh, S. Ando, A. Hatakeyama, K. Hatanaka, K. Imai, T. Murakami, H.S. Nataraj, Y. Shimizu, T. Sato, T. Wakasa, H.P. Yoshida, Y. Sakemi, 2016 *Appl. Opt.* **55** (5) 1164-1169.

[2] K Harada , T Aoki, K Kato, H Kawamura, T Inoue, T Aoki, A Uchiyama, K Sakamoto, S Ito, M Itoh, T Hayamizu, A Hatakeyama, K Hatanaka, T Wakasa and Y Sakemi, 2016 *J. Phys. Conf. Ser.* **691** 012017.

[3] Y. Sakemi, T. Aoki, R. Calabrese, H. Haba, K. Harada, T. Hayamizu, Y. Ichikawa, K. Jungmann6, A. Kastberg, Y. Kotaka, Y. Matsuda, Y. Matsuo, H. Nagahama, K. Nakamura, M. Otsuka, N. Ozawa1, K. S. Tanaka, A. Uchiyama, H. Ueno, and L. Willmann, 2021 *AIP Conference Proceedings* 2319, 080020.

[4] H. Nagahama, K. Nakamura, N. Ozawa, Y. Kotaka, H. Haba, T. Hayamizu,M. Ohtsuka, K. S. Tanaka, H. Ueno, Y. Ichikawa, 149 Y. Sakemi, 2020 *Accel. Prog. Rep.* **53**, 57.

[5] N. Ozawa H. Nagahama, T. Hayamizu, K. Nakamura, M. Sato, S. 153 Nagase, Y. Kotaka, K. Kamakura, K. S. Tanaka, M.154 Ohtsuka, Y. Ichikawa, A. Takamine, H. Haba, H. Ueno, and Y. Sakemi, 2021 *Accel. Prog. Rep.* **54**, 117.

[6] T Hayamizu, K Nakamura, T Aoki, Y Kotaka, H Nagahama, S Nagase, M Ohtsuka, N Ozawa, M Sato, A Takamine, K. S. Tanaka, H Haba, Y Sakemi, 2021 *Accel. Prog. Rep.* **54**, 116.

[7] T Hayamizu, H Haba, K Nakamura, T Aoki, H Nagahama, K. S. Tanaka, N Ozawa, M Ohtsuka, Y Sakemi, 2022 *Few-Body Systems* **63**, 11.

[8] N. Shitara, N. Yamanaka, B.K. Sahoo, T. Watanabe, B.P. Das, CP violating effects in 210Fr and prospects for new physics beyond the Standard Model. 2021 JHEP **2021**, 124.

[9] V. Andreev, D.G. Ang, D. DeMille, J.M. Doyle, G. Gabrielse, J. Haefner, N.R. Hutzler, Z. Lasner, C. Meisenhelder, B.R. O' Leary, C.D. Panda, A.D. West, E.P. West, X. Wu, A.C.M.E. Collaboration, 2018 *Nature* **562**, 355.

[10]J. E. Simsarian, A. Ghosh, G. Gwinner, L. A. Orozco, G. D. Sprouse, and P. A. Voytas, 1996 *Phys. Rev. Lett.* **76**, 3522-3525.

[11] S. N. Atutov, R. Calabrese, V. Guidi, B. Mai, E. Scansani, G. Stancari, L. Tomassetti, L. Corradi, A. Dainelli, V. Biancalana, A. Burchianti, C. Marinelli, E. Mariotti, L. Moi, and S. Veronesi, "Cooling and trapping of radioactive atoms: the Legnaro francium magneto-optical trap," 2003 *J. Opt. Soc. Am.* B **20**, 953-959.

[12] W. Z. Zhao, J. E. Simsarian, L. A. Orozco, and G. D. Sprouse, "A computer-based digital feedback control of frequency drift of multiple lasers," 1998 *Rev. Sci. Instrum.* **69**, 3737-3740 .

[13] Y Hisai, D Akamatsu, T Kobayashi, S Okubo, H Inaba, K Hosaka, M Yasuda, and Feng-Lei Hong, 2019 *Opt. Express* **27**, 6404.

[14] K. Saleh, J. Millo, A. Didier, Y. Kersale, and C. Lacroute, 2015 *Appl. Opt.* **54**, 9446.

[15] L. Couturier, I. Nosske, F. Hu, C. Tan, C. Qiao Y. H. Jiang, Peng Chen, and M Weidemuller, 2018 *Rev. Sci. Instrum.* **89**, 043103.
[16] König, K., Imgram, P., Krämer, B Maaß1 · K Mohr, T Ratajczyk, F Sommer, W. Nörtershäuser, 2020 *Appl. Phys.* B **126**, 86.

Effects Beyond Center-of-Mass Separability in a Trapped Bosonic Mixture: Exact Results

O E Alon[1,2] and L S Cederbaum[3]

[1] Department of Mathematics, University of Haifa, Haifa, Israel
[2] Haifa Research Center for Theoretical Physics and Astrophysics, University of Haifa, Haifa, Israel
[3] Theoretical Chemistry, Physical Chemistry Institute, Heidelberg University, Heidelberg, Germany

E-mail: ofir@research.haifa.ac.il

Abstract. An exactly solvable model mimicking demixing of two Bose-Einstein condensates at the many-body level of theory is devised. Various properties are expressed in closed form along the demixing pathway and investigated. The connection between the center-of-mass coordinate and in particular the relative center-of-mass coordinate and demixing is explained. The model is also exactly solvable at the mean-field level of theory, allowing thereby comparison between many-body and mean-field properties. Applications are briefly discussed.

1. Introduction

Demixing of Bose-Einstein condensates has drawn an extensive attention and usually studied numerically, either at the mean-field level or at the many-body level of theory, see, e.g., [1-29]. Spatial inhomogeneity makes the problem analytically almost intractable. Here we introduce a solvable model which enables one to emulate demixing, or mixing, of two Bose-Einstein condensates at the many-body as well as at mean-field levels of theory. With a solvable model we can investigate analytically various properties along the pathway of demixing, such as the energetics, spatial overlap of the bosonic clouds, and entanglement between the two species to list a few. To this end, we extend the harmonic-interaction model for mixtures [30-32] to treat demixing. Our work builds on and naturally goes beyond [33-35]. The harmonic-interaction model for bosons, fermions, and mixtures has drawn much attention, see, e.g., [36-51].

2. Theory and Properties

Consider two Bose-Einstein condensates which consist of species 1 and species 2 bosons, respectively. Condensate 1 is placed in an harmonic potential localized at the origin and condensate 2 is held in an harmonic potential centered at a distance L from the first. All particle-particle interactions are harmonic. The many-particle Hamiltonian hence reads

$$\hat{H}(x_1,\ldots,x_{N_1},y_1,\ldots,y_{N_2}) = \\ = \sum_{j=1}^{N_1}\left(-\frac{1}{2m_1}\frac{\partial^2}{\partial x_j^2}+\frac{1}{2}m_1\omega^2 x_j^2\right)+\sum_{j=1}^{N_2}\left[-\frac{1}{2m_2}\frac{\partial^2}{\partial y_j^2}+\frac{1}{2}m_2\omega^2\left(y_j-L\right)^2\right]+ \\ +\lambda_1\sum_{1\le j<k}^{N_1}\left(x_j-x_k\right)^2+\lambda_2\sum_{1\le j<k}^{N_2}\left(y_j-y_k\right)^2+\lambda_{12}\sum_{j=1}^{N_1}\sum_{k=1}^{N_2}\left(x_j-y_k\right)^2. \tag{1}$$

Throughout this investigation we work in one spatial dimension and take $\hbar = 1$. There are N_1 bosons of type 1 and N_2 bosons of type 2, the corresponding masses are m_1 and m_2, the intra-species interaction strengths are λ_1 and λ_2, and the inter-species interaction strength is λ_{12}. The total number of bosons is denoted by $N = N_1 + N_2$.

Expressed in terms of the Jacoby coordinates of the mixture, $Q_k = \frac{1}{\sqrt{k(k+1)}}\sum_{j=1}^{k}(x_{k+1} - x_j)$, $1 \le k \le N_1 - 1$; $Q_{N_1-1+k} = \frac{1}{\sqrt{k(k+1)}}\sum_{j=1}^{k}(y_{k+1} - y_j)$, $1 \le k \le N_2 - 1$; $Q_{N-1} = \sqrt{N_1N_2}\,(X_{N_1} - Y_{N_2})$; and $Q_N = \frac{m_1N_1}{M}X_{N_1} + \frac{m_2N_2}{M}Y_{N_2}$, where $X_{N_1} = \frac{1}{N_1}\sum_{j=1}^{N_1} x_j$ and $Y_{N_2} = \frac{1}{N_2}\sum_{j=1}^{N_2} y_j$ are the center-of-mass coordinates of the individual species, the Hamiltonian (1) is diagonalized and takes on the form

$$\hat{H}(Q_1,\ldots,Q_N) = \sum_{k=1}^{N_1-1}\left(-\frac{1}{2m_1}\frac{\partial^2}{\partial Q_k^2}+\frac{1}{2}m_1\Omega_1^2Q_k^2\right)+\sum_{k=N_1}^{N-2}\left(-\frac{1}{2m_2}\frac{\partial^2}{\partial Q_k^2}+\frac{1}{2}m_2\Omega_2^2Q_k^2\right)+ \\ +\left[-\frac{1}{2M_{12}}\frac{\partial^2}{\partial Q_{N-1}^2}+\frac{1}{2}M_{12}\Omega_{12}^2\left(Q_{N-1}-Q_{N-1}^0\right)^2\right]+\left[-\frac{1}{2M}\frac{\partial^2}{\partial Q_N^2}+\frac{1}{2}M\omega^2\left(Q_N-Q_N^0\right)^2\right]+ \\ +\frac{1}{2}\frac{m_1N_1m_2N_2}{M}\omega^2L^2\left(1-\frac{\omega^2}{\Omega_{12}^2}\right), \tag{2}$$

where $M_{12} = \frac{m_1m_2}{M}$ and $M = N_1m_1 + N_2m_2$ are the relative center-of-mass and total masses, respectively. The shifts in the relative center-of-mass and center-of-mass coordinates

$$Q_{N-1}^0 = -\sqrt{N_1N_2}\frac{\omega^2}{\Omega_{12}^2}L, \qquad Q_N^0 = \frac{m_2N_2}{M}L \tag{3}$$

emerge from completing the squares but actually govern, as we shall discuss below, the demixing of the two condensates. In particular, unless $L \neq 0$, demixing cannot occur in our model. The specific case $L = 0$, which cannot exhibit demixing, has been analyzed in [32]. Finally, the interaction-dressed frequencies are given by

$$\Omega_{12} = \sqrt{\omega^2+2\left(\frac{N_1}{m_2}+\frac{N_2}{m_1}\right)\lambda_{12}}, \\ \Omega_1 = \sqrt{\omega^2+\frac{2}{m_1}\left(N_1\lambda_1+N_2\lambda_{12}\right)}, \quad \Omega_2 = \sqrt{\omega^2+\frac{2}{m_2}\left(N_2\lambda_2+N_1\lambda_{12}\right)} \tag{4}$$

and seen to be independent of L. For attractive inter-species interaction $\frac{\Omega_{12}}{\omega} > 1$ and for repulsive interaction $0 < \frac{\Omega_{12}}{\omega} < 1$. Correspondingly, $|Q_{N-1}^0| \to 0$ with increasing inter-species attraction and $|Q_{N-1}^0| \to \infty$ with increasing repulsion. From a different perspective, for an attractive inter-species interaction the last term in the Hamiltonian (2) and hence the total energy, see below, decrease as $L \to 0$ whereas for repulsive inter-species interaction the total energy decreases as $L \to \infty$, as is expected for attractive and repulsive forces. The center-of-mass of the mixture is localized at Q^0_{N-1} independently of any interaction.

With diagonalization of the Hamiltonian (1) to (2), the wavefunction and energy of the ground state are readily given by

$$\Psi(Q_1,\ldots,Q_N) = \left(\frac{m_1\Omega_1}{\pi}\right)^{\frac{N_1-1}{4}} \left(\frac{m_2\Omega_2}{\pi}\right)^{\frac{N_2-1}{4}} \left(\frac{M_{12}\Omega_{12}}{\pi}\right)^{\frac{1}{4}} \left(\frac{M\omega}{\pi}\right)^{\frac{1}{4}} \times$$
$$\times e^{-\frac{1}{2}\left\{m_1\Omega_1 \sum_{k=1}^{N_1-1} Q_k^2 + m_2\Omega_2 \sum_{k=N_1}^{N-2} Q_k^2 + M_{12}\Omega_{12}\left[Q_{N-1}-Q^0_{N-1}\right]^2 + M\omega\left[Q_N - Q^0_N\right]^2\right\}} \quad (5)$$

and

$$E = \frac{1}{2}\left[(N_1-1)\sqrt{\omega^2 + \frac{2}{m_1}(N_1\lambda_1 + N_2\lambda_{12})} + (N_2-1)\sqrt{\omega^2 + \frac{2}{m_2}(N_2\lambda_2 + N_1\lambda_{12})} + \right.$$
$$\left. + \sqrt{\omega^2 + 2\left(\frac{N_1}{m_2} + \frac{N_2}{m_1}\right)\lambda_{12}} + \omega + \frac{m_1N_1m_2N_2}{M}\omega^2 L^2\left(1 - \frac{\omega^2}{\Omega_{12}^2}\right)\right]. \quad (6)$$

It is instructive to compare the structures of the wavefunction (5) and energy (6) to the solution of the problem for $L = 0$ [32]. In particular, the center-of-mass and relative center-of-mass are not centered anymore at their origins and a term proportional to L^2 is added to the energy. We shall analyze the implications of these structures shortly.

To translate the wavefunction to the laboratory frame, we plug in the mixture's Jacoby coordinates explicitly into (5). Furthermore, the shifts of the relative center-of-mass and center-of-mass coordinates (3) have to be 'translated' to corresponding coordinates' shifts x_0 and y_0 of the two species in the laboratory frame. Using the inverse relations $X_{N_1} = Q_N + \frac{m_2}{M}\sqrt{\frac{N_2}{N_1}}Q_{N-1}$ and $Y_{N_2} = Q_N - \frac{m_1}{M}\sqrt{\frac{N_1}{N_2}}Q_{N-1}$ the final result for the wavefunction is given by

$$\Psi(x_1,\ldots,x_{N_1},y_1,\ldots,y_{N_2}) = \left(\frac{m_1\Omega_1}{\pi}\right)^{\frac{N_1-1}{4}} \left(\frac{m_2\Omega_2}{\pi}\right)^{\frac{N_2-1}{4}} \left(\frac{M_{12}\Omega_{12}}{\pi}\right)^{\frac{1}{4}} \left(\frac{M\omega}{\pi}\right)^{\frac{1}{4}} \times$$
$$\times e^{-\frac{\alpha_1}{2}\sum_{j=1}^{N_1}(x_j-x_0)^2 - \beta_1\sum_{1\le j<k}^{N_1}(x_j-x_0)(x_k-x_0)} e^{-\frac{\alpha_1}{2}\sum_{j=1}^{N_2}(y_j-y_0)^2 - \beta_2\sum_{1\le j<k}^{N_2}(y_j-y_0)(y_k-y_0)} \times$$
$$\times e^{+\gamma\sum_{j=1}^{N_1}\sum_{k=1}^{N_2}(x_j-x_0)(y_k-y_0)}, \quad (7)$$

where the coefficients α, β, and γ are collected in Appendix A. The shifts of the coordinates in the laboratory frame are given by the expectation values of the individual species' center-of-mass operators

$$x_0 = \langle\Psi|\hat{X}_{N_1}|\Psi\rangle = \frac{m_2N_2}{M}\left(1 - \frac{\omega^2}{\Omega_{12}^2}\right)L, \quad y_0 = \langle\Psi|\hat{Y}_{N_2}|\Psi\rangle = \left[1 - \frac{m_1N_1}{M}\left(1 - \frac{\omega^2}{\Omega_{12}^2}\right)\right]L. \quad (8)$$

We can now discuss the meaning of repulsion and attraction between the two condensates which becomes transparent within our model. When $\lambda_{12} = 0$, species 1 is localized at $x_0 = 0$ and species 2 at $y_0 = L$. For repulsive inter-species interaction the two species grow apart: x_0 decreases and y_0 increases; whereas for attractive inter-species interaction the two species come closer together: x_0 increases and y_0 decreases. Side by side, the term added to the energy [last term in (6)] takes on an appealing form as a function of the individual species' center-of-mass expectation values (8),

$$\frac{1}{2}\frac{m_1N_1m_2N_2}{M}\omega^2L^2\left(1 - \frac{\omega^2}{\Omega_{12}^2}\right) = \frac{1}{2}m_1N_1\omega^2x_0^2 + \frac{1}{2}m_2N_2\omega^2(y_0 - L)^2 + \lambda_{12}N_1N_2\left(x_0 - y_0\right)^2, \quad (9)$$

also see Appendix B. The added energy term (9) can be interpreted as the 'classical potential energy' of a N_1 particles of mass m_1 shifted by x_0 in a harmonic potential localized at the origin, N_2 particles of mass m_2 shifted by $y_0 - L$ in a harmonic potential localized at L, and the corresponding energy of their mutual harmonic interaction which scales like the square of their distance $(x_0 - y_0)^2$. We emphasis that all terms of the mixture's energy (6), except that originating from the center-of-mass, depend on the inter-species interaction λ_{12} and hence vary along the demixing pathway.

Let us proceed to many-body quantities derived from the many-particle wavefunction of the mixture (7). The all-particle density matrix of the mixture is defined by (here normalized to unity):

$$\rho_{12}^{(N)}(x_1,\ldots,x_{N_1},y_1,\ldots,y_{N_2},x'_1,\ldots,x'_{N_1},y'_1,\ldots,y'_{N_2}) =$$
$$= \left(\frac{m_1\Omega_1}{\pi}\right)^{\frac{N_1-1}{2}} \left(\frac{m_2\Omega_2}{\pi}\right)^{\frac{N_2-1}{2}} \left(\frac{M_{12}\Omega_{12}}{\pi}\right)^{\frac{1}{2}} \left(\frac{M\omega}{\pi}\right)^{\frac{1}{2}} \times$$
$$\times e^{-\frac{\alpha_1}{2}\sum_{j=1}^{N_1}\left[(x_j-x_0)^2+(x'_j-x_0)^2\right]-\beta_1\sum_{1\le j<k}^{N_1}\left[(x_j-x_0)(x_k-x_0)+(x'_j-x_0)(x'_k-x_0)\right]} \times$$
$$\times e^{-\frac{\alpha_2}{2}\sum_{j=1}^{N_2}\left[(y_j-y_0)^2+(y'_j-y_0)^2\right]-\beta_2\sum_{1\le j<k}^{N_2}\left[(y_j-y_0)(y_k-y_0)+(y'_j-y_0)(y'_k-y_0)\right]} \times$$
$$\times e^{+\gamma\sum_{j=1}^{N_1}\sum_{k=1}^{N_2}\left[(x_j-x_0)(y_k-y_0)+(x'_j-x_0)(y'_k-y_0)\right]}. \quad (10)$$

The integration of (10) to the intra-species and inter-species reduced density matrices [52] follows the lines of [32, 34] and are not reproduced here. The final results for the lowest-order intra-species and inter-species reduced density matrices are given by

$$\rho_1^{(1)}(x,x') = N_1\left(\frac{\alpha_1+C_{1,0}}{\pi}\right)^{\frac{1}{2}} e^{-\frac{\alpha_1}{2}\left[(x-x_0)^2+(x'-x_0)^2\right]} e^{-\frac{1}{4}C_{1,0}\left[(x-x_0)+(x'-x_0)\right]^2},$$

$$\rho_2^{(1)}(y,y') = N_2\left(\frac{\alpha_1+C'_{0,1}}{\pi}\right)^{\frac{1}{2}} e^{-\frac{\alpha_2}{2}\left[(y-y_0)^2+(y'-y_0)^2\right]} e^{-\frac{1}{4}C'_{0,1}\left[(y-y_0)+(y'-y_0)\right]^2},$$

$$\rho_{12}^{(2)}(x,x',y,y') = N_1N_2\left[\frac{(\alpha_1+C_{1,1})(\alpha_2+C'_{1,1})-D_{1,1}^2}{\pi^2}\right]^{\frac{1}{2}} e^{-\frac{\alpha_1}{2}\left[(x-x_0)^2+(x'-x_0)^2\right]} \times$$
$$\times e^{-\frac{\alpha_2}{2}\left[(y-y_0)^2+(y'-y_0)^2\right]} \times e^{-\frac{1}{4}C_{1,1}\left[(x-x_0)+(x'-x_0)\right]^2} e^{-\frac{1}{4}C'_{1,1}\left[(y-y_0)+(y'-y_0)\right]^2} \times$$
$$\times e^{+\frac{1}{2}D_{1,1}\left[(x-x_0)+(x'-x_0)\right]\left[(y-y_0)+(y'-y_0)\right]} e^{+\frac{1}{2}D'_{1,1}\left[(x-x_0)-(x'-x_0)\right]\left[(y-y_0)-(y'-y_0)\right]}, \quad (11)$$

where the various coefficients $C_{1,0}$, $C'_{0,1}$, $C_{1,1}$, $C'_{1,1}$, $D_{1,1}$, and $D'_{1,1}$ are results of coupled recursive relations [32] and prescribed in Appendix A. The respective densities, i.e., the diagonals of the reduced density matrices (11), are given by

$$\rho_1^{(1)}(x) = N_1\left(\frac{\alpha_1+C_{1,0}}{\pi}\right)^{\frac{1}{2}} e^{-(\alpha_1+C_{1,0})(x-x_0)^2},$$

$$\rho_2^{(1)}(y) = N_2\left(\frac{\alpha_2+C'_{0,1}}{\pi}\right)^{\frac{1}{2}} e^{-(\alpha_2+C'_{0,1})(y-y_0)^2},$$

$$\rho_{12}^{(2)}(x,y) = N_1N_2\left[\frac{(\alpha_1+C_{1,1})(\alpha_2+C'_{1,1})-D_{1,1}^2}{\pi^2}\right]^{\frac{1}{2}} e^{-(\alpha_1+C_{1,1})(x-x_0)^2} \times$$
$$\times e^{-(\alpha_2+C'_{1,1})(y-y_0)^2} e^{+2D_{1,1}(x-x_0)(y-y_0)}. \quad (12)$$

Clearly, the intra-species densities are localized around x_0 or y_0 whereas the inter-species density is localized both at x_0 and y_0. Since all these many-body quantities are given analytically, we can evaluate them explicitly at each point along the demixing pathway.

The extent to which the two condensates mix along the demixing pathway can be quantified using the spatial overlap between their corresponding densities (12). Explicitly, we consider the overlap of the square roots of the one-particle densities per particle,

$$S_{12} = \int_{-\infty}^{+\infty} dx \sqrt{\frac{\rho_1^{(1)}(x)}{N_1}} \sqrt{\frac{\rho_2^{(1)}(x)}{N_2}} =$$
$$= \left(\frac{\sqrt{(\alpha_1 + C_{1,0})(\alpha_2 + C'_{0,1})}}{\frac{1}{2}[(\alpha_1 + C_{1,0}) + (\alpha_2 + C'_{0,1})]} \right)^{\frac{1}{2}} e^{-\frac{(\alpha_1+C_{1,0})(\alpha_2+C'_{0,1})}{2[(\alpha_1+C_{1,0})+(\alpha_2+C'_{0,1})]}(x_0-y_0)^2} =$$
$$= \left(\frac{\sqrt{(\alpha_1 + C_{1,0})(\alpha_2 + C'_{0,1})}}{\frac{1}{2}[(\alpha_1 + C_{1,0}) + (\alpha_2 + C'_{0,1})]} \right)^{\frac{1}{2}} e^{-\frac{(\alpha_1+C_{1,0})(\alpha_2+C'_{0,1})}{2[(\alpha_1+C_{1,0})+(\alpha_2+C'_{0,1})]}\frac{\omega^4}{\Omega_{12}^4}L^2}. \quad (13)$$

Definition (13) seems natural since it reduces to unity when the two densities are equal. Generally, when the distance between the two condensates $y_0 - x_0 = \frac{\omega^2}{\Omega_{12}^2} L$ increases, i.e., for $\frac{\Omega_{12}}{\omega} < 1$, their overlap decreases and vice versa. Of course, expression (13) gives the precise value of the two condensates' overlap as a function of all parameters; the masses, interaction strengths, and the number of particles of each species. Below, we obtain the analogous expression evaluated at the mean-field level of theory, in which the interplay between the intra-species and inter-species interactions can be straightforwardly analyzed.

Let us move to the mean-field solution for demixing which is obtained analytically as follows. The derivation generalizes that in [32, 34, 41]. The Gross-Pitaevskii ansatz for the mixture's wavefunction is

$$\Phi(x_1, \ldots, x_{N_1}, y_1, \ldots, y_{N_2}) = \prod_{j=1}^{N_1} \phi_1(x_j) \prod_{k=1}^{N_2} \phi_2(y_k). \quad (14)$$

The orbitals $\phi_1(x)$ and $\phi_2(y)$ have to be determined self consistently. Sandwiching the many-body Hamiltonian (1) with the mean-field ansatz (14) one gets the Gross-Pitaevskii energy functional. Minimizing the latter with respect to the shapes of the two normalized orbitals $\phi_1(x)$ and $\phi_2(y)$, one obtains the coupled nonlinear integro-differential equations

$$\left\{ -\frac{1}{2m_1}\frac{\partial^2}{\partial x^2} + \frac{1}{2} m_1 \omega^2 x^2 + \Lambda_1 \int dx' |\phi_1(x')|^2 (x - x')^2 + \right.$$
$$\left. +\Lambda_{21} \int dy |\phi_2(y)|^2 (x - y)^2 \right\} \phi_1(x) = \mu_1 \phi_1(x),$$
$$\left\{ -\frac{1}{2m_2}\frac{\partial^2}{\partial y^2} + \frac{1}{2} m_2 \omega^2 (y - L)^2 + \Lambda_2 \int dy' |\phi_2(y')|^2 (y - y')^2 + \right.$$
$$\left. +\Lambda_{12} \int dx |\phi_1(x)|^2 (x - y)^2 \right\} \phi_2(y) = \mu_2 \phi_2(y), \quad (15)$$

where μ_1 and μ_2 stand for the respective chemical potentials and the mean-field interaction parameters are given by $\Lambda_1 = \lambda_1(N_1 - 1)$, $\Lambda_2 = \lambda_1(N_2 - 1)$, $\Lambda_{12} = \lambda_{12} N_1$, and $\Lambda_{21} = \lambda_{12} N_2$.

Recall that within the Gross-Pitaevskii treatment of demixing only the interaction parameters Λ_1, Λ_2, Λ_{12}, and Λ_{21} appear.

The coupled Gross-Pitaevskii equations (15) admit an analytic solution. This is intriguing in itself, as we are not aware of other analytical mean-field solutions for demixing of two spatially-inhomogeneous Bose-Einstein condensates. The final result for the orbitals reads

$$\phi_1(x) = \left(\frac{m_1}{\pi}\sqrt{\omega^2 + \frac{2}{m_1}(\Lambda_1+\Lambda_{21})}\right)^{\frac{1}{4}} e^{-\frac{m_1}{2}\sqrt{\omega^2+\frac{2}{m_1}(\Lambda_1+\Lambda_{21})}(x-x_0)^2},$$
$$\phi_2(y) = \left(\frac{m_2}{\pi}\sqrt{\omega^2 + \frac{2}{m_2}(\Lambda_2+\Lambda_{12})}\right)^{\frac{1}{4}} e^{-\frac{m_2}{2}\sqrt{\omega^2+\frac{2}{m_2}(\Lambda_2+\Lambda_{12})}(y-y_0)^2}, \qquad (16)$$

where further details are collected in Appendix B. Finally, the Gross-Pitaevskii energy of the mixture takes on the following form, expressed as a function of the interaction parameters only:

$$\varepsilon^{GP} = \frac{E^{GP}}{N} = \frac{1}{2}\frac{\Lambda_{12}\sqrt{\omega^2+\frac{2}{m_1}(\Lambda_1+\Lambda_{21})}+\Lambda_{21}\sqrt{\omega^2+\frac{2}{m_2}(\Lambda_2+\Lambda_{12})}}{\Lambda_{12}+\Lambda_{21}} +$$
$$+\frac{1}{2}\frac{\Lambda_{12}m_1\omega^2x_0^2+\Lambda_{21}m_2\omega^2(y_0-L)^2}{\Lambda_{12}+\Lambda_{21}} + \frac{\Lambda_{12}\Lambda_{21}(x_0-y_0)^2}{\Lambda_{12}+\Lambda_{21}}. \qquad (17)$$

Indeed, the first line in (17) is the mean-field energy as if the two harmonic traps overlap ($L = 0$), see [32], and the second line is precisely the 'potential-energy-and-interaction' term added at the many-body level of theory to describe the demixing for $L \neq 0$, see (6) and (9).

Let us intermediately summarize. We have put forward an exactly-solvable model for demixing of two Bose-Einstein condensates, whose many-body and mean-field ground-state solutions are given in closed and analytical forms. We can now ask further questions, on energetics, condensation, correlations, and on other properties, first at the many-body level of theory and than at the mean-field level of theory, and investigate the respective differences. From what we have depicted so far above, the model (1) looks sufficiently rich such that a detailed account can only find sufficient room beyond the present paper. We hence proceed with exploration of just two additional selected quantities.

We return to the degree of mixing of the two condensates characterized by their spatial overlap (13), but now at the mean-field level. The Gross-Pitaevskii densities per particle are nothing but $\phi_1^2(x)$ and $\phi_2^2(y)$ [the orbitals (16) are real-valued functions]. Consequently, we readily find

$$S_{12}^{GP} = \int_{-\infty}^{+\infty} dx\phi_1(x)\phi_2(x) = \left(\frac{\sqrt{m_1m_2\sqrt{\omega^2+\frac{2}{m_1}(\Lambda_1+\Lambda_{21})}\sqrt{\omega^2+\frac{2}{m_2}(\Lambda_2+\Lambda_{12})}}}{\frac{1}{2}\left[m_1\sqrt{\omega^2+\frac{2}{m_1}(\Lambda_1+\Lambda_{21})}+m_2\sqrt{\omega^2+\frac{2}{m_2}(\Lambda_2+\Lambda_{12})}\right]}\right)^{\frac{1}{2}} \times$$
$$\times e^{-\frac{m_1m_2\sqrt{\omega^2+\frac{2}{m_1}(\Lambda_1+\Lambda_{21})}\sqrt{\omega^2+\frac{2}{m_2}(\Lambda_2+\Lambda_{12})}}{2\left[m_1\sqrt{\omega^2+\frac{2}{m_1}(\Lambda_1+\Lambda_{21})}+m_2\sqrt{\omega^2+\frac{2}{m_2}(\Lambda_2+\Lambda_{12})}\right]}(x_0-y_0)^2}, \qquad (18)$$

where $x_0 - y_0 = -\frac{\omega^2}{\Omega_{12}^2}L$ just like in the many-body case, see Appendix B. Equation (18) shows in a transparent manner the dependence of the spatial overlap between the two condensates on the masses, m_1 and m_2, and intra-species Λ_1, Λ_2 and inter-species Λ_{12}, Λ_{21} interaction parameters. One can push the analysis further in the case the parameters of both species are equal, i.e., $m_2 = m_1$, $\Lambda_2 = \Lambda_1$, and $\Lambda_{12} = \Lambda_{21}$. Then, $S_{12}^{GP} = e^{-\frac{m_1}{4}\sqrt{\omega^2+\frac{2}{m_1}(\Lambda_1+\Lambda_{21})}\frac{\omega^4}{\Omega_{12}^4}L^2}$. For a given intra-species interaction Λ_1, attraction or repulsion, the overlap decreases monotonously with inter-species repulsion and increases monotonously with inter-species attraction. Furthermore, for a given inter-species interaction Λ_{21}, attraction or repulsion, the overlap increases monotonously with intra-species repulsion and decreases monotonously with intra-species attraction. The latter reflects the common wisdom that it is more difficult to spatially-separate condensates when they are made of repulsive species.

Perhaps, the most obvious difference between the many-body and mean-field wavefunctions of the mixture, equations (7) and (14), respectively, is that in the many-body treatment the two species are entangled whereas in the mean-field description, using the separable product state, the two species are obviously not entangled. Thus, finally, we move to the Schmidt decomposition of the many-body wavefunction, thereby generalizing recent results in the specific case of a symmetric mixture [35]. We begin from and employ Meher's formula which can be written as follows:

$$\left(\frac{s}{\pi}\right)^{\frac{1}{2}} e^{-\frac{1}{2}\frac{(1+\rho^2)s}{1-\rho^2}(x^2+y^2)} e^{+\frac{2\rho s}{1-\rho^2}xy} = \\ = \sum_{n=0}^{\infty} \sqrt{1-\rho^2}\rho^n \frac{1}{\sqrt{2^n n!}}\left(\frac{s}{\pi}\right)^{\frac{1}{4}} H_n(\sqrt{s}x)e^{-\frac{1}{2}sx^2} \frac{1}{\sqrt{2^n n!}}\left(\frac{s}{\pi}\right)^{\frac{1}{4}} H_n(\sqrt{s}y)e^{-\frac{1}{2}sy^2}, \quad (19)$$

where the parameters $s > 0$, $1 > \rho \geq 0$ for Schmidt decomposition of the wavefunction, and H_n are the Hermite polynomials. The wavefunction (7) is rewritten in terms of the Jacoby coordinates of each of the species, including the shifts of the coordinates x_0 and y_0,

$$\Psi(\bar{X}_1,\ldots,\bar{X}_{N_1},\bar{Y}_1,\ldots,\bar{Y}_{N_2}) = \left(\frac{m_1\Omega_1}{\pi}\right)^{\frac{N_1-1}{4}} \left(\frac{m_2\Omega_2}{\pi}\right)^{\frac{N_2-1}{4}} \left(\frac{M_{12}\Omega_{12}}{\pi}\right)^{\frac{1}{4}} \left(\frac{M\omega}{\pi}\right)^{\frac{1}{4}} \times \\ \times e^{-\frac{1}{2}\left(m_1\Omega_1\sum_{k=1}^{N_1-1}\bar{X}_k^2 + m_2\Omega_2\sum_{k=1}^{N_2-1}\bar{Y}_k^2\right)} \times \\ \times e^{-\frac{1}{2}\left(m_1\frac{m_2N_2\Omega_{12}+m_1N_1\omega}{M}\bar{X}_{N_1}^2 + m_2\frac{m_1N_1\Omega_{12}+m_2N_2\omega}{M}\bar{Y}_{N_2}^2\right)} e^{+\frac{m_1m_2\sqrt{N_1N_2}}{M}(\Omega_{12}-\omega)\bar{X}_{N_1}\bar{Y}_{N_2}}, \quad (20)$$

where we denote for brevity here and hereafter $\bar{X}_k = \frac{1}{\sqrt{k(k+1)}}\sum_{j=1}^{k}[(x_{k+1}-x_0)-(x_j-x_0)]$, $1 \leq k \leq N_1-1$; $\bar{Y}_k = \frac{1}{\sqrt{k(k+1)}}\sum_{j=1}^{k}[(y_{k+1}-y_0)-(y_j-y_0)]$, $1 \leq k \leq N_2-1$; $\bar{X}_{N_1} = \frac{1}{N_1}\sum_{j=1}^{N_1}(x_j-x_0)$; and $\bar{Y}_{N_2} = \frac{1}{N_2}\sum_{j=1}^{N_2}(y_j-y_0)$. Furthermore, in (20) it is convenient to treat first the case of mixing, i.e., of attractive inter-species interaction $\Omega_{12} > \omega$; the slight modifications in the treatment of demixing, i.e., for repulsive inter-species interaction $\Omega_{12} < \omega$, are put forward below. Clearly, the wavefunction (20) boils down to that of the symmetric mixture when the parameters of species 1 and species 2 bosons are equal and for $L = 0$. On the other hand, unlike the Schmidt decomposition of the symmetric mixture [35] and before Mehler's formula (19) can be applied, equation (20) would require a squeeze mapping of $\bar{X}_{N_1}$ and $\bar{Y}_{N_2}$. Thus, defining

$$\widetilde{X}_{N_1}^2 \equiv \bar{X}_{N_1}^2\sqrt{\frac{m_1\left(m_2N_2\Omega_{12}+m_1N_1\omega\right)}{m_2\left(m_1N_1\Omega_{12}+m_2N_2\omega\right)}}, \quad \widetilde{Y}_{N_2}^2 \equiv \bar{Y}_{N_2}^2\sqrt{\frac{m_2\left(m_1N_1\Omega_{12}+m_2N_2\omega\right)}{m_1\left(m_2N_2\Omega_{12}+m_1N_1\omega\right)}} \quad (21)$$

(satisfying $\widetilde{X}_{N_1}\widetilde{Y}_{N_2} = \bar{X}_{N_1}\bar{Y}_{N_2}$), the last row of the wavefunction (20) transforms and reads

$$e^{-\frac{1}{2}\frac{m_1m_2\sqrt{N_1N_2}}{M}\sqrt{(\Omega_{12}+\omega)^2+\frac{(m_1N_1-m_2N_2)^2}{m_1N_1m_2N_2}\Omega_{12}\omega}\left(\widetilde{X}_{N_1}^2+\widetilde{Y}_{N_2}^2\right)} e^{+\frac{m_1m_2\sqrt{N_1N_2}}{M}(\Omega_{12}-\omega)\widetilde{X}_{N_1}\widetilde{Y}_{N_2}}. \quad (22)$$

Now, with equation (22), Mehler's formula can be directly applied. The final result for the

Schmidt decomposition reads

$$\Psi(\bar{X}_1,\ldots,\widetilde{X}_{N_1},\bar{Y}_1,\ldots,\widetilde{Y}_{N_2}) = \sum_{n=0}^{\infty}\sqrt{1-\rho^2}\rho^n\Phi_{1,n}(\bar{X}_1,\ldots,\widetilde{X}_{N_1})\Phi_{2,n}(\bar{Y}_1,\ldots,\widetilde{Y}_{N_2}),$$

$$\Phi_{1,n}(\bar{X}_1,\ldots,\widetilde{X}_{N_1}) = \left(\frac{m_1\Omega_1}{\pi}\right)^{\frac{N_1-1}{4}} e^{-\frac{1}{2}m_1\Omega_1\sum_{k=1}^{N_1-1}\bar{X}_k^2}\frac{1}{\sqrt{2^n n!}}\left(\frac{s}{\pi}\right)^{\frac{1}{4}} H_n\left(\sqrt{s}\widetilde{X}_{N_1}\right)e^{-\frac{1}{2}s\widetilde{X}_{N_1}^2},$$

$$\Phi_{2,n}(\bar{Y}_1,\ldots,\widetilde{Y}_{N_2}) = \left(\frac{m_2\Omega_2}{\pi}\right)^{\frac{N_2-1}{4}} e^{-\frac{1}{2}m_2\Omega_2\sum_{k=1}^{N_2-1}\bar{Y}_k^2}\frac{1}{\sqrt{2^n n!}}\left(\frac{s}{\pi}\right)^{\frac{1}{4}} H_n\left(\sqrt{s}\widetilde{Y}_{N_2}\right)e^{-\frac{1}{2}s\widetilde{Y}_{N_2}^2}, \quad (23)$$

where the Schmidt parameters are

$$\rho = \frac{\left[\frac{\sqrt{\left(\frac{\Omega_{12}}{\omega}+1\right)^2+\frac{(m_1N_1-m_2N_2)^2}{m_1N_1m_2N_2}\frac{\Omega_{12}}{\omega}}+\left(\frac{\Omega_{12}}{\omega}-1\right)}{\sqrt{\left(\frac{\Omega_{12}}{\omega}+1\right)^2+\frac{(m_1N_1-m_2N_2)^2}{m_1N_1m_2N_2}\frac{\Omega_{12}}{\omega}}-\left(\frac{\Omega_{12}}{\omega}-1\right)}\right]^{+\frac{1}{2}}-1}{\left[\frac{\sqrt{\left(\frac{\Omega_{12}}{\omega}+1\right)^2+\frac{(m_1N_1-m_2N_2)^2}{m_1N_1m_2N_2}\frac{\Omega_{12}}{\omega}}+\left(\frac{\Omega_{12}}{\omega}-1\right)}{\sqrt{\left(\frac{\Omega_{12}}{\omega}+1\right)^2+\frac{(m_1N_1-m_2N_2)^2}{m_1N_1m_2N_2}\frac{\Omega_{12}}{\omega}}-\left(\frac{\Omega_{12}}{\omega}-1\right)}\right]^{+\frac{1}{2}}+1}, \qquad s = \sqrt{m_1m_2\omega\Omega_{12}}. \quad (24)$$

Finally, the case of repulsive inter-species interaction $\Omega_{12} < \omega$ implies the assignments of, e.g., $\bar{Y}_{N_2} \to -\bar{Y}_{N_2}$ in (20) and $\rho \to -\rho$ in (24), similarly to [35].

Equation (24) quantifies precisely as a function of the mixture's parameters the entanglement between the two species along the demixing pathway. It is instrumental to discuss a few limiting cases. Without inter-species interaction, i.e., when $\Omega_{12} = \omega$, one has $\rho = 0$ and the two species are, of course, not entangled; The condensates themselves can possess strong intra-species interactions though. When Ω_{12} is very large (strong inter-species attraction) or very small (strong inter-species repulsion), ρ increases more and more towards unity and a high degree of entanglement emerges. Hence, within our model, a high degree of entanglement accompanies both mixing and demixing. Last but not least, when a very large asymmetry between the two species exists, explicitly, say, $m_1N_1 \gg m_2N_2$, ρ decreases more and more towards zero, implying the entanglement diminishes further and further, see in this context the situation for $L = 0$ [31]. As can be expected, for symmetric mixtures and $L = 0$ some of the above-obtained results boil down to those in [35]. This is a suitable place to stop the current investigation.

3. Concluding Remarks

In the present work a solvable model mimicking demixing of two Bose-Einstein condensates at the many-body level of theory is devised and investigated. The wavefunction, energy, reduced density matrices, densities, spatial overlap, and entanglement between the two condensates expressed as the Schmidt decomposition of the many-particle wavefunction are given in closed form along the demixing pathway. The connection between the center-of-mass and in particular relative center-of-mass coordinates and demixing is elucidated, within our model. Furthermore, the model is also solved analytically at the mean-field level of theory, and the above-computed properties are expressed in closed form also at the mean-field level of theory. A short discussion on the differences between properties computed at the many-body and mean-field levels of theory along the demixing pathway is made.

There are several research directions the present investigation can lead to, of which we list the following three. An immediate study would be a comprehensive comparison between many-body and mean-field descriptions of demixing at the limit of an infinite number of particles. Therein, some properties, like the energy per particle and densities per particle, would exactly coincide and other properties, like variances per particle of many-particle observables and the overlap between the many-body and mean-field wavefunctions, can differ substantially [53-65]. For finite mixtures, the fragmentation [66] along the demixing pathway would be instrumental to follow. Another research venue that is worth pursuing is benchmarking multiconfigurational time-dependent Hartree methods [67-70] and other numerical approaches for bosonic mixtures along the demixing pathway. Finally, a more distant but rewarding challenge would be the emulation and subsequent investigation of many-body effects when scattering attractive bosonic clouds from a potential barrier, or off each other, fully analytically [71-74].

Acknowledgments
This research was supported by the Israel Science Foundation (Grant No. 1516/19). We thank Alexej I. Streltsov for motivating discussions.

Appendix A. Coefficients of the inter-species and intra-species reduced density matrices

It can be shown that the parameters of the wavefunction (7) in the laboratory frame entering the reduced density matrices are those of the wavefunction without the coordinates' shifts x_0 and y_0, and therefore are given by [32]:

$$\begin{aligned}
\alpha_1 &= m_1\Omega_1 + \beta_1, \qquad \beta_1 = m_1\left[-\Omega_1\frac{1}{N_1} + (m_2N_2\Omega_{12} + m_1N_1\omega)\frac{1}{MN_1}\right], \\
\alpha_2 &= m_2\Omega_2 + \beta_2, \qquad \beta_2 = m_2\left[-\Omega_2\frac{1}{N_2} + (m_1N_1\Omega_{12} + m_2N_2\omega)\frac{1}{MN_2}\right], \\
\gamma &= \frac{m_1m_2}{M}(\Omega_{12} - \omega).
\end{aligned} \tag{A.1}$$

Correspondingly, since the various integrations of the all-particle density matrix (10) are taken along the variables $(x'_j - x_0) = (x_j - x_0)$ and $(y'_k - y_0) = (y_k - y_0)$, and hence, upon these integrations, the coordinates' shifts x_0 and y_0 can be eliminated, the coefficients of the inter-species reduced density matrix (11) are also those of the corresponding reduced density matrix without coordinates' shifts [32]:

$$\begin{aligned}
\alpha_1 + C_{1,1} &= (\alpha_1 - \beta_1)\frac{[(\alpha_1-\beta_1) + N_1\beta_1][(\alpha_2-\beta_2) + (N_2-1)\beta_2] - \gamma^2 N_1(N_2-1)}{[(\alpha_1-\beta_1) + (N_1-1)\beta_1][(\alpha_2-\beta_2) + (N_2-1)\beta_2] - \gamma^2(N_1-1)(N_2-1)}, \\
\alpha_2 + C'_{1,1} &= (\alpha_2 - \beta_2)\frac{[(\alpha_2-\beta_2) + N_2\beta_2][(\alpha_1-\beta_1) + (N_1-1)\beta_1] - \gamma^2 N_2(N_1-1)}{[(\alpha_1-\beta_1) + (N_1-1)\beta_1][(\alpha_2-\beta_2) + (N_2-1)\beta_2] - \gamma^2(N_1-1)(N_2-1)}, \\
D_{1,1} &= \gamma\frac{(\alpha_1-\beta_1)(\alpha_2-\beta_2)}{[(\alpha_1-\beta_1) + (N_1-1)\beta_1][(\alpha_2-\beta_2) + (N_2-1)\beta_2] - \gamma^2(N_1-1)(N_2-1)}, \\
D'_{1,1} &= \gamma.
\end{aligned} \tag{A.2}$$

Analogously, the coefficients of the intra-species reduced density matrices (11) are given by [32]:

$$\begin{aligned}
\alpha_1 + C_{1,0} &= (\alpha_1 - \beta_1)\frac{[(\alpha_1-\beta_1) + N_1\beta_1][(\alpha_2-\beta_2) + N_2\beta_2] - \gamma^2 N_1N_2}{[(\alpha_1-\beta_1) + (N_1-1)\beta_1][(\alpha_2-\beta_2) + N_2\beta_2] - \gamma^2(N_1-1)N_2}, \\
\alpha_2 + C'_{0,1} &= (\alpha_2 - \beta_2)\frac{[(\alpha_1-\beta_1) + N_1\beta_1][(\alpha_2-\beta_2) + N_2\beta_2] - \gamma^2 N_1N_2}{[(\alpha_2-\beta_2) + (N_2-1)\beta_2][(\alpha_1-\beta_1) + N_1\beta_1] - \gamma^2(N_2-1)N_1}.
\end{aligned} \tag{A.3}$$

Of course, all these coefficients depend explicitly on the masses m_1, m_2, interaction strengths λ_1, λ_2, λ_{12}, and the numbers of particles N_1, N_2, and vary along the demixing pathway.

Appendix B. Further details of the solution of the coupled Gross-Pitaevskii equations and the coordinates' shifts

It is useful to rewrite (15) in terms of shifts of the coordinates x_0 and y_0, which themselves have to be determined explicitly within mean-field theory. Making use of the normalization of $\phi_1(x)$ and $\phi_2(y)$ and that they are even functions with respect to x_0 and y_0 [see the obtained self-consistent solution (16) given in the main text], one finds

$$\begin{aligned}
&\left\{-\frac{1}{2m_1}\frac{\partial^2}{\partial x^2}+\frac{1}{2}m_1\omega^2(x-x_0)^2+\Lambda_1\int dx'|\phi_1(x')|^2\left[(x-x_0)-(x'-x_0)\right]^2+\right.\\
&\left.+\Lambda_{21}\int dy|\phi_2(y)|^2\left[(x-x_0)-(y-y_0)\right]^2\right\}\phi_1(x)=\left[\mu_1-\frac{1}{2}m_1x_0^2-\Lambda_{21}(x_0-y_0)^2\right]\phi_1(x),\\
&\left\{-\frac{1}{2m_2}\frac{\partial^2}{\partial y^2}+\frac{1}{2}m_2\omega^2(y-y_0)^2+\Lambda_2\int dy'|\phi_2(y')|^2\left[(y-y_0)-(y'-y_0)\right]^2+\right.\\
&\left.+\Lambda_{12}\int dx|\phi_1(x)|^2\left[(x-x_0)-(y-y_0)\right]^2\right\}\phi_2(y)=\left[\mu_2-\frac{1}{2}m_2(y_0-L)^2-\Lambda_{12}(x_0-y_0)^2\right]\phi_2(y),
\end{aligned} \tag{B.1}$$

where, for the terms linear in $(x - x_0)$ and $(y - y_0)$ to drop out, x_0 and y_0 must obey

$$\begin{aligned}
&m_1\omega^2x_0+2\Lambda_{21}(x_0-y_0)=0,\\
&m_2\omega^2(y_0-L)-2\Lambda_{12}(x_0-y_0)=0.
\end{aligned} \tag{B.2}$$

The solution of the linear system (B.2) is $x_0 = \frac{2\Lambda_{21}}{m_1\Omega_{12}^2}L$ and $y_0 = \left(1-\frac{2\Lambda_{12}}{m_2\Omega_{12}^2}\right)L$, where $\Omega_{12} = \sqrt{\omega^2+2\left(\frac{\Lambda_{12}}{m_2}+\frac{\Lambda_{21}}{m_1}\right)}$, hence $x_0 - y_0 = \left[2\left(\frac{\Lambda_{12}}{m_2}+\frac{\Lambda_{21}}{m_1}\right)\frac{1}{\Omega_{12}^2}-1\right]L$. These are exactly the same values found within the many-body solution, see (8). Intriguingly, the frequency of the relative center-of-mass Jacoby coordinate in the many-body treatment is obtained (in the mean-field treatment) from re-expressing the mean-field equations using the coordinates' shifts x_0 and y_0.

Consequently, the Gross-Pitaevskii solution for the demixing scenario ($L \neq 0$) can be related to the solution of the coupled equations without the shifts of coordinates ($L = 0$). The self-consistent orbitals are given in the main text, see (16), and the respective chemical potentials read

$$\begin{aligned}
\mu_1&=\frac{1}{2}\left(\sqrt{\omega^2+\frac{2}{m_1}(\Lambda_1+\Lambda_{21})}+\frac{\Lambda_1}{\sqrt{\omega^2+\frac{2}{m_1}(\Lambda_1+\Lambda_{21})}}+\frac{\Lambda_{21}}{\sqrt{\omega^2+\frac{2}{m_2}(\Lambda_2+\Lambda_{12})}}\right)+\\
&+\frac{1}{2}m_1x_0^2+\Lambda_{21}(x_0-y_0)^2,\\
\mu_2&=\frac{1}{2}\left(\sqrt{\omega^2+\frac{2}{m_2}(\Lambda_2+\Lambda_{12})}+\frac{\Lambda_2}{\sqrt{\omega^2+\frac{2}{m_2}(\Lambda_2+\Lambda_{12})}}+\frac{\Lambda_{12}}{\sqrt{\omega^2+\frac{2}{m_1}(\Lambda_1+\Lambda_{21})}}\right)+\\
&+\frac{1}{2}m_2(y_0-L)^2+\Lambda_{12}(x_0-y_0)^2.
\end{aligned} \tag{B.3}$$

The total Gross-Pitaevskii energy (17) for demixing is equivalently obtained from the chemical potentials and the interaction energy, $E^{GP} = N_1\mu_1 + N_2\mu_2 - \frac{N_1}{2}\Big[\Lambda_1 \int dxdx'|\phi_1(x)|^2|\phi_1(x')|^2(x-$

Journal of Physics: Conference Series **2249** (2022) 012011 doi:10.1088/1742-6596/2249/1/012011

$x')^2 + \Lambda_{21} \int dxdy|\phi_1(x)|^2|\phi_2(y)|^2(x - y)^2\Big] - \frac{N_2}{2}\Big[\Lambda_2 \int dydy'|\phi_2(y)|^2|\phi_2(y')|^2(y - y')^2 + \Lambda_{12} \int dxdy|\phi_1(x)|^2|\phi_2(y)|^2(x - y)^2\Big]$.

Finally and for completeness, it is instructive to re-express the Hamiltonian in the laboratory frame (1) using the shifts of the coordinates x_0 and y_0,

$$\hat{H}(x_1,\ldots,x_{N_1},y_1,\ldots,y_{N_2}) = \sum_{j=1}^{N_1}\left[-\frac{1}{2m_1}\frac{\partial^2}{\partial x_j^2} + \frac{1}{2}m_1\omega^2\left(x_j - x_0\right)^2\right] + \\ + \sum_{j=1}^{N_2}\left[-\frac{1}{2m_2}\frac{\partial^2}{\partial y_j^2} + \frac{1}{2}m_2\omega^2\left(y_j - y_0\right)^2\right] + \lambda_1 \sum_{1\le j<k}^{N_1}\left[(x_j - x_0) - (x_k - x_0)\right]^2 + \\ + \lambda_2 \sum_{1\le j<k}^{N_2}\left[(y_j - y_0) - (y_k - y_0)\right]^2 + \lambda_{12}\sum_{j=1}^{N_1}\sum_{k=1}^{N_2}\left[(x_j - x_0) - (y_k - y_0)\right]^2 + \\ + \frac{1}{2}m_1N_1\omega^2x_0^2 + \frac{1}{2}m_2N_2\omega^2(y_0 - L)^2 + \lambda_{12}N_1N_2\left(x_0 - y_0\right)^2. \qquad \text{(B.4)}$$

Equation (B.4) admits an appealing physical interpretation of demixing and its energetics, within our model, as discussed in the main text.

References

[1] Ho T-L and Shenoy V B 1996 *Phys. Rev. Lett.* **77** 3276
[2] Esry B D, Greene C H, Burke J P Jr and Bohn J L 1997 *Phys. Rev. Lett.* **78** 3594
[3] Pu H and Bigelow N P 1998 *Phys. Rev. Lett.* **80** 1130
[4] Timmermans E 1998 *Phys. Rev. Lett.* **81** 5718
[5] Esry B D and Greene C H 1999 *Phys. Rev. A* **59** 1457
[6] Alon O E, Streltsov A I and Cederbaum L S 2006 *Phys. Rev. Lett.* **97** 230403
[7] Liu X P 2007 *Phys. Rev. A* **76** 053615
[8] Sakhel A R, DuBois J L and Glyde H R 2008 *Phys. Rev. A* **77** 043627
[9] Zöllner S, Meyer H-D and Schmelcher P 2008 *Phys. Rev. A* **78** 013629
[10] Oleś B and Sacha K 2008 *J. Phys. A* **41** 145005
[11] Papp S B, Pino J M and Wieman C E 2008 *Phys. Rev. Lett.* **101** 040402
[12] Navarro R, Carretero-González R and Kevrekidis P G 2009 *Phys. Rev. A* **80** 023613
[13] Hao Y and Chen S 2009 *Phys. Rev. A* **80** 043608
[14] Malet F, Kavoulakis G M and Reimann S M 2010 *Phys. Rev. A* **81** 013630
[15] Girardeau M D and Astrakharchik G E 2010 *Phys. Rev. A* **81** 043601
[16] Gautam S and Angom D 2010 *J. Phys. B* **43** 095302
[17] Mason P and Aftalion A 2011 *Phys. Rev. A* **84** 033611
[18] Sabbatini J, Zurek W H and Davis M J 2011 *Phys. Rev. Lett.* **107** 230402
[19] Wen L, Liu W M, Cai Y, Zhang J M and Hu J 2012 *Phys. Rev. A* **85** 043602
[20] García-March M Á and Busch T 2013 *Phys. Rev. A* **87** 063633
[21] Gautam S and Adhikari S K 2014 *Phys. Rev. A* **90** 043619
[22] Mistakidis S I, Katsimiga G C, Kevrekidis P G and Schmelcher P 2018 *New J. Phys.* **20** 043052
[23] Han J and Tsubota M 2019 *Phys. Rev. A* **99** 033607
[24] Richaud A, Zenesini A and Penna V 2019 *Sci. Rep.* **9** 6908
[25] Richaud A and Penna V 2019 *Phys. Rev. A* **100** 013609
[26] Chen T, Shibata K, Eto Y, Hirano T and Saito H 2019 *Phys. Rev. A* **100** 063610
[27] Sowiński T and García-March M Á 2019 *Rep. Prog. Phys.* **82** 104401
[28] Hejazi S S S, Polo J, Sachdeva R and Busch T 2020 *Phys. Rev. A* **102** 053309
[29] Warner C, Lam A Z, Bigagli N, Liu H C, Stevenson I and Will S 2021 *Phys. Rev. A* **104** 033302
[30] Osadchii M S and Muraktanov V V 1991 *Int. J. Quant. Chem.* **39** 173
[31] Bouvrie P A, Majtey A P, Tichy M C, Dehesa J S and Plastino A R 2014 *Eur. Phys. J. D* **68** 346
[32] Alon O E 2017 *J. Phys. A* **50** 295002
[33] Armstrong J R, Volosniev A G, Fedorov D V, Jensen A S and Zinner N T 2015 *J. Phys. A* **48** 085301
[34] Alon O E 2020 *Entropy* **22** 1342

[35] Alon O E 2021 *Atoms* **9** 92
[36] Pruski S, Maćkowiak J and Missuno O 1972 *Rep. Math. Phys.* **3** 227
[37] Pruski S, Maćkowiak J and Missuno O 1972 *Rep. Math. Phys.* **3** 241
[38] Robinson P D 1977 *J. Chem. Phys.* **66** 3307
[39] Hall R L 1978 *J. Phys. A* **11** 1227
[40] Hall R L 1978 *J. Phys. A* **11** 1235
[41] Cohen L and Lee C 1985 *J. Math. Phys.* **26** 3105
[42] Załuska-Kotur M A, Gajda M, Orłowski A and Mostowski J 2000 *Phys. Rev. A* **61** 033613
[43] Yan J 2003 *J. Stat. Phys.* **113** 623
[44] Gajda M 2006 *Phys. Rev. A* **73** 023603
[45] Armstrong J R, Zinner N T, Fedorov D V and Jensen A S 2011 *J. Phys. B* **44** 055303
[46] Armstrong J R, Zinner N T, Fedorov D V and Jensen A S 2012 *Phys. Rev. E* **86** 021115
[47] Kościk P and Okopińska A 2013 *Few-Body Syst.* **54** 1637
[48] Schilling C 2013 *Phys. Rev. A* **88** 042105
[49] Benavides-Riveros C L, Toranzo I V and Dehesa J S 2014 *J. Phys. B* **47** 195503
[50] Schilling C and Schilling R 2016 *Phys. Rev. A* **93** 021601(R)
[51] Klaiman S, Streltsov A I and Alon O E 2017 *Chem. Phys.* **482** 362
[52] Coleman A J and Yukalov V I 2000 *Reduced Density Matrices: Coulson's Challenge (Lectures Notes in Chemistry vol 72)* (Berlin: Springer)
[53] Castin Y and Dum R 1998 *Phys. Rev. A* **57** 3008
[54] Lieb E H, Seiringer R and Yngvason J 2000 *Phys. Rev. A* **61** 043602
[55] Lieb E H and Seiringer R 2002 *Phys. Rev. Lett.* **88** 170409
[56] Yukalov V I 2011 *Phys. Part. Nucl.* **42** 460
[57] Klaiman S and Alon O E 2015 *Phys. Rev. A* **91** 063613
[58] Klaiman S and Cederbaum L S 2016 *Phys. Rev. A* **94** 063648
[59] Anapolitanos I, Hott M and Hundertmark D 2017 *Rev. Math. Phys.* **29** 1750022
[60] Michelangeli A and Olgiati A 2017 *Anal. Math. Phys.* **7** 377
[61] Cederbaum L S 2017 *Phys. Rev. A* **96** 013615
[62] Klaiman S, Streltsov A I and Alon O E 2018 *J. Phys.: Conf. Ser.* **999** 012013
[63] Sakmann K and Schmiedmayer J 2018 arXiv:1802.03746v2 [cond-mat.quant-gas]
[64] Alon O E 2019 *Symmetry* **11** 1344
[65] Michelangeli A, Nam P T and Olgiati A 2019 *Rev. Math. Phys.* **31** 1950005
[66] Mueller E J, Ho T L, Ueda M and Baym G 2006 *Phys. Rev. A* **74** 033612
[67] Alon O E, Streltsov A I and Cederbaum L S 2007 *Phys. Rev. A* **76** 062501
[68] Krönke S, Cao L, Vendrell O and Schmelcher P 2013 *New J. Phys.* **15** 063018
[69] Cao L, Krönke S, Vendrell O and Schmelcher P 2013 *J. Chem. Phys.* **139** 134103
[70] Lévêque C and Madsen L B 2018 *J. Phys. B* **51** 155302
[71] Weiss C and Castin Y 2009 *Phys. Rev. Lett.* **102** 010403
[72] Streltsov A I, Alon O E and Cederbaum L S 2009 *Phys. Rev. A* **80** 043616
[73] Cederbaum L S, Streltsov A I, Band Y B and Alon O E 2007 *Phys. Rev. Lett.* **98** 110405
[74] Gertjerenken B, Billam T P, Blackley C L, Le Sueur C R, Khaykovich L, Cornish S L and Weiss C 2013 *Phys. Rev. Lett.* **111** 100406

Calculating critical temperature and critical exponents by self-similar approximants

V.I. Yukalov[1,2] and E.P. Yukalova[3]

[1]*Bogolubov Laboratory of Theoretical Physics, Joint Institute for Nuclear Research, Dubna 141980, Russia*

[2]*Instituto de Fisica de São Carlos, Universidade de São Paulo, CP 369, São Carlos 13560-970, São Paulo, Brazil*

[3]*Laboratory of Information Technologies, Joint Institute for Nuclear Research, Dubna 141980, Russia*

E-mails: *yukalov@theor.jinr.ru*, *yukalova@theor.jinr.ru*

Abstract

Self-similar approximation theory allows for defining effective sums of asymptotic series. The method of self-similar factor approximants is applied for calculating the critical temperature and critical exponents of the $O(N)$-symmetric φ^4 field theory in three dimensions by summing asymptotic ε expansions. This method is shown to be essentially simpler than other summation techniques involving complicated numerical calculations, while enjoying comparable accuracy.

Keywords: Self-similar factor approximants; asymptotic series; critical exponents

1 Introduction

Asymptotic expansions are widely employed in physics and applied mathematics [1, 2]. The expansion parameters are usually not small, because of which one needs to involve some kind of effective summation of divergent series. For this purpose, one often uses the method of Padé approximants [3], Borel summation [4] and its variants, such as Padé-Borel summation and Borel summation with conformal mapping. One also uses the methods of renormalization group, conformal bootstrap, Monte Carlo simulations, and other methods (see [5]) requiring quite heavy numerical calculations.

In this communication, we advocate another approach based on self-similar approximation theory [6–9]. The idea of this theory is to represent the transition from one approximation term to another as the motion of a dynamical system, with the approximation order playing the role of discrete time. Then the sequence of approximation terms is bijective to the dynamical-system trajectory and the limit of the sequence is bijective to the fixed point of the trajectory. In the vicinity of a fixed point, the evolution equation acquires the form of a self-similar relation, which justifies the name of the self-similar approximation theory. Mathematical details can be found in the review articles [10, 11]. Here we only give the main ideas of the general theory in Sec. 2. In Sec. 3, we formulate the method of self-similar factor approximants. We do not plunge into the foundations of the theory, but just give the scheme of the method application. We apply the method for calculating the critical temperature shift for the $O(N)$ -symmetric φ^4 field theory in three dimensions in Sec. 4 and the critical exponents of this model in Sec. 5. We show that the method of self-similar factor approximants provides the accuracy comparable with other elaborate techniques involving heavy numerical calculations, while being essentially simpler. We compare our results with the most accurate values obtained by other methods summarized in Refs. [5,12–14]. We compare our results with Monte Carlo simulations, conformal bootstrap, hypergeometric Meijer summation, Borel summation, Borel summation with conformal mapping, and the method of nonperturbative renormalization group. This comparison shows good agreement of the self-similar factor approximants with the results obtained by other methods. Sec. 6 concludes.

2 Self-similar approximation theory

Let the function of interest be presented as an asymptotic expansion over a parameter x assumed to be small,

$$f_k(x) = \sum_{n=0}^{k} a_n x^n \qquad (x \to 0) . \tag{1}$$

For concreteness, we keep in mind a real function of a real variable. Although the expansion in powers of x requires that the latter be asymptotically small, in reality this variable is finite, and sometimes even is rather large. It is the standard situation that series (1) diverges, as k increases, for any finite value of x. Hence, it is necessary to have in hands a method for ascribing to the divergent series an effective sum. This can be done by resorting to self-similar approximation theory that is based on the following ideas (see reviews [10,11]).

First, the given series has to be transformed by implanting in it control parameters,

$$\{f_k(x)\} \mapsto \{F_k(x, u_k)\} . \tag{2}$$

These parameters are converted into control functions $u_k = u_k(x)$ whose role is to make convergent the series $\{F_k(x, u_k(x))\}$.

Then the transition between the terms F_k is reorganized so that to represent a motion of a dynamical system, where the approximation order plays the role of discrete time,

$$F_k \mapsto F_{k+1} \mapsto \ldots \mapsto F^* . \tag{3}$$

The implantation of control functions can be done in different ways. For example, by incorporating control parameters into the initial step of perturbation theory. Thus, if the considered problem is characterized by a Hamiltonian H, one defines the Hamiltonian

$$H_\varepsilon = H_0(u) + \varepsilon[\, H - H(u)\,] \qquad (\varepsilon \to 1) , \tag{4}$$

with a dummy parameter ε. Then perturbation theory is accomplished with respect to this dummy parameter.

Another way of implanting control parameters is by a change of the variable, with a control parameter entering this change, and the following re-expansion in powers of a new variable. For instance, one can substitute a change of the variable $x = x(z, u)$ into series (1) and then re-expand it in powers of z obtaining another series,

$$\sum_{n=0}^{k} a_n x^n \mapsto \sum_{n=0}^{k} b_n(u) z^n . \tag{5}$$

Also, it is possible to transform the given series so that the transformed series would contain control parameters,

$$\hat{T}[\, u \,]\; f_k(x) = F_k(x, u_k) \; . \tag{6}$$

The convertion of control parameters into control functions is based on the Cauchy criterion of convergence, according to which the series converges if and only if for any positive ε there exists $k^* = k^*(\varepsilon)$, such that

$$|\; F_{k+p}(x, u_{k+p}) - F_k(x, u_k) \;| < \varepsilon \tag{7}$$

for all $k > k^*$ and $p > 0$. The difference

$$C(F_{k+p}, F_k) \equiv F_{k+p}(x, u_{k+p}) - F_k(x, u_k) \tag{8}$$

is termed the Cauchy difference. The control functions are to be the minimizers of the convergence cost functional:

$$\min_u \; \frac{1}{2} \sum_k C^2(F_{k+p}, F_k) \;\; \mapsto \;\; u_k(x) \; . \tag{9}$$

The sequence of terms $F_k(x, u_k(x))$ is transformed into the trajectory of a dynamical system in the following way. We impose the rheonomic constraint

$$F_0(x, u_k(x)) = f \tag{10}$$

that defines the expansion function

$$x = x_k(f) \; . \tag{11}$$

Let us introduce the endomorphism

$$y_k(f) : \; \mathbb{Z}_+ \times \mathbb{R} \;\; \mapsto \;\; \mathbb{R} \; , \tag{12}$$

where

$$y_k(f) \equiv F_k(x_k(f), u_k(x_k(f))) \; . \tag{13}$$

The family of these endomorphsims forms a dynamical system in discrete time

$$\{ y_k(f) : \; \mathbb{Z}_+ \times \mathbb{R} \;\; \mapsto \;\; \mathbb{R} \} \; , \tag{14}$$

with the initial condition

$$y_0(f) = f \; . \tag{15}$$

By construction, the dynamical system trajectory $\{y_k(f)\}$ is bijective to the sequence of approximants $\{F_k(x, u_k(x))\}$. If the limit of the approximation sequence exists, it is bijective to the trajectory fixed point

$$y_k(y^*(f)) = y^*(f) \; . \tag{16}$$

Since in practical applications we deal with finite approximation orders k, we obtain the approximants

$$F_k^*(x) = F_k(x, u_k(x)) \; . \tag{17}$$

In the vicinity of a fixed point, the self-similar relation

$$y_{k+p}(f) = y_k(y_p(f)) \tag{18}$$

Journal of Physics: Conference Series **2249** (2022) 012012 doi:10.1088/1742-6596/2249/1/012012

holds [10, 11]. The family $\{y_k(f)\}$, with relation (18), composes an approximation cascade. The approximation cascade can be embedded into an approximation flow,

$$\{y_k(f) : \mathbb{Z}_+ \times \mathbb{R} \mapsto \mathbb{R}\} \subset \{y(t,f) : \mathbb{R}_+ \times \mathbb{R} \mapsto \mathbb{R}\}, \tag{19}$$

which implies that the flow, which is a dynamical system in continuous time, enjoys the self-similar relation

$$y(t+t', f) = y(t, y(t', f)) \tag{20}$$

and the continuous flow trajectory passes through all points of the cascade trajectory,

$$y(t,f) = y_k(f) \qquad (t = k) \tag{21}$$

starting at the same initial point

$$y(0,f) = f \; . \tag{22}$$

The self-simlar relation (20) can be represented as the Lie equation

$$\frac{\partial}{\partial t}\, y(t,f) = v(y(t,f)) \tag{23}$$

whose integration gives the evolution integral

$$\int_{F_k}^{F_k^*} \frac{dy}{v_k(y)} = t_k \; . \tag{24}$$

The entering here velocity can be defined by means of the Euler discretization

$$v_k(y) = y_{k+1}(f) - y_k(f) \; . \tag{25}$$

The upper limit of the evolution integral gives the self-similar approximant

$$f_k^*(x) = F_k^*(x, u_k(x)) \; . \tag{26}$$

The procedure stability is characterized by calculating the local map multipliers

$$\mu_k(f) \equiv \frac{\partial y_k(f)}{\partial f} \tag{27}$$

showing that the procedure is stable provided that

$$\mu_k < 1 \; , \qquad \mu_k \equiv \sup_f |\, \mu_k(f)\, | \; . \tag{28}$$

Respectively, the found fixed point is stable, when the fixed-point multiplier

$$\mu_k^*(f) \equiv \frac{\partial y_k^*(f)}{\partial f} \tag{29}$$

satisfies the condition

$$\mu_k^* < 1 \; , \qquad \mu_k^* \equiv \sup_f |\, \mu_k^*(f)\, | \; . \tag{30}$$

Instead of local multipliers, one can consider the local Laypunov exponents

$$\lambda_k(f) = \frac{1}{k} \ln \mid \mu_k(f) \mid \tag{31}$$

in terms of which the stability condition reads as

$$\lambda_k < 0 \; , \qquad \lambda_k \equiv \sup_f \lambda_k(f) \; . \tag{32}$$

A particular example of transformation (2) is the fractal transformation

$$F_k(x,s) = x^s f_k(x) = \sum_{n=0}^{k} a_n x^{n+s} \; , \tag{33}$$

with the scaling exponent s playing the role of a control parameter, and satisfying the scaling relation

$$\frac{F_k(bx,s)}{f_k(bx)} = \frac{F_k(x,s)}{f_k(x)} \, b^s \; . \tag{34}$$

The rheonomic contsraint (10) becomes

$$F_0(x,s) = x^s = f \tag{35}$$

giving the expansion function

$$x(f) = f^{1/s} \; . \tag{36}$$

Endomorphism (13) acquires the form

$$y_k(f) = \sum_{n=0}^{k} a_n f^{1+n/s} \; . \tag{37}$$

The cascade velocity (25) reduces to

$$v_k(f) = y_k(f) - y_{k-1}(f) = a_k f^{1+n/s} \; . \tag{38}$$

Finding the fixed point $y_k^*(f)$ and accomplishing the inverse fractal transformation yields the self-similar approximant

$$f_k^*(x) = x^{-s} y_k^*(x^s) \; . \tag{39}$$

3 Self-similar factor approximants

Employing the ideas described above, a convenient type of approximants has been derived, called self-similar factor approximants [10,11,15,16]. Let an asymptotic expansion for a real function be given:

$$f_k(x) = f_0(x) \left(1 + \sum_{n=1}^{k} a_n x^n \right) \tag{40}$$

where a real parameter x is assumed to be asymptotically small. However, we need to find the value of the function at a finite value of the parameter or even for $x \to \infty$. In what follows, we assume that the sought function is sign-defined. Without the loss of generality, it can be treated as positive (non-negative). This is because a negative function f can always be replaced by a positive function $-f$.

The self-similar factor approximants, obtained from series (40), read as

$$f_k^*(x) = f_0(x) \prod_{j=1}^{N_k} (1 + A_j x)^{n_j} \; , \tag{41}$$

where the number of factors is

$$N_k = \left\{ \begin{array}{ll} k/2, & k = 2, 4, 6, \ldots \\ (k+1)/2, & k = 3, 5, 7, \ldots \end{array} \right. . \tag{42}$$

The parameters A_j and n_j are uniquely determined by the accuracy-through-order procedure, by equating the like-order terms in the expansions at small x,

$$f_k^*(x) \simeq f_k(x) \qquad (x \to 0) \; . \tag{43}$$

This procedure gives the equations

$$\sum_{j=1}^{N_k} n_j A_j^n = D_n \qquad (n = 1, 2, \ldots, k) \; , \tag{44}$$

in which

$$D_n \equiv \frac{(-1)^{n-1}}{(n-1)!} \lim_{x \to 0} \frac{d^n}{dx^n} \ln \left(1 + \sum_{m=1}^{n} a_m x^m \right) \; . \tag{45}$$

For even orders k, Eq. (44) consists of k equations uniquely defining the $k/2$ parameters A_j and $k/2$ parameters n_j. For odd orders k, the system of k equations (44) contains $k+1$ unknowns, where one of the parameters A_j, say A_1, is arbitrary. It is possible to use the normalization $A_1 = 1$, which makes the system of equations (44) self-consistent and all parameters uniquely defined [10, 17]. If the parameters lead to a complex-valued approximant, it is replaced by the nearest real-valued approximant. The final result is given by the average between the last two approximants $[f_k^*(x) + f_{k-1}^*(x)]/2$ and the error bar is defined as the half-difference between the last two different approximants $[f_k^*(x) - f_{k-1}^*(x)]/2$.

4 Critical temperature

The interaction strength of weakly interacting particles can be characterized by the coupling parameter, also called gas parameter,

$$\gamma \equiv \rho^{1/3} a_s \; , \tag{46}$$

where ρ is average density and a_s, scattering length. Many-body systems are often modeled by $O(N)$-symmetric φ^4 field theory in three dimensions. When diminishing temperature, these systems exhibit a phase transition with spontaneous symmetry breaking.

An interesting problem that has attracted much attention is the dependence of the phase transition temperature on the interaction strength (see reviews [18–20]). One usually considers the critical temperature shift

$$\frac{\Delta T}{T_0} \equiv \frac{T_c(\gamma) - T_0}{T_0} \tag{47}$$

describing the relative difference between the critical temperature $T_c(\gamma)$, as a function of the gas parameter, and the critical temperature T_0 of noninteracting particles, when $\gamma = 0$. At asymptotically weak interactions, the critical temperature shift behaves as

$$\frac{\Delta T}{T_0} \simeq c_1 \gamma \qquad (\gamma \to 0) . \tag{48}$$

The problem is to find the coefficient c_1.

This coeficient has been calculated [21–23] in the loop expansion representing c_1 as an asymptotic series

$$c_1 = c_1(x) \simeq \sum_{n=1}^{5} a_n x^n \qquad (x \to 0) \tag{49}$$

in powers of the variable

$$x = (N+2)\,\frac{\lambda_{eff}}{\sqrt{\mu_{eff}}} , \tag{50}$$

where λ_{eff} is effective interaction strength and μ_{eff} is effective chemical potential. The difficulty arises at the phase transition temperature, where the chemical potential μ_{eff} tends to zero, hence the expansion variable x tends to infinity. Thus one comes to the necessity of giving to the value $c_1(\infty)$ some meaning.

The problem can be solved by using self-similar factor approximants [20, 24]. Then series (49) is transformed into the expression

$$c_1^*(x) = a_1 x \prod_{i=1}^{N_k} (1 + A_i x)^{n_i} , \tag{51}$$

with the parameters A_i and n_i defined by the accuracy-through-order procedure. At large x, expression (51) behaves as

$$c_1^*(x) \simeq B_k x^{\beta_k} \qquad (x \to \infty) , \tag{52}$$

with the amplitude and exponent

$$B_k = a_1 \prod_{i=1}^{N_k} A_i^{n_i} , \qquad \beta_k = 1 + \sum_{i=1}^{N_k} n_i . \tag{53}$$

The required finiteness of c_1 implies that β_k equals zero, which yields

$$c_1^*(\infty) = B_k \qquad (\beta_k = 0) . \tag{54}$$

This procedure can be complemented by a correction accelerating convergence [20, 24].

The described scheme has been applied to expansion (49) for the number of components $N = 0, 1, 2, 3, 4$. The coefficients a_n of the loop expansion (49) can be found in [20, 23]. The results for c_1 are presented in Table 1 and compared with the available Monte Carlo simulations for $N = 1$ [25], for $N = 2$ [26, 27], and for $N = 4$ [25].

5 Critical exponents

Critical exponents can be represented in the form of ε-expansions in powers of $\varepsilon = 4 - d$, where d is space dimensionality. For $O(N)$-symmetric φ^4 field theory in three dimensions, the five-loop expansions can be found in the book [28]. The summation of the five-loop expansions by means of self-similar approximants was considered in [29] for all N. It has been shown that for $N = -2$ and $N \to \infty$ self-similar approximants reproduce the exact values of the exponents. The results are very accurate for large $N \gg 1$, with the errors decreasing as $1/N$ with increasing N.

Here we demonstrate that the accuracy of self-similar factor approximants for the $O(N)$ - symmetric φ^4 theory in three dimensions can be essentially improved by using the seven-loop ε-expansions that are known for $N = 1$ [30] and that have been derived, using the seven-loop coupling-parameter expansions [31], for $N = 0, 2, 3, 4$ [32].

For $N = 0$, the expansions are

$$\nu^{-1} = 2 - 0.25\,\varepsilon - 0.08594\,\varepsilon^2 + 0.11443\,\varepsilon^3 - 0.28751\,\varepsilon^4 + 0.95613\,\varepsilon^5 - 3.8558\,\varepsilon^6 + 17.784\,\varepsilon^7\,, \quad (55)$$

$$\eta = 0.015625\,\varepsilon^2 + 0.016602\,\varepsilon^3 - 0.0083675\,\varepsilon^4 + 0.026505\,\varepsilon^5 - 0.09073\,\varepsilon^6 + 0.37851\,\varepsilon^7\,, \quad (56)$$

$$\omega = \varepsilon - 0.65625\,\varepsilon^2 + 1.8236\,\varepsilon^3 - 6.2854\,\varepsilon^4 + 26.873\,\varepsilon^5 - 130.01\,\varepsilon^6 + 692.1\,\varepsilon^7\,. \quad (57)$$

For $N = 1$, the ε expansions read as

$$\nu^{-1} = 2 - 0.333333\,\varepsilon - 0.11728\,\varepsilon^2 + 0.12453\,\varepsilon^3 - 0.30685\,\varepsilon^4 + 0.95124\,\varepsilon^5 - 3.5726\,\varepsilon^6 + 15.287\,\varepsilon^7\,, \quad (58)$$

$$\eta = 0.018519\,\varepsilon^2 + 0.01869\,\varepsilon^3 - 0.0083288\,\varepsilon^4 + 0.025656\,\varepsilon^5 - 0.081273\,\varepsilon^6 + 0.31475\,\varepsilon^7\,, \quad (59)$$

$$\omega = \varepsilon - 0.62963\,\varepsilon^2 + 1.6182\,\varepsilon^3 - 5.2351\,\varepsilon^4 + 20.75\,\varepsilon^5 - 93.111\,\varepsilon^6 + 458.74\,\varepsilon^7\,. \quad (60)$$

The ε expansions for $N = 2$ are

$$\nu^{-1} = 2 - 0.4\,\varepsilon - 0.14\,\varepsilon^2 + 0.12244\,\varepsilon^3 - 0.30473\,\varepsilon^4 + 0.87924\,\varepsilon^5 - 3.103\,\varepsilon^6 + 12.419\,\varepsilon^7\,, \quad (61)$$

$$\eta = 0.02\,\varepsilon^2 + 0.019\,\varepsilon^3 - 0.0078936\,\varepsilon^4 + 0.023209\,\varepsilon^5 - 0.068627\,\varepsilon^6 + 0.24861\,\varepsilon^7\,, \quad (62)$$

$$\omega = \varepsilon - 0.6\,\varepsilon^2 + 1.4372\,\varepsilon^3 - 4.4203\,\varepsilon^4 + 16.374\,\varepsilon^5 - 68.777\,\varepsilon^6 + 316.48\,\varepsilon^7\,. \quad (63)$$

The ε expansions for $N = 3$read as

$$\nu^{-1} = 2 - 0.45455\,\varepsilon - 0.1559\,\varepsilon^2 + 0.11507\,\varepsilon^3 - 0.2936\,\varepsilon^4 + 0.78994\,\varepsilon^5 - 2.6392\,\varepsilon^6 + 9.9452\,\varepsilon^7\,, \quad (64)$$

$$\eta = 0.020661\,\varepsilon^2 + 0.018399\,\varepsilon^3 - 0.0074495\,\varepsilon^4 + 0.020383\,\varepsilon^5 - 0.057024\,\varepsilon^6 + 0.19422\,\varepsilon^7\,, \quad (65)$$

$$\omega = \varepsilon - 0.57025\,\varepsilon^2 + 1.2829\,\varepsilon^3 - 3.7811\,\varepsilon^4 + 13.182\,\varepsilon^5 - 52.204\,\varepsilon^6 + 226.02\,\varepsilon^7\,. \quad (66)$$

The ε expansions for $N = 4$ take the form

$$\nu^{-1} = 2 - 0.5\,\varepsilon - 0.16667\,\varepsilon^2 + 0.10586\,\varepsilon^3 - 0.27866\,\varepsilon^4 + 0.70217\,\varepsilon^5 - 2.2337\,\varepsilon^6 + 7.9701\,\varepsilon^7\,, \quad (67)$$

$$\eta = 0.020833\,\varepsilon^2 + 0.017361\,\varepsilon^3 - 0.0070852\,\varepsilon^4 + 0.017631\,\varepsilon^5 - 0.047363\,\varepsilon^6 + 0.15219\,\varepsilon^7\,, \quad (68)$$

$$\omega = \varepsilon - 0.54167\,\varepsilon^2 + 1.1526\,\varepsilon^3 - 3.2719\,\varepsilon^4 + 10.802\,\varepsilon^5 - 40.567\,\varepsilon^6 + 166.26\,\varepsilon^7\,. \quad (69)$$

We calculate the corresponding self-similar factor approximants $f_k^*(\varepsilon)$, as is explained in Sec. 3, and set $\varepsilon = 1$. The results of the calculations for the exponents are shown in Table 2 for

$N = 0$, Table 3 for $N = 1$, Table 4 for $N = 2$, Table 5 for $N = 3$, and Table 6 for $N = 4$. We compare the results obtained by means of Factor approximants (FA) with the results of other methods: Monte Carlo simulations (MC) [33–38], Conformal bootstrap (CB) [39–43], Hypergeometric Meijer summation (HGM) [32], Borel summation complimented by additional conjectures on the behavior of coefficients (BAC) [44], Borel summation with conformal mapping (BCM) [45], and Nonperturbative renormalization group (NPRG) [5, 46–48].

Other exponents, α, β, γ, and δ, are calculated using the relations

$$\alpha = 2 - 3\nu \; , \qquad \beta = \frac{\nu}{2}\,(1+\eta) \qquad \gamma = \nu(2-\eta) \; , \qquad \delta = \frac{5-\eta}{1+\eta} \; . \tag{70}$$

Table 7 summarizes these results.

6 Conclusion

The method of self-similar factor approximants is a very simple and convenient tool for the summation of asymptotic series. The basis of the method is the consideration of the transfer from one approximation term to another as a motion of a dynamical system, with the approximation order playing the role of time. The motion in the vicinity of a fixed point is described by a self-similar relation. The fixed point of the evolution equation defines the effective limit of the sequence. The method can be applied to a large class of problems, as can be inferred from the review articles [10, 11] and the recent papers [49, 50].

The method of self-similar factor approximants is used for the summation of ε expansions for the $O(N)$- symmetric field theory in three dimensions. The series of seventh order in ε are employed. The method is shown to produce accurate approximations, at the same time being very simple. The results are compatible with other known methods of summation.

Author Contributions

Both the authors, V.I. Yukalov and E.P. Yukalova, equally contributed to this work.

This research did not receive any specific grant from funding agencies in the public, commercial, or not-for-profit sectors.

References

[1] Giacaglia G E 1972 *Perturbation Methods in Non-Linear Systems* (New York: Springer)

[2] Nayfeh A H 1973 *Perturbation Methods* (New York: Wiley)

[3] Baker G A and Graves-Moris P 1996 *Padé Approximants* (Cambridge: Cambridge University)

[4] Kleinert H 2004 *Path Integrals* (Singapore: World Scientific)

[5] Dupuis N, Canet L, Eichhorn A, Metzner W, Pawlowski J, Tissier M, Wschebor N 2021 *Phys. Rep.* **910** 1

[6] Yukalov V I 1990 *Phys. Rev. A* **42** 3324

[7] Yukalov V I 1990 *Physica A* **167** 833

[8] Yukalov V I 1991 *J. Math. Phys.* **32** 1235

[9] Yukalov V I 1992 *J. Math. Phys.* **33** 3994

[10] Yukalov V I 2019 *Phys. Part. Nucl.* **50** 141

[11] Yukalov V I and Yukalova E P 2021 *Physics* **3** 829

[12] Pelissetto A and Vicari E 2002 *Phys. Rep.* **368** 549

[13] Abhignan V and Sankaranarayanan R 2021 *J. Stat. Phys.* **183** 4

[14] Shalaby A M 2021 *Eur. Phys. J. C* **81** 87

[15] Yukalov V I, Gluzman S and Sornette D 2003 *Physica A* **328** 409

[16] Gluzman S, Yukalov V I and Sornette D 2003 *Phys. Rev. E* **67** 026109

[17] Yukalov V I and Yukalova E P 2007 *Phys. Lett. A* **368** 341

[18] Andersen J O 2004 *Rev. Mod. Phys.* **76** 599

[19] Yukalov V I 2004 *Laser Phys. Lett.* **1** 435

[20] Yukalov V I and Yukalova E P 2017 *Laser Phys. Lett.* **14** 073001

[21] Kastening B 2004 *Laser Phys.* **14** 586

[22] Kastening B 2004 *Phys. Rev. A* **69** 043613

[23] Kastening B 2004 *Phys. Rev. A* **70** 043621

[24] Yukalov V I and Yukalova E P 2017 *Eur. Phys. J. Web Conf.* **138** 03011

[25] Sun X 2003 *Phys. Rev. E* **67** 066702

[26] Kashurnikov V A, Prokof'ev N and Svistunov B 2001 *Phys. Rev. Lett.* **87** 120402

[27] Arnold P and Moore G 2001 *Phys. Rev. Lett.* **87** 120401

[28] Kleinert H and Schulte-Frohlinde V 2006 *Critical Properties of φ^4-Theories* (Singapore: World Scientific)

[29] Yukalov V I and Yukalova E P 2007 *Eur. Phys. J. B* **55** 93

[30] Ryttov T A 2020 *J. High Energy Phys.* **04** 072

[31] Schnetz O 2018 *Phys. Rev. D* **97** 085018

[32] Shalaby A M 2021 *Eur. Phys. J. C* **81** 87

[33] Hasenbusch M 2001 *J. Phys. A* **34** 8221

[34] Hasenbusch M 2010 *Phys. Rev. B* **82** 174433

[35] Hasenbusch M and Vicari E 2011 *Phys. Rev. B* **84** 125136

[36] Clisby N and Dunweg B 2016 *Phys. Rev. E* **94** 052102

[37] Clisby N 2017 *J. Phys. A* **50** 264003

[38] Hasenbusch M 2019 *Phys. Rev. B* **100** 224517

[39] El-Showk S, Paulos M F, Poland D, Rychkov S, Simmons-Duffin D and Vichi A 2014 *J. Stat. Phys.* **157** 869

[40] Shimada H and Hikami S 2016 *J. Stat. Phys.* **165** 1006

[41] Kos F, Poland D, Simmons-Duffin D and Vichi A 2016 *J. High Energy Phys.* **08** 036

[42] Echeverri A C, von Harling B and Serone M 2016 *J. High Energy Phys.* **09** 097

[43] Simmons-Duffin D 2017 *J. High Energy Phys.* **03** 086

[44] Kompaniets M V and Wiese K J 2020 *Phys. Rev. E* **101** 012104

[45] Kompaniets M V and Panzer E 2017 *Phys. Rev. D* **96** 036016

[46] Hasselmann N 2012 *Phys. Rev. E* **86** 041118

[47] Rose F and Dupuis N 2018 *Phys. Rev. B* **97** 174514

[48] De Polsi G, Balog I, Tissier M and Wschebor N 2020 *Phys. Rev. E* **101** 042113

[49] Yukalov V I and Yukalova E P 2021 *Phys. Rev. D* **103** 076019

[50] Yukalov V I, Yukalova E P 2022 *Phys. Lett. A* **425** 127899

Journal of Physics: Conference Series **2249** (2022) 012012 doi:10.1088/1742-6596/2249/1/012012

Table 1: Coefficient c_1 of the critical temperature shift for different numbers of the field components N, found by means of self-similar factor approximants, and compared with Monte Carlo simulations.

N	c_1	*Monte Carlo*
0	0.77 ± 0.03	
1	1.06 ± 0.05	1.09 ± 0.09 ([25])
2	1.29 ± 0.07	1.29 ± 0.05 ([26]) 1.32 ± 0.02 ([27])
3	1.46 ± 0.08	
4	1.60 ± 0.09	1.60 ± 0.10 ([25])

Table 2: Critical exponents for $N = 0$, found by different methods: Self-similar factor approximants (FA), Monte Carlo simulations (MC), Conformal bootstrap (CB), Hypergeometric Meijer summation (HGM), Borel summation with additional conjectures on the behaviour of coefficients (BAC), Borel summation with conformal mapping (BCM), and Nonperturbative renormalization group (NPRG).

Method	ν	η	ω
FA	0.5877 (2)	0.0301 (2)	0.821 (15)
MC	0.5875970 (4)	0.031043 (3)	0.899 (12)
CB	0.5877 (12)	0.0282 (4)	–
HGM	0.5877 (2)	0.0312 (7)	0.8484 (17)
BAC	0.5874 (2)	0.0304 (2)	0.846 (15)
BCM	0.5874 (3)	0.0310 (7)	0.841 (13)
NPRG	0.5876 (2)	0.0312 (9)	0.901 (24)

Table 3: Critical exponents for $N = 1$, found by different methods listed in Table 2.

Method	ν	η	ω
FA	0.6300 (3)	0.0353 (3)	0.808 (9)
MC	0.63002 (10)	0.03627 (10)	0.832 (6)
CB	0.62999 (5)	0.03631 (3)	0.830 (2)
HGM	0.6298 (2)	0.0365 (7)	0.8231 (5)
BAC	0.6296 (3)	0.0355 (3)	0.827 (13)
BCM	0.6292 (5)	0.0362 (6)	0.820 (7)
NPRG	0.63012 (16)	0.0361 (11)	0.832 (14)

Table 4: Critical exponents for $N = 2$, found by different methods listed in Table 2.

Method	ν	η	ω
FA	0.6710 (3)	0.0372 (4)	0.809 (11)
MC	0.67169 (7)	0.03810 (8)	0.789 (4)
CB	0.67175 (10)	0.0385 (6)	0.811 (10)
HGM	0.6708 (4)	0.0381 (6)	0.789 (13)
BAC	0.6706 (2)	0.0374 (3)	0.808 (7)
BCM	0.6690 (10)	0.0380 (6)	0.804 (3)
NPRG	0.6716 (6)	0.0380 (13)	0.791 (8)

Table 5: Critical exponents for $N = 3$, found by different methods listed in Table 2.

Method	ν	η	ω
FA	0.7099 (1)	0.0372 (4)	0.7919 (3)
MC	0.7116 (10)	0.0378 (3)	0.773
CB	0.7121 (28)	0.0386 (12)	0.791 (22)
HGM	0.7091 (2)	0.0381 (6)	0.764 (18)
BAC	0.70944 (2)	0.0373 (3)	0.794 (4)
BCM	0.7059 (20)	0.0378 (5)	0.795 (7)
NPRG	0.7114 (9)	0.0376 (13)	0.796 (11)

Table 6: Critical exponents for $N = 4$, found by different methods listed in Table 2.

Method	ν	η	ω
FA	0.7459 (2)	0.0361 (4)	0.7913 (8)
MC	0.750 (2)	0.0360 (3)	0.765 (30)
CB	0.751 (3)	0.0378 (32)	0.817 (30)
HGM	0.7443 (3)	0.0367 (4)	0.7519 (13)
BAC	0.7449 (4)	0.0363 (2)	0.7863 (9)
BCM	0.7397 (35)	0.0366 (4)	0.794 (9)
NPRG	0.7478 (9)	0.0360 (12)	0.761 (12)

Table 7: Critical exponents for the three-dimensional $O(N)$-symmetric φ^4 field theory, calculated by means of self-similar factor approximants

N	α	β	γ	δ	ν	η	ω
0	0.2369	0.3027	1.15771	4.8247	0.5877	0.0301	0.821
1	0.1100	0.3261	1.23776	4.7954	0.6300	0.0353	0.808
2	−0.0130	0.3480	1.31704	4.7848	0.6710	0.0372	0.809
3	−0.1297	0.3682	1.39339	4.7848	0.7099	0.0372	0.792
4	−0.2377	0.3864	1.46487	4.7910	0.7459	0.0361	0.791

Journal of Physics: Conference Series **2249** (2022) 012013 doi:10.1088/1742-6596/2249/1/012013

Coherent and incoherent superposition of transition matrix elements of the squeezing operator.

Sándor Varró[1,2]

1) Wigner Research Centre for Physics, Eötvös Loránd Research Network, Budapest, Hungary

2) ELI-ALPS Attosecond Light Pulse Source, ELI-HU Non-Profit Ltd., Szeged, Hungary

Abstract. We discuss the general matrix elements of the squeezing operator between number eigenstates of a harmonic oscillator (which may also represent a quantized mode of the electromagnetic radiation). These matrix elements have first been used by Popov and Perelomov (1969) long ago, in their thorough analysis of the parametric excitation of harmonic oscillators. They expressed the matrix elements in terms of transcendental functions, the associated Legendre functions. In the present paper we will show that these matrix elements can also be derived in an different form, expressed by the classical Gegenbauer polynomials. This new expression makes it possible to determine coherent and incoherent superpositions of these matrix elements in closed analytic forms. As an application, we describe multiphoton transitions in the system "charged particle + electromagnetic radiation", induced by a (strong) coherent field or by a black-body radiation component (with a Planck-Bose photon number distribution). The exact results are compared with the semi-classical ones. We will show that in case of interaction with a thermal field, the semi-classical result (with a Gaussian stochastic field amplitude) yields an acceptable approximation only in the Rayleigh-Jeans limit, however, in the Wien limit it completely fails.

1. Introduction

The „non-classical states of light", in particular, „squeezed light" have been a subject of intensive theoretical and experimental research for many decades (Stoler 1970, 1971, Yuen 1976, Walls 1979, Loudon and Knight 1987, Dodonov 2002, Dodonov and Man'ko 2003, Andersen et al 2016), and have become standard subjects in any textbooks on quantum optics (Scully and Zubairy 1997, Loudon 2000, Schleich 2001). The phenomenon „squeezing of the amplitudes" (of a mechanical oscillator, or of a component of electromagnetic radiation) often appears in parametric processes (Husimi 1953, Louisell, Yariv and Siegman 1961, Mollow and Glauber 1967, Popov and Perelomov 1969, Malkin, Man'ko and Trifonov 1970, Milburn and Walls 1983, Kiss, Janszky and Ádám 1994, Dodonov 2003, Fedorov *et al* 2008, Straupe *et al* 2011). Concerning the oscillatory behaviour of the photon number distribution in highly-squeezed coherent states, see Wheeler (1985), Schleich and Wheeler (1987), Schleich, Walls and Wheeler (1988), Schleich (2001). The squeezing operator is the generator of the Bogoljubov transformation, which has played an important role in the theory of superconductivity (Bogoljubov 1958, Valatin 1958). The quantum dynamics of charged particles in time-varying external (electro)magnetic fields also shows interesting phenomena, related to non-classical states of linear oscillators (Malkin, Man'ko and Trifonov 1970, Varró 1984, Varró and Ehlotzky 1985, 1987, Dodonov 2018, Dodonov and Horovits 2021). Recently the interest in entangled states is more pronounced then in nonclassical pure states. Anyway, undoubtely, the „classic non-classical states" are the squeezed coherent states, which have been the ones most extensively studied in the last decades. It is lesser known by the quantum optics community that such states of the radiation field naturally appear as exact stationary solutions in the simplest (and most fundamental) system of quantum electrodynamics, namely in the interaction of a free electron with a quantized radiation field (see Berson 1969, Fedorov and Kazakov 1973, Bergou and Varró 1981a-b, Gazazyan and Sherman 1989, Varró 2021a). In the theory of intereactions of strong laser radiation with electrons such exact solutions of the Schrödinger or Dirac equations are of outstanding importance in treating the multiphoton processes non-perturbatively (though the laser field is mostly taken as an external field).

Our motivation for the present study has been to calculate the coherent and incoherent superposition of multiphoton transition amplitudes, which we have derived in the frame of non-relativistic quantum electrodynamics, for considering high-order harmonic generation (Varró 2021a). In the non-perturbative treatment of such processes one neccesarily encounters matrix elements between photon number eigenstates of the displacement and squeezing operators (which are standard objects in quantum optics). It is very remarkable that, though the photon number distributions of squeezed coherent states (Hermite polynomials), or displaced number states (Laguerre polynomials) have long been known, the matrix elements S_{mn}=<m|S|n> of the squeezing operator alone have not been published in a classical polynomial form. More precisely, S_{mn} has been expressed by Bargmann (1947) in his paper on the irreducible representation of the Lorentz group, in terms of hypergeometric functions (see also Sannikov 1965). In the context of parametric processes, Popov and Perelomov (1969) and Malkin, Man'ko and Trifonov (1970) derived a form of S_{mn}, expressed in terms of associated Legendre functions. Later S_{mn} has been related also to Jacobi polynomials (see, in particular, the very thorough recent studies of Wünsche 2017a-b), however a compact, classical polynomial expression has not been derived (see also Tanabe 1973, Rashid 1975, Satyanarayana 1985 and Mendaš and Popović 1995). Recently we have derived a compact classical polynomial expression, in term of Gegenbauer polynomials (Varró 2021b), and the larger part of the present paper will be devoted to the derivation and the main properties of the matrix element S_{mn} of the squeezing operator.

In Section 2 we summarize the main steps of the derivation of the Gegenbauer polynomial expression of the matrix elements of the squeezing operator between photon number eigenstates. In order to have the paper self-contained, we have included an Appendix which contains all the necessary details of this derivation. We shall also present some numerical illustrations to display the main fetures of the photon number distribution in a squeezed number state. In Section 3 we apply the new formula for the matrix elements for calculating the coherent and incoheren superpositions of multiphoton squeezing transitions.

2. The matrix elements of the squeezing operator between photon number eigenstates.

In the present section we summarize the basic steps of the calculation of the matrix elements between the number eigenstates of the squeezing operator, $S_{m,n} = \langle m|S(\xi)|n\rangle$, where

$$S(\xi) = \exp[\tfrac{1}{2}\xi(a^+)^2 - \tfrac{1}{2}\xi^*(a)^2], \quad S(\xi) = \exp(\xi K_+ - \xi^* K_-), \quad \xi = |\xi| e^{i\varphi}, \tag{1}$$

with $K_+ = \frac{1}{2}(a^+)^2$ and $K_- = \frac{1}{2}(a)^2$. Here a and a^+ are the photon absorption and emission operators, respectively. (In a more general context, they are the quantum amplitudes associated to the decrease and increase by 1 of the excitation index of the system, like a mechanical linear oscillator.) The a and a^+ satisfy the commutation relation $[a, a^+] = aa^+ - a^+a = I$, where I is the unit operator of the complex Hilbert space $\mathscr{H}$, which models the system. The effects of a and a^+ on the Fock states $|n\rangle$ (which are the photon number eigenstates, satisfying the eigenvalue equation $a^+a|n\rangle = n|n\rangle$) are expressed by equations $a|n\rangle=\sqrt{n}|n-1\rangle$ and $a^{+}|n\rangle=\sqrt{n+1}|n+1\rangle$. For the ground state $a|0\rangle=\vec{0}$, where $\vec{0}$ is the zero vector of $\mathscr{H}$. The Fock states $|n\rangle$ form a complete orthogonal set ($\langle m|n\rangle=\delta_{m,n}$) in $\mathscr{H}$. By introducing $K_0=\frac{1}{4}(aa^{+}+a^{+}a)$, we have the closed set of generators of a special Lie algebra of the $SU(1,1)$ group, having the commutation relations $[K_0,K_{\pm}]=\pm K_{\pm}$, and $[K_-,K_+]=2K_0$. In Appendix A we summarize some basic properties of this group, on the basis of which we derive the normally ordered form of $S(\xi)$. The details of the derivations are presented in Appendix A, in the present section we just outline the basic steps, leading to the final result.

According to Eq. (A.5), the normal form of $S(\xi)$ becomes

$$S(\xi)=\exp[\tfrac{1}{2}\zeta(a^{+})^2]\exp[-\eta\tfrac{1}{2}(a^{+}a+\tfrac{1}{2})]\exp[-\tfrac{1}{2}\zeta^{*}(a)^2],\quad \eta=2\log\cosh|\xi|,\quad \zeta=\mathrm{e}^{i\varphi}\tanh|\xi|. \tag{2}$$

By consecutively applying the relation $a^2|n\rangle=\sqrt{n(n-1)}|n-2\rangle$, we have the finite sum

$$\exp(-\tfrac{1}{2}\zeta^{*}a^2)|n\rangle=\sum_{k=0}^{[n/2]}\frac{(-\zeta^{*}/2)^k}{k!}\sqrt{n(n-1)(n-2)\cdots(n-2k+1)}|n-2k\rangle, \tag{3}$$

as the right-hand factor in the scalar product $S_{m,n}=\langle m|S(\xi)|n\rangle$. Similarly, the left-hand factor becomes also a finite sum, running up to $[m/2]$. The operator in the middle of the normal form in Eq. (2) results in a multiplication for any $|k\rangle$ by a number, namely, $\exp[-\eta\frac{1}{2}(a^{+}a+\frac{1}{2})]|k\rangle=\exp[-\eta\frac{1}{2}(k+\frac{1}{2})]|k\rangle$. Owing to the orthogonality of the Fock states, the scalar product, as a double sum, reduces to a single sum, as is shown in Eqs. (A.7) and (A.8) for $m\geq n$ and $m\leq n$, respectively. Only those matrix elements are non-vanishing in which m and n have the same parity, so $\frac{1}{2}(m-n)$ must be an integer. The finite sums in Eqs. (A.7) and (A.8) are expressed by hypergeometric functions in Eqs. (A.9) and (A.10). Here we display the result in the case $m\geq n$,

$$\langle m|S(\xi)|n\rangle=\exp[-\tfrac{1}{2}(n+\tfrac{1}{2})\eta](\zeta/2)^{(m-n)/2}\sqrt{\frac{m!}{n!}}\frac{1}{\Gamma(\alpha+1)}F\left(-\frac{n}{2},\frac{1-n}{2};\frac{m-n}{2}+1;z\right)\qquad (m\geq n), \tag{4}$$

where $z=-\sinh^2|\xi|$ and $\exp(\eta/2)=\cosh|\xi|$. A similar expression comes out in the other case $m\leq n$, as is shown in Eq. (A.10). This formula has been first published by Bargmann (1947), who used the analytic function representation in the Fock-Bargmann space. It was rederived by Popov and Perelomov (1969), by performing the scalar products in the L^2 representation, and using generating functions.

The crucial step in our derivation is that now we apply formula 8.962.1 of Gradshteyn and Ryzhik (2000), to connect the hypergeometric functions with the Jacobi polynomials $P_{n/2}^{(\alpha,-1/2)}(x)$, and $P_{(n-1)/2}^{(\alpha,+1/2)}(x)$, where $\alpha=\frac{1}{2}(m-n)\geq 0$, and $n/2$ and $(n-1)/2$ are integers (corresponding to even or odd n, respectively). This is shown in Eqs. (A.11) and (A.12) in the Appendix. The problem with these expressions is that they contain different Jacobi polynomials with different parameters, depending on the parity. This means that the functional forms of the matrix elements are not the same for even-even and odd-odd transitions. Fortunately, we can get rid off this severe assymmetry by using the formulas 22.5.22 and 22.5.21 of Abramowitz and Stegun (1970) in Eqs. (A.11) and (A.12), to be applied for the even-even and odd-odd transitions, respectively. These formulas, Eqs. (A.13a-b), connect the Jacobi polynomials of the type $P_k^{(\alpha,\pm 1/2)}$ with the Gegenbauer polynomials $C_n^{\alpha+\frac{1}{2}}$. In this way we receive a *unifying formula* (in the sub-case $m\geq n$), *valid for both even and odd indeces*, as is expressed in Eq. (A.14). A similar unifying formula (which is also valid for both even and odd paritys) results in the other sub-case $m \le n$ as is shown in Eq. (A.15). Some additional details of the calculation are explained in Appendix A. On the basis of these considerations, the final result can be summarized in the form

$$\langle m|S(\xi)|n\rangle = \mathrm{e}^{-\eta/4}(2\mathrm{e}^{i\varphi}\tanh|\xi|)^{\alpha}\sqrt{\frac{n!}{m!}}\frac{\Gamma(\alpha+\frac{1}{2})}{\sqrt{\pi}}C_n^{\alpha+\frac{1}{2}}\left(\frac{1}{\cosh|\xi|}\right), \quad \alpha=\tfrac{1}{2}(m-n), \quad (m \ge n), \tag{5}$$

$$\langle m|S(\xi)|n\rangle = \mathrm{e}^{-\eta/4}(-2\mathrm{e}^{-i\varphi}\tanh|\xi|)^{\lambda}\sqrt{\frac{m!}{n!}}\frac{\Gamma(\lambda+\frac{1}{2})}{\sqrt{\pi}}C_m^{\lambda+\frac{1}{2}}\left(\frac{1}{\cosh|\xi|}\right), \quad \lambda=\tfrac{1}{2}(n-m), \quad (m \le n). \tag{6}$$

The formulas in Eq. (5) and Eq. (6) are valid for both even-even and odd-odd transitions, so they are valid in the whole Hilbert space $\mathscr{H} = \mathscr{H}_{1/4} \oplus \mathscr{H}_{3/4}$. The order ν of the Gegenbauer polynomials $C_n^{\nu}(x)$ in Eqs. (7) and (8) satisfy the condition $\nu > -1/2$, and this means that the matrix elements are expressed by the classical orthogonal polynomials.

In the simplest special case, when $m = n$, the Gegenbauer polynomials reduce to the Legendre polynomials, according to the relation $C_n^{\frac{1}{2}}(x) = P_n(x)$ (see e.g. Gradshteyn and Ryzhik (2000)), i.e.

$$\langle n|S(\xi)|n\rangle = \sqrt{x}P_n(x), \quad x = 1/\cosh r, \quad r = |\xi| \qquad (m = n). \tag{7}$$

In the general case the new formulas for the matrix elements $S_{mn} = \langle m|S(\xi)|n\rangle$, Eqs. (5) and (6), can be brought to another equivalent form, expressed in terms of the associated Legendre functions $P_{\mu}^{\nu}(x)$. By using the formula 11.4 (10) of Erdélyi (1953), the Gegenbauer polynomials can be interrelated with the associated Legendre functions,

$$\frac{1}{\sqrt{\pi}}\Gamma(\mu+\tfrac{1}{2})C_{\nu-\mu}^{\mu+\frac{1}{2}}(x) = (-1)^{\mu}2^{-\mu}(1-x^2)^{-\frac{1}{2}\mu}P_{\nu}^{\mu}(x),$$

and we have

$$\langle m|S(\xi)|n\rangle = (-1)^k\sqrt{\frac{n!}{m!}}\sqrt{x}P_l^k(x)\mathrm{e}^{ik\varphi}, \qquad k=\tfrac{1}{2}(m-n), \qquad (m \ge n), \tag{8}$$

$$\langle m|S(\xi)|n\rangle = \sqrt{\frac{m!}{n!}}\sqrt{x}P_l^{k'}(x)\mathrm{e}^{-ik'\varphi}, \qquad k'=\tfrac{1}{2}(n-m), \qquad (m \le n), \tag{9}$$

where $x = 1/\cosh|\xi|$ and $l = \frac{1}{2}(m+n)$ .The transition probability then has the compact form

$$w_{m,n} = |\langle m|S(\xi)|n\rangle|^2 = \frac{n_<!}{n_>!}x\,|P_{(m+n)/2}^{|m-n|/2}(x)|^2, \tag{10}$$

where $n_< = \min(m,n)$, $n_> = \max(m,n)$, $x = 1/\cosh r$ (see Popov and Perelomov (1969), Malkin, Man'ko and Trifonov (1970)). The result in Eq. (10) directly shows the symmerty of the transition probalbilities with respect to the interchange of n and m. Besides, the matrix elements $S_{mn} = \langle m|S(\xi)|n\rangle$ are formally related to the representation of spatial rotations (see e.g. Aronson, Malkin and Man'ko (1974), Witschel (1975)). By introducing the non-normalized spherical harmonics $X_l^k(\vartheta,\varphi)$,

$$X_l^k(\vartheta,\varphi) = \sqrt{\frac{4\pi}{(2l+1)}}Y_l^k(\vartheta,\varphi) = (-1)^k\left[\frac{(l-k)!}{(l+k)!}\right]^{1/2}P_l^k(\cos\vartheta)\mathrm{e}^{ik\varphi}, \tag{11}$$

Eqs. (8) and (9) can be summarized in a compact formula for the general matrix elements of the squeezing operator. On the basis of Eqs. (8), (9) and (11), we have

$$\langle m|S(\xi)|n\rangle = \sqrt{\cos\theta}\,X_l^k(\theta,\varphi)\,,\quad \cos\theta = 1/\cosh r\,,\quad l = \tfrac{1}{2}(m+n)\,,\quad k = \tfrac{1}{2}(m-n)\,,\quad (m,n \geq 0\,), \tag{12}$$

where $0 < \theta < \pi/2$, $l = 0,1,2,\ldots$, $k = -l,-l+1,\ldots,l-1,l$. In the special case $n = 0$, the order equals to $k = l = m/2$, and we encounter with the well-known special formula for spherical harmonics,

$$X_l^l(\theta,\varphi) = (-1)^l\sqrt{\frac{(2l)!}{2^{2l}(l!)^2}}(\sin\theta)^l\,\mathrm{e}^{il\varphi}\,,\quad l = m/2\,,\quad \sin\theta = \tanh r\,. \tag{13}$$

By multiplying this with $\sqrt{\cos\theta}$ we receive the photon number distribution amplitude of a *squeezed vacuum state* $S(\xi)|0\rangle$. We note that, in contrast to our Gegenbauer polynomial expressions, Eqs. (5) and (6), in the other equivalent formulas, given by Eqs. (8), (9) and (12), the degree $l = \frac{1}{2}(m+n)$ of the functions is a combination of the initial and final indeces, n and m, respectively. This circumstance would severely complicate the calculation for building up the coherent and incoherent superpositions, to be discussed in Section 3.

In the rest of the present section we illustrate some main features of the photon number distribution of a squeezed number state, determined by the Gegenbauer distribution corresponding to Eqs. (5) and (6). Figure 1 shows the distribution for the lowest n-values with different parities.

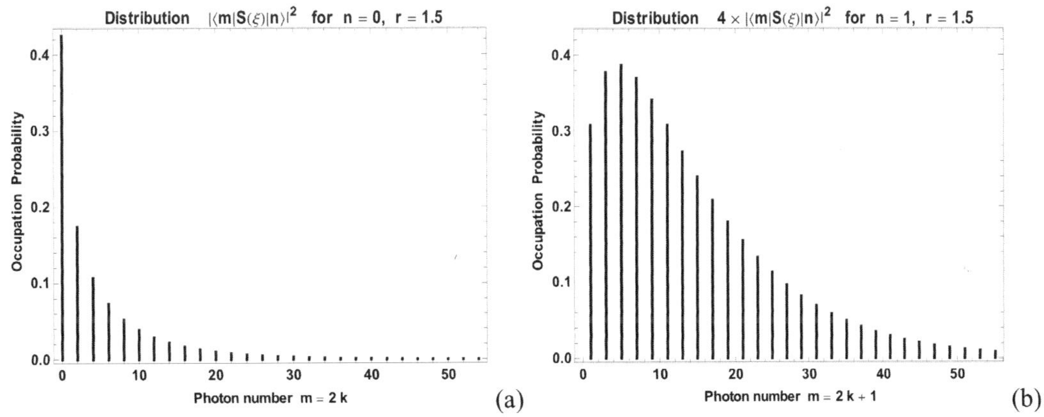

Figure 1. Shows the photon number distributions $p_m(n) = |\langle m|S(\xi)|n\rangle|^2$, according to Eqs. (5-6), for the lowest n-values with different parities. In both cases $|\xi| \equiv r = 1.5$, and the squeezing parameter $s = \mathrm{e}^r = 4.48$. **(a)**: $p_m(0)$ for a squeezed vacuum state $S(\xi)|0\rangle$, in which only even photon number states are occupied; $m = 2k$ (i.e. $m = 0,2,4,\ldots$). This distribution is determined by Eq. (13). The expectation value of the dimensionless energy is $\langle a^+a + 1/2\rangle = 0.5\cosh 2r = 5.03$. **(b)**: $p_m(1)$ for a squeezed one-photon state $S(\xi)|1\rangle$, in which only odd photon number states are occupied; $m = 2k+1$ (i.e. $m = 1,3,5,\ldots$). The expectation value of the dimensionless energy is $1.5\cosh 2r = 15.1$. For a better comparison with **(a)**, in **(b)** we have plotted $4 \times p_m(1)$.

In Fig. 2 the photon number distribution $p_m(5)$ for a squeezed 5-photon state $S(\xi)|5\rangle$ have been plotted, for $|\xi| \equiv r = 1.5$. In this case $n/\cosh^2 r < 1$, and it is justified to use the asymptotic formula 8.936 in Gradshteyn and Ryzhik (2000),

$$\lim_{\nu\to\infty} \nu^{-n/2} C_n^{\nu/2}[t(2/\nu)^{1/2}] = (1/2^{n/2} n!) H_n(t), \tag{14}$$

which yields $p_m(n) \approx |\psi_n(t_m)|^2 / t_m$, where $\psi_n(x)$ is the usual hermite function for the oscillator eigenfunction in L^2 representation, so

$$\langle m|S(\xi)|n\rangle \approx \frac{e^{i\varphi(m-n)/2}}{(\cosh r)\sqrt{t_m}} \psi_n(t_m), \quad \psi_n(x) = \frac{1}{\sqrt{\pi^{1/2} 2^n n!}} H_n(x) \exp(-\tfrac{1}{2}x^2), \quad t_m = \sqrt{\frac{m}{2\cosh^2 r}}. \tag{15}$$

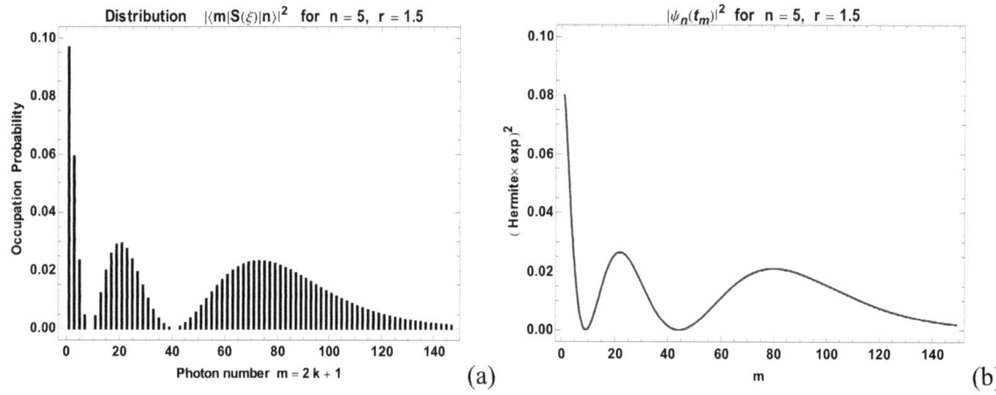

Figure 2. Shows the photon number distributions $p_m(5)$, according to Eqs. (5) and (6). In both cases $|\xi| \equiv r = 1.5$, $s = e^r = 4.48$ and $\tanh^2 r = 0.82$. **(a)**: the exact $p_m(5)$ distribution for a squeezed 5-photon state $S(\xi)|5\rangle$, with average energy is $5.5\cosh 2r = 55.4$. **(b)**: the approximate $p_m(5)$ for the same squeezed 5-photon state. The discrete point are connected for a better visibility of the shape of the distribution Here $n^2 \tanh^2 r = 20.5$ large, but the parameter $n(1-\tanh^2 r) = 0.9$ is smaller than unity; the distribution is result of that the largely stretched state (in 'coordinate representation') is swept by the narrow original wavefunctions $\psi_m(x)$, whose main contribution comes around the points t_m of Eq. (15).

Fig. 2 (b) has been plotted by using the approximate formula, Eq. (15), for a comparison. The agreement with Fig. 2(a) is quite good. It is very remarkable, that the approcimation in Eq. (15), is derived from the exact expressions Eqs. (5) and (6), which are the result of an exclusively abstract, algebraic procedure. So, we would even say, that we 'did not know at the beginning that wave functions in L^2 exist, at all.' Still, we find that a quite faithful „coarse-grained picture" of the modulus square of the wave function has been created from the photon number distribution. We note that Popov and Perelomov (1969) have also found this approximation (see their note at the very end of their paper), however they a priori calculated the matrix element in the L^2 coordinate representation.

In Fig. 3 the characteristic quantum oscillations are nicely seen in this parameter regime $n/\cosh^2 r >> 1$. Such oscillation have also been analysed in details by Popov and Perelomov (1969).

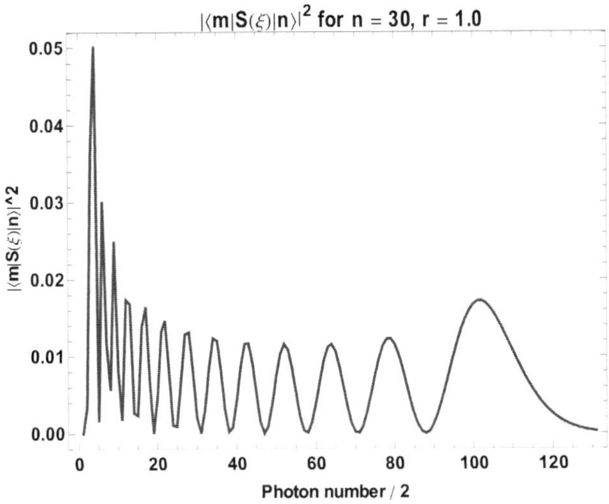

Figure 3. Shows the photon number distribution $p_m(30) = |\langle m|S|30\rangle|^2$ for a squeezed 30-photon state $S(\xi)|30\rangle$, with $|\xi| \equiv r = 1$. The discrete point are connected for a better visibiliti of the shape of the distribution. In this regime the quantum oscillations are nicely seen.

3. Coherent and incoherent superposition of the matrix elements of squeezing transitions.

In the present Section we determine the coherent and incoherent superpositions of the matrix elements of the squeezing operator, on the basis of the newly found Gegenbauer polynomial expressions in Eqs. (5) and (6). First we note that in the semi-classical description, when the photon field is considered as an external field $F_0 \sin \omega t$, for a free electron in a Volkov state (see e.g. Varró 2021a), the oscillating phase factor $\exp[-i(U_p / 4\hbar\omega)\sin 2\omega t]$ appears, which comes from the A^2-interaction. U_p is the ponderomotive energy of the electron which is proportional to the intensity of the (laser) field. So, in the semiclassical approximation this factor contributes to the double-photon emission and absorption processes by the Fourier components, governed by the Bessel functions,

$$\exp[-iz \sin 2\omega t] = \sum_{n=-\infty}^{\infty} J_k(z) e^{-i2k\omega \cdot t} . \tag{16}$$

The corresponding coherent superposition of the transition amplitudes can be calculated on the basis of the generating formula for the Gegenbauer polynomials (see Gradshteyn and Ryzhik 2000),

$$\Gamma(k+\tfrac{1}{2}) \frac{2^k}{\sqrt{\pi}} \sum_{n=0}^{\infty} C_n^{k+\frac{1}{2}}(x) \frac{z^n}{\Gamma(2\lambda+n)} = \mathrm{e}^{z\cos\theta} \frac{1}{(z\sin\theta)^k} J_k(z\sin\theta) .$$

By applying this formula for the (fixed) $2k$-order emission processes, taking Eq. (5), we have

$$\begin{aligned} &\sum_{n=0}^{\infty} \frac{(\beta^*)^{n+2k}}{\sqrt{(n+2k)!}} \mathrm{e}^{-\frac{1}{2}|\beta|^2} \frac{\alpha^n}{\sqrt{n!}} \mathrm{e}^{-\frac{1}{2}|\alpha|^2} \langle n+2k|S(\xi)|n\rangle = \\ &= \exp(-\tfrac{1}{2}|\beta|^2 - \tfrac{1}{2}|\alpha|^2 + \beta^*\alpha\cos\theta) \mathrm{e}^{-\eta/4} \mathrm{e}^{ik\varphi} \left(\frac{\beta^*}{\alpha}\right)^k J_k(\beta^*\alpha \tanh|\xi|) \end{aligned} . \tag{17}$$

In Eq. (17) α and β are the (optionally highly-excited) initial and final coherent states, which represent the laser field, and the argument of the Bessel function can be related to average photon number, which is connected to the classical intensity. For a correct probability interpretation of this expression, the initial and final coherent states can be considered as von Neumann lattice coherent states (Neumann 1932), as is explained in our recent works in Gombkötő *et al* (2016, 2020), Földi *et al* (2021) and Varró (2021a). By comparing Eq. (17) and the Bessel function amplitudes in Eq. (16), we see that if $\alpha=\beta$, the semiclassical result reproduces the result obtained with the quatized description of the field excitations. However, in the quantal description there is always a chance for α and β being different, which expresses the effect of the back-reaction of the electron. So, according to the exact result, Eq. (17), if this back-reaction is not negligible, then the emission (absorption) probability distribution of the $2k$ -order processe can be qualitatively different from the semi-classical one.

In the case of the interaction with a spectral component of black-body radiation, or with some chaotic radiation, in the semi-classical description we use an averaging with respect to the field amplitude. In case of a Gaussian distribution we need the average of $J_n^2(z)$ in Eq. (16). We have

$$\frac{1}{I_0}\int_0^\infty dI e^{-I/I_0} J_k^2(\gamma I) = \frac{1}{\pi\gamma I_0} Q_{k-\frac{1}{2}}(1+1/2\gamma^2 I_0^2) \qquad (k=0,1,2,\ldots), \tag{18}$$

where I_0 is the average intensity of the field, and $Q_{k-\frac{1}{2}}(z)$ is a Legendre function of second kind (here it is a so-called toroidal function, see Gradshteyn and Ryzhik, 2000). For the absorption processes we receive the same expression.

We have also calculated the incoheren superpositions of the Gegenbauer amplitudes in Eqs. (5), by using Planck-Bose weights,

$$p_n = (1-b)b^n, \quad b = e^{-h\nu/k_B T} = \frac{\bar{n}}{1+\bar{n}}, \quad p_n = \frac{\bar{n}^n}{(1+\bar{n})^{1+n}}, \quad \bar{n} = \frac{1}{e^{h\nu/k_B T}-1}, \tag{19}$$

where k_B is the Boltzmann constant, and T is the absolute temperature of the black-body radiation. For $2k$ -order emission processes we have the result

$$\sum_{n=0}^\infty p_n \left|\langle n+2k|S(\xi)|n\rangle\right|^2 = \frac{\cos\theta}{\pi\sqrt{\bar{n}(1+\bar{n})\sin^2\theta}} b^{-k} Q_{k-\frac{1}{2}}(z), \quad z = 1+\frac{1}{2\bar{n}(1+\bar{n})\sin^2\theta}, \quad k=0,1,2\ldots. \tag{20}$$

For $2k$ -order absorptions we have recived the average probability

$$\sum_{n=2l}^\infty p_n \left|\langle n-2l|S(\xi)|n\rangle\right|^2 = \frac{\cos\theta}{\pi\sqrt{\bar{n}(1+\bar{n})\sin^2\theta}} b^{+l} Q_{l-\frac{1}{2}}(z), \quad z = 1+\frac{1}{2\bar{n}(1+\bar{n})\sin^2\theta}, \quad l=0,1,2\ldots. \tag{21}$$

In Eqs. (20) and (21) $\sin^2\theta = \tanh^2 r$. The semi-classical result, Eq. (18), approximates quite well the exact quantum result, calculated on the basis of Eqs. (5) and (6), if the prefactors $b^l = e^{-l\cdot h\nu/k_B T}$ are close to unity, i.e. if $h\nu/k_B T << 1$, which just corresponds to the Rayleigh-Jeans limit of the Planck distribution. In the opposite case, in the Wien-limit, when $h\nu/k_B T >> 1$, the two expressions are quatitatively different.

4. Summary

In Section 2 (and in the Appendix) we have proved that matrix elements of the squeezing operator (Bogoliubov transformation) can be expressed in terms of the classical Gegenbauer polynomials. On the basis of this new formula, we have given few illustrative examples for the photon number distribution of a squeezed number state. This new expression makes it possible to determine coherent and incoherent superpositions of these matrix elements in closed analytic forms. In Section 3 we have described multiphoton transitions in the system "charged particle + electromagnetic radiation", induced by a (strong) coherent field or by a spectral component of black-body radiation. The exact results have been compared with the semi-classical ones (the latter are based on non-perturbative matrix elements with c-number radiation fields). It has been found that in the case of interaction with a thermal field, the semi-classical result (calculated with a c-number, Gaussian stochastic field amplitude) yields an acceptable approximation only in the Rayleigh-Jeans limit ($h\nu/kT \ll 1$). It has been explicitely shown that in the Wien limit ($h\nu/kT \gg 1$) the semiclassical formula for the multiphoton absorption and emission probabilities cannot even 'mimic' its quantum counterpart, because it has a differen functional form.

Acknowledgments
The author thanks Prof. Dr. Wolfgang Schleich of Universität Ulm for the illuminating discussions on the subject of the present paper, we had during his recent visit in Budapest. Support by the ELI-ALPS project is acknowledged. The ELI-ALPS Project No. GINOP 2.3.6-15 is supported by the European Union and co-financed by the European Regional Development Fund.

Appendix A. Algebraic derivation of the matrix elements of the squeezing operator between photon number eigenstates.

In the present Appendix A we give the details of the calculation of the matrix elements between the number eigenstates of the unitary squeezing operator, $S_{m,n} = \langle m|S(\xi)|n\rangle$, where

$$S(\xi) = \exp[\tfrac{1}{2}\xi(a^+)^2 - \tfrac{1}{2}\xi^*(a)^2], \quad S(\xi) = \exp(\xi K_+ - \xi^* K_-), \quad \xi = |\xi|\, e^{i\varphi}, \tag{A.1}$$

with $K_+ = \frac{1}{2}(a^+)^2$ and $K_- = \frac{1}{2}(a)^2$. By introducing $K_0 = \frac{1}{4}(aa^+ + a^+a)$, we have the closed set of generators of a special Lie algebra of the $SU(1,1)$ group, having the commutation relations $[K_0, K_\pm] = \pm K_\pm$, and $[K_-, K_+] = 2K_0$. The Casimir operator $\hat{C}_2 = K_0^2 - \frac{1}{2}(K_+K_- + K_-K_+)$ commutes with all the generators, hence it is proportional with the unit operator, $C_2 = \kappa(\kappa-1)\hat{I}$, where κ is called the Bargmann index. In the special case under discussion, we have $C_2 = -(3/16)\hat{I}$, which yields $\kappa = 1/4$ or $\kappa = 3/4$. For $\kappa = 1/4$, the states $|n\rangle$ with even n form a basis of the unitary irreducible representation space of the group $SU(1,1)$, and the states $|n\rangle$ with odd n form a basis corresponding to $\kappa = 3/4$. In the 'even representation space' $K_0|2k\rangle = (k+1/4)|2k\rangle$ ($k = 0,1,2,\ldots$), and in the 'odd representation space' $K_0|2l+1\rangle = (l+3/4)|2l+1\rangle$ ($l = 0,1,2,\ldots$). Each equation can be written in the form $K_0|\psi_\nu\rangle = \nu|\psi_\nu\rangle$, where $\nu - \kappa = 0,1,2,\ldots$. Of course, the complete Hilbert space is the direct sum of these sub-spaces, $\mathscr{H} = \mathscr{H}_{1/4} \oplus \mathscr{H}_{3/4}$. In the present and in similar contexts, the properties of the $SU(1,1)$ group have been used by Popov and Perelomov (1969), Malkin, Man'ko and Trifonov (1970), Gilmore (1974), Perelomov (1977), Kelemen (1975), Witschel (1975), Fisher, Nieto and Sandberg (1984) or Schumaker and Caves (1985), Wünsche (1999, 2003, 2017a-b).

Since neither the commutation relations, nor the parameters depend on the Bargmann index, the ordering (factorization) of $S(\xi)$ can be performed in any concrete representation, of course. Perhaps the simplest way to derive the normal (and anti-normal) form of $S(\xi)$ is to use the 'spinor representation' of the generators (see e.g. Gilmore (1974), Fisher, Nieto and Sandberg (1984) or Schumaker and Caves (1985)). It is an easy matter to check that the following 2×2 matrices

$$-\sigma_+ = \begin{pmatrix} 0 & -1 \\ 0 & 0 \end{pmatrix}, \quad \sigma_- = \begin{pmatrix} 0 & 0 \\ 1 & 0 \end{pmatrix}, \quad \tfrac{1}{2}\sigma_3 = \tfrac{1}{2}\begin{pmatrix} 1 & 0 \\ 0 & -1 \end{pmatrix}$$

satisfy the same commutation rules as K_+ and K_- and K_0, respectively, thus $-\sigma_+$, σ_- and $\frac{1}{3}\sigma_3$ also form an $SU(1,1)$ algebra . The ordering can be performed by expanding the exponential expressions of the 2×2 matrix representants (which are also 2×2 matrices), and comparing the parameters of the original exponential expression with the parameters of the new exponential factors. From this we can determine the coefficients of $(-\sigma_+)$, σ_- and $\frac{1}{2}\sigma_3$ in the new exponential factors, and these coefficients must be identified with that of K_+ and K_- and K_0 in the normal (anti-normal) form. In the 'spinor representation' we have

$$S(\xi) = \exp(\xi K_+ - \xi^* K_-) \to \exp\left[-\begin{pmatrix} 0 & \xi \\ \xi^* & 0 \end{pmatrix}\right] = \begin{pmatrix} \cosh|\xi| & -e^{i\varphi}\sinh|\xi| \\ -e^{-i\varphi}\sinh|\xi| & \cosh|\xi| \end{pmatrix}, \tag{A.2}$$

$$\exp(\zeta K_+)\exp(-\eta K_0)\exp(\zeta' K_-) \to \begin{pmatrix} 1 & -\zeta \\ 0 & 1 \end{pmatrix}\begin{pmatrix} e^{-\eta/2} & 0 \\ 0 & e^{\eta/2} \end{pmatrix}\begin{pmatrix} 1 & 0 \\ \zeta' & 1 \end{pmatrix} = \begin{pmatrix} e^{-\eta/2} - \zeta\zeta' e^{\eta/2} & -\zeta e^{\eta/2} \\ e^{\eta/2}\zeta' & e^{\eta/2} \end{pmatrix}. \tag{A.3}$$

By equating the components of the right hand sides,

$$\cosh|\xi| = e^{\eta/2}, \text{ i.e. } \eta = 2\log\cosh|\xi|, \quad -e^{i\varphi}\sinh|\xi| = -\zeta e^{\eta/2} = -\zeta\cosh|\xi|, \text{ i.e. } \zeta = e^{i\varphi}\tanh|\xi|,$$
$$-e^{-i\varphi}\sinh|\xi| = \zeta' e^{\eta/2} = \zeta'\cosh|\xi|, \text{ i.e. } \zeta' = -e^{-i\varphi}\tanh|\xi| = -\zeta^*. \tag{A.4}$$

Thus, the normal form of $S(\xi)$ becomes

$$S(\xi) = \exp(\xi K_+ - \xi^* K_-) = \exp(\zeta K_+)\exp(-\eta K_0)\exp(-\zeta^* K_-). \tag{A.5}$$

On the basis of (A.4), the new parameters are determinedto be $\eta = 2\log\cosh|\xi|$, $\zeta = e^{i\varphi}\tanh|\xi|$. We note that the anti-normal form can also be obtained similarly,

$$\hat{S}(\xi) = \exp(\xi K_+ - \xi^* K_-) = \exp(-\zeta^* K_-)\exp(\eta K_0)\exp(\zeta K_+). \tag{A.6}$$

With the help of the normal form in (A.5) the calculation of the matrix element $\langle m|S(\xi)|n\rangle$ is now straightforward. The effect of $\exp(-\zeta^* \frac{1}{2}a^2)$ on $|n\rangle$ yields a finite sum, in which the highest power of $-\zeta^*$ is $[n/2]$, where $[x]$ denotes the integer part of x (i.e. the smallest integer, which is larger of equal to x). The factor in the middle of the expression in (A.5) is diagonal, i.e. $\exp(-\eta K_0)|k\rangle = \exp[-(\eta/2)(k+\frac{1}{2})]|k\rangle$ for any k. The factor $\langle m|\exp(\zeta K_+) = (\exp(\zeta^* \frac{1}{2}a^2)|m\rangle)^+$ on the left yields also a finite sum, in which the highest power of ζ is $[m/2]$. The scalar product of the two sums terminates at the smaller maximum summation index $\min([m/2],[n/2])$. Any term like $\langle m|(a^{+2})^l(a^2)^k|n\rangle$ is proportional to $\langle m|n-2k+2l\rangle = \delta_{m,n-2k+2l}$, thus, only those matrix elements are non-vanishing in which m and n have the same parity, and this means that their difference $m-n$ must be an even number, $m-n=2\alpha$, where α is an integer. By taking all these considerations into account, a straightforward calculation leads to the explicit result,

$$\langle m|S(\xi)|n\rangle = \exp[-\tfrac{1}{2}(n+\tfrac{1}{2})\eta](\zeta/2)^{(m-n)/2}\sqrt{m!}\sqrt{n!}\sum_{k=0}^{[n/2]}\frac{(-|\zeta|^2 e^{\eta}/4)^k}{[\frac{1}{2}(m-n)+k]!k!(n-2k)!}, \quad (m \ge n), \tag{A.7}$$

$$\langle m|S(\xi)|n\rangle = \exp[-\tfrac{1}{2}(m+\tfrac{1}{2})\eta](-\zeta^*/2)^{(n-m)/2}\sqrt{m!}\sqrt{n!}\sum_{l=0}^{[m/2]}\frac{(-|\zeta|^2 e^{\eta}/4)^l}{[\frac{1}{2}(n-m)+l]!\,l!\,(m-2l)!} \quad (m \le n). \quad \text{(A.8)}$$

These expressions (A.7) and (A.8) are equivalent to the ones derived long ago by Husimi (1953), who used two-variable generating functions of S_{mn}, and the L^2 representation (Hermite functions) for the oscillator basis states (see also Popov and Perelomov 1969). For a real ζ Eq. (A.7) reduces to the equation (27) in Kelemen's (1975) paper. See also Tanabe (1973), Rashid (1975), Satyanarayana (1985) and Mendaš and Popović (1995).

Explicit form of the matrix elements of the squeezing operator between photon number eigenstates, expressed in terms of Gegenbauer polynomials.

Now, our task is to express the finite sums (A.7) and (A.8) in terms of known functions, which makes it possible to perform the analytic calculations shown in the main text. Let us consider the sum on the right hand side of (A.3), which refers to the sub-case $m \ge n$. We observe that the first factorial in the denominator can be expressed as

$$[\tfrac{1}{2}(m-n)+k]! = [\tfrac{1}{2}(m-n)]!\left(\tfrac{1}{2}(m-n)+1\right)_k, \qquad (a)_k = (a)\cdot(a+1)\cdots(a+k-1) = \frac{\Gamma(a+k)}{\Gamma(a)},$$

where we have introduced the Pochhammer's symbol $(a)_k$, which can also be expressed in terms of the gamma functions (Abramowitz and Stegun, 1970). Besides, from the explicit form of the factorials, one can show

$$\frac{n!}{(n-2k)!} = 2^{2k}\left(-\frac{n}{2}\right)_k\left(\frac{1-n}{2}\right)_k .$$

These manipulations allow us to bring into consideration the hypergeometric series $F(a,b;c;z)$, which may be used to represent orthogonal polynomials,

$$F(a,b;c;z) = 1 + \frac{a\cdot b}{c\cdot 1}z + \frac{a(a+1)b(b+1)}{c(c+1)\cdot 1\cdot 2}z^2 + \ldots = \sum_{k=0}^{\infty}\frac{(a)_k(b)_k}{k!(c)_k}z^k .$$

In our formula (A.7) this series terminates when the summation index takes on the value $k=[n/2]$ (and in (A.8) the sum terminates at $k=[m/2]$). The expressions (A.3) and (A.4) for the matrix elements can then be brought to the equivalent forms, expressed in terms of the hypergeometric functions,

$$\langle m|S(\xi)|n\rangle = \exp[-\tfrac{1}{2}(n+\tfrac{1}{2})\eta](\zeta/2)^{(m-n)/2}\sqrt{\frac{m!}{n!}}\frac{1}{\Gamma(\alpha+1)}F\left(-\frac{n}{2},\frac{1-n}{2};\frac{m-n}{2}+1;z\right) \quad (m \ge n), \quad \text{(A.9)}$$

$$\langle m|S(\xi)|n\rangle = \exp[-\tfrac{1}{2}(m+\tfrac{1}{2})\eta](-\zeta^*/2)^{(n-m)/2}\sqrt{\frac{n!}{m!}}\frac{1}{\Gamma(\lambda+1)}F\left(-\frac{m}{2},\frac{1-m}{2};\frac{n-m}{2}+1;z\right) \quad (m \le n), \quad \text{(A.10)}$$

where $z = -|\zeta|^2 e^{\eta} = -\sinh^2|\xi|$, and $\alpha = \tfrac{1}{2}(m-n)$ in (A.9) and $\lambda = \tfrac{1}{2}(n-m) = |\alpha|$ in (A.10) are non-negative integers. It can be seen that Equations (A.9) and (A.10) go over into each other if we exchange n and m, take the complex conjugate of the resulting expression, and multiply this by $(-1)^{(m-n)/2}$. Thus, it is enough to detail the forthcoming derivation only for one of these equations, e.g. (A.9). In addition we note that Eqs. (A.9) and (A.10) are equivalent to equations (10.28a-b) of Bargmann (1947) (see also Sciarrino and Toller (1967)).

Consider the transitions between *even* photon number states, i.e., assume that $n=2k$ where $k=0,1,2\ldots$ (and $m=2q$ with $q=0,1,2\ldots$). Then, on the right hand side of equation (A.9) we have $F(-k,-k-\beta;\alpha+1;z)$, where $\beta=-1/2$ and $\alpha=\tfrac{1}{2}(m-n)=k-q$ is the same non-negative integer, as has been defined above. We apply the formula 8.962.1 of Gradshteyn and Ryzhik (2000), p. 990, which gives the connection between the Jacobi polynomials and the hypergeometric functions,

$$P_k^{(\alpha,\beta)}(x) = \frac{\Gamma(k+\alpha+1)}{k!\Gamma(\alpha+1)}\left(\frac{x+1}{2}\right)^k F\left(-k,-k-\beta;\alpha+1;\frac{x-1}{x+1}\right), \quad \frac{x-1}{x+1} = z, \tag{*}$$

where $P_k^{(\alpha,\beta)}(x)$ is a Jacobi polynomial of degree $k = n/2$, and $\beta = -1/2$ in the present case. The argument of the Jacobi polynomial can simply be expressed as $x = (1+z)/(1-z)$, where $z = -\sinh^2|\xi|$. By taking this connection (*) into account in Equation (A.9), we receive

$$\langle m|S(\xi)|n\rangle = e^{-\eta/4}(\zeta/2)^\alpha \frac{\sqrt{m!}\,(\frac{1}{2}n)!}{\sqrt{n!}\,(\frac{1}{2}m)!} P_{n/2}^{(\alpha,-1/2)}(x), \quad x = \frac{1+z}{1-z}, \tag{A.11}$$

($m \geq n$, even m and even $n \geq 0$).

Consider now the transitions between *odd* photon number states, i.e., assume that $n = 2k+1$ where $k = 0,1,2\ldots$ (and $m = 2q+1$ with $q = 0,1,2\ldots$). We take into account that the hypergeometric functions are symmetric in the first two parameters; $F(a,b;c;z) = F(b,a;c;z)$. Then, on the right hand side of equation (A.5) we have $F(-k,-k-\beta';\alpha+1;z)$, but now $\beta' = +1/2$. As before, we again apply the same formula 8.962.1 of Gradshteyn and Ryzhik (2000), denoted by (*) above, for the connection of the Jacobi polynomials and the hypergeometric functions, and receive

$$\langle m|S(\xi)|n\rangle = e^{-\eta/4}(\zeta/2)^\alpha \frac{\sqrt{m!}\,[\frac{1}{2}(n-1)]!}{\sqrt{n!}\,[\frac{1}{2}(m-1)]!}\sqrt{\frac{1+x}{2}} P_{(n-1)/2}^{(\alpha,+1/2)}(x), \quad x = \frac{1+z}{1-z}, \tag{A.12}$$

($m \geq n$, odd m and odd $n \geq 1$).

The argument of the Jacobi polynomial is $x = (1+z)/(1-z)$ (where $z = -\sinh^2|\xi|$), and the other parameters have been defined in (A.4); $\zeta = \mathrm{e}^{i\varphi}\tanh|\xi|$, and $\mathrm{e}^{\eta/2} = \cosh|\xi|$.

Finally, we use the formulas 22.5.22 and 22.5.21 of Abramowitz and Stegun (1970) in Eq. (A.11) and Eq. (A.12), referring to the even-even and odd-odd transitions, respectively,

$$P_k^{(\alpha,-1/2)}(x) = \frac{(\frac{1}{2})_k}{(\alpha+\frac{1}{2})_k} C_{2k}^{\alpha+\frac{1}{2}}\left(\sqrt{\frac{1+x}{2}}\right), \quad P_k^{(\alpha,+1/2)}(x) = \frac{(\frac{1}{2})_{k+1}}{(\alpha+\frac{1}{2})_{k+1}\sqrt{\frac{1+x}{2}}} C_{2k+1}^{\alpha+\frac{1}{2}}\left(\sqrt{\frac{1+x}{2}}\right), \tag{A.13a}$$

$$x = \frac{1+z}{1-z}, \quad \frac{1+x}{2} = \frac{1}{1-z}, \quad z = -\sinh^2|\xi|, \quad \sqrt{\frac{1+x}{2}} = \frac{1}{\cosh|\xi|}. \tag{A.13b}$$

Here $C_n^\nu(x)$ are the Gegenbauer polynomials (see e.g. Gradshteyn and Ryzhik (2000)). Inserting the first and the second equations of (A.13a) in Eqs. (A.11) and (A.12), respectively, and taking (A.13b) into account, we receive a *unifying formula* (in the sub-case $m \geq n$), *valid for both even and odd indeces,*

$$\langle m|S(\xi)|n\rangle = \mathrm{e}^{-\eta/4}(2\mathrm{e}^{i\varphi}\tanh|\xi|)^\alpha \sqrt{\frac{n!}{m!}}\frac{\Gamma(\alpha+\frac{1}{2})}{\sqrt{\pi}} C_n^{\alpha+\frac{1}{2}}\left(\frac{1}{\cosh|\xi|}\right), \quad \alpha = \tfrac{1}{2}(m-n), \quad (m \geq n). \tag{A.14}$$

A similar unifying formula (which is also valid for both paritys) results in the other sub-case $m \leq n$,

$$\langle m|S(\xi)|n\rangle = \mathrm{e}^{-\eta/4}(-2\mathrm{e}^{-i\varphi}\tanh|\xi|)^\lambda \sqrt{\frac{m!}{n!}}\frac{\Gamma(\lambda+\frac{1}{2})}{\sqrt{\pi}} C_m^{\lambda+\frac{1}{2}}\left(\frac{1}{\cosh|\xi|}\right), \quad \lambda = \tfrac{1}{2}(n-m), \quad (m \leq n). \tag{A.15}$$

In deriving Equation (A.14) from Eqs. (A.11) and (A.12), we have also taken into account the so-called doubling formula $\Gamma(2x) = \pi^{-1/2}2^{2x-1}\Gamma(x)\Gamma(x+\frac{1}{2})$ of the gamma function (see e.g. Gradshteyn and Ryzhik 2000), and the explicit form of the parameter $\eta = 2\log\cosh|\xi|$. According to the functional relation $C_n^{\frac{1}{2}}(x) = P_n(x)$, if $m = n$, then, both in (A.14) and (A.15), the Gegenbauer polynomials reduce to the Legendre polynomial $P_n(x)$,

$$\langle n|S(\xi)|n\rangle = \sqrt{x}P_n(x), \quad x = 1/\cosh r, \quad r = |\xi| \qquad (m = n). \tag{A.16}$$

We also note that, in contrast to Eqs. (A.11) and (A.12) being separately valid in the sub-spaces $\mathscr{H}_{1/4}$ and $\mathscr{H}_{3/4}$, respectively, the unifying formulas (A.14) (and also (A.15)) are valid in the whole Hilbert space $\mathscr{H} = \mathscr{H}_{1/4} \oplus \mathscr{H}_{3/4}$. On "unifying formula" we mean that the functional form of the expressions are the same for both the $\mathscr{H}_{1/4}$ (even) and the $\mathscr{H}_{3/4}$ (odd) subspaces. This is not the case for Eqs. (A.11) and (A.12), because they contain different Jacobi polynomials $P_{n/2}^{(\alpha,-1/2)}(x)$, and $P_{(n-1)/2}^{(\alpha,+1/2)}(x)$ with different parameters, respectively.

References.

Abramowitz M and Stegun I A 1965 (Editors) *Handbook of Mathematical Functions* (Dover Publications, Inc., New York, 1965)
Andersen U L, Gehring T, Marquardt C and Leuchs G 2016, 30 years of squeezed light generation. *Phys. Scr.* **91**, 053001 (2016).
Aronson E B, Malkin I A and Man'ko V I 1974 Matrix elements of Bogolubov canonical transformations. *Lett. Nuovo Cimento* **11**, 44-46 (1974).
Bargmann V 1947 Irreducible unitary representation of the Lorentz group. *Ann. Math.*, **48** (3), 568-640 (1947).
Bergou J and Varró S 1981a Nonlinear scattering processes in the presence of a quantized radiation field: I. Nonrelativistic treatment. *J. Phys. A: Math. Gen.* **14**, 1469-1482 (1981).
Bergou J and Varró S 1981b Nonlinear scattering processes in the presence of a quantized radiation field: II. Relativistic treatment. *J. Phys. A: Math. Gen.* **14**, 2281-2303 (1981).
Berson I Y 1969 Electron in the quantized field of a monochromatic electromagnetic wave. *Zh. Eksp. Teor. Fiz.* **56**, 1627–1633 (1969), [*Sov. Phys. JETP* 1969, **29**, 871-874.]
Bogoljubov N N 1958 On a new method in the theory of superconductivity. *Nuovo Cim.* **7**, 794-805 (1958).
Dodonov V V 2002 'Nonclassical' states in quantum optics: a 'squeezed' review of the first 75 years. *J. Opt. B: Quantum Semiclass. Opt.* **4**, R1-R33 (2002).
Dodonov V V and Man'ko V I 2003 (Eds.) *Theory of Nonclassical States of Light* (Taylor & Francis: London, UK; New York, NY, USA, 2003).
Dodonov V V, Parametric excitation and generation of nonclassical states in linear media. In Dodonov V V and Man'ko V I (Eds.) *Theory of Nonclassical States of Light*; Taylor & Francis: London, UK; New York, NY, USA, 2003) pp 153-218.
Dodonov V V 2018 Coherent states and their generalizations for a charged particle in a magnetic field. In *Coherent States and Their Applications: A Contemporary Panorama*; Springer Proc. in Physics; Antoine J-P, Bagarello F and Gazeau J-P, (Eds.)Volume 205, pp. 311–338. (Springer: Berlin, Germany, 2018)
Dodonov V V and Horovits M B 2021 Energy and magnetic moment of a quantum charged particle in time-dependent magnetic and electric fields of circular and plane solenoids. *Entropy* **23**, 1579 (54 pages).
Erdélyi A 1953 (Editor) *Higher transcendental functions. Vol. II.* (McGraw-Hill Book Company, Inc., N. Y., 1953).
Fedorov M V and Kazakov A E 1973 An electron in a quantized plane wave and in a constant magnetic field *Z. Phys.* **261**, 191-202 (1973).
Fedorov M V, Efremov M A, Volkov P A, Moreva E V, Straupe S S and Kulik S P 2008 Spontaneous parametric down-conversion: Anisotropical and anomalously strong narrowing of biphoton momentum correlation distributions. Phys. Rev. A 77, 032336 (2008).
Fisher R A, Nieto M M and Sandberg V D 1984 Impossibility of naively generalizing squeezed coherent states. *Phys. Rev. D* **29**, 013418. (1984).
Földi P, Magashegyi I, Gombkötő Á and Varró S 2021 Describing high-order harmonic generation using quantum optical models. *Photonics* **8**, 263 (2021). https://doi.org/10.3390/photonics8070263
Gazazyan A D and Sherman B G 1989 Effects of quantum coherence of electromagnetic radiation in interactions with electrons. *Zh. Eksp. Teor. Fiz.* **96**, 71-84 (1989) [*Sov. Phys. JETP* **69** (1), 38-45 (1989)]
Gilmore R 1974 Baker-Campbell-Hausdorff formulas. *J. Math. Phys.* 15, 2090-2092 (1974).

Gombkötő Á, Czirják A, Varró S and Földi P 2016 Quantum-optical model for the dynamics of high-order-harmonic generation. *Phys. Rev. A* **94**, 013853. (2016).
Gombkötő Á, Varró S, Mati P and Földi P 2020 High-order harmonic generation as induced by a quantized field: Phase-space picture. *Phys. Rev. A* **101**, 013418. (2020).
Gradshteyn I S and Ryzhik I M 2000 *Table of integrals, series and products* (Academic Press, San Diego, 2000).
Husimi K 1953 Miscellanea in elementary quantum mechanics. II. *Progr. Theor. Phys.* **9** (4) 381-402.
Kelemen A 1975 New proof of closed formulas for Bogoljubov boson transformations. *Z. Phys. A* **274**, 109-111.
Kiss T, Janszky J and Adam P 1994 Time evolution of harmonic oscillators with time-dependent parameters: A step-function approximation. *Phys. Rev. A* **49**, 4935–4942 (1994).
Loudon R 2000 The *Quantum Theory of Light*; Clarendon Press: Oxford, UK, 2000. (1st Edition 1973.)
Loudon, R.; Knight, P.L. Squeezed light. *J. Mod. Opt.* **34**, 709–759 (1987).
Louisell W H , Yariv A and Siegman A E 1961 Quantum fluctuations and noise in parametric processes. I. *Phys. Rev.* **124**, 1646–1654 (1961).
Malkin I A, Man'ko V I and Trifonov D A 1970 Coherent states and transition probabilities in a time-dependent electromagnetic field. *Phys. Rev. D* **2** (2), 1371-1385 (1970).
Mendaš I and Popović D B 1995 Number-phase uncertainty product for generalized squeezed states arising from the Pegg-Barnett Hermitian phase operator formalism. *Phys. Rev. A* **52**, 4356-4364 (1995).
Milburn G J and Walls D F 1983 Squeezed states and intensity fluctuations in degenerate parametric oscillation. *Phys. Rev. A* **27**, 392-394 (1983).
Mollow B R and Glauber R J 1967 Quantum theory of parametric amplification. I. *Phys. Rev.* **160**, 1076 (1967).
Neumann J von 1932 *Mathematische Grundlagen der Quantenmechanik*; (Springer: Berlin, Germany).
Perelomov A M 1977 Generalized coherent states and some of their applications. *Usp. Fiz. Nauk* 123, 23-55 (1977).
Popov V S and Perelomov A M 1969 Parametric excitation of a quantum oscillator. *Zh. Eksp. Teor Fiz.* **56** (4), 1375-1390 (1969). [*Sov. Phys. JETP* **29** (4), 738-745 (1969).]
Rashid M A, Simple expression for the linear boson transformation coefficients. II. *J. Math. Phys.* **16**, 378-380 (1975).
Sannikov S S 1965 Non-compact symmertry group of a quantum oscillator. *Zh. Eksp. Teor. Fiz.* **49** (6), 1913-1922 (1965). [*Sov. Phys. JETP* **22** (6), 1306-1311 (1966).]
Satyanarayana M V, Generalized coherent states and generalized squeezed coherent states. Phys. Rev. D **32**, 400-404 (1985).]
Schleich W and Wheeler J A 1987 Oscillations in photon distribution of squeezed states and interference in phase space. *Nature* **326**, 574-577 (1987).
Schleich W, Walls D F and Wheeler J A 1988 Area of overlap and interference in phase spave versus Wigner pseudoprobabilities. *Phys. Rev. A* **38** (3), 1177-1186 (1988).
Schleich W P 2001 *Quantum Optics in Phase Space.* (Wiley-VCH:Weinheim, Germany, 2001).
Sciarrino A and Toller M 1967 Decomposition of the unitary irreducible representations of the group SL(2C) restricted to the subgroup SU(1,1). *J. Math. Phys.* **8** (6), 1252-1265 (1967).
Scully M O and Zubairy M S 1997 *Quantum Optics* (Cambridge University Press: Cambridge, UK).
Schumaker B L and Caves C M 1985 New formalism for two-photon quantum optics. II. Mathematical foundation and compact notation. *Phys. Rev. A* **31**, 3093-3111 (1985).
Stoler D 1970 Equivalence classes of minimum uncertainty packets. *Phys. Rev. D* **1**, 3217-3219 (1970).
Stoler D 1971 Generalized coherent states. *Phys. Rev. D* **4**, 2309-2312 (1971).
Straupe S S, Ivanov D P, Kalinkin A A, Bobrov I B and Kulik S P 2011 Angular Schmidt modes in spontaneous parametric down-conversion. *Phys. Rev. A* **83**, 060302(R) (2011).
Tanabe K, The explicit determination of the linear boson transformation coefficients. *J. Math. Phys.* **14**, 618-622 (1973).
Valatin J G 1958 Comments on the theory of superconductivity. *Nuovo Cim.* **7**, 843-857 (1958).
Varró S 1984 Coherent states of an electron in a homogeneous constant magnetic field and the zero magnetic field limit. *J. Phys. A Math. Gen.* **17**, 1631–1638 (1984).

Varró S and Ehlotzky F 1985 Classical limit of Compton scattering and electron scattering in external fields. *J. Phys. B: At. Mol. Phys.* **18**, 3395-3406 (1985).

Varró S and Ehlotzky 1987 Generalized coherent states for electrons in external fields and application to potential scattering. *Phys. Rev. A* **36**, 497-509 (1987).

Varró S 2015 Regular phase operator and SU(1,1) coherent states of the harmonic oscillator. *Physica Scripta* **90** (7) (2015) 074053 (18pp).

Varró S 2021a Quantum optical aspects of high-harmonic generation. *Photonics* **8**, 269 (2021).

Varró S 2021b Coherent and incoherent superposition of the transition matrix elements of the squeezing operator. S7.5.3, presented at Seminar 7 of LPHYS'21; 29th International Laser Physics Workshop, 19-23 July 2021.

Walls D F 1979 Evidence for the quantum nature of light. *Nature* **280**, 451-454 (1979).

Wheeler J A 1985 Franck-Condon effect and squeezed-state physics as double-source interference phenomena. *Lett. Math. Phys*. **10**, 201-206 (1985).

Witschel W 1975 On the general linear (Bogoliubov-) transformation for bosons. *Z. Phys. B* **21**, 313-318 (1975).

Wünsche A 1999 Ordered operator expansions and reconstruction from ordered moments. *J. Opt. B: Quantum Semiclass.* **1**, 264-288 (1999).

Wünsche A 2003 Squeezed states. pp 95-152.; In Dodonov V V and Man'ko V I (Eds.) *Theory of Nonclassical States of Light* (Taylor & Francis: London, UK; New York, NY, USA, 2003)

Wünsche A 2017a Operator methods and SU(1,1) symmetry in the theory of Jacobi and of ultraspherical polynomials. *Adv. Pure Math.* **7**, 213-261 (2017).

Wünsche A 2017b Squeezed coherent states in non-unitary approach and relation to sub- and super-Poissonian statistics. *Adv. Pure Math.* **7**, 706-757 (2017).

Yuen H P 1976 Two-photon coherent states of the radiation field. *Phys. Rev. A*. **13**, 2226-2243 (1976).

Phase Transitions in Quantum Complex Networks

A.P. Alodjants, A. Yu. Bazhenov and M. M. Nikitina

National Research University for Information Technology, Mechanics and Optics (ITMO), St. Petersburg, 197101, Russia

E-mail: b.a.y@mail.ru

Abstract. In this work we examine a superradiant (SR) and/or ferromagnetic (FM) - paramagnetic (PM) phase transitions problem in quantum materials which may be established by Barabási-Albert (BA) scale-free network that possesses power law degree distribution and specific degree correlations. We represent quantum material by means of Dicke-Ising model, that describes the interaction between a spin-1/2 (two-level) system and external classical (magnetic) and quantized (transverse) fields. To describe PM-FM and SR phase transitions we introduce three order parameters: the total (topologically) weighted as well as un-weighted z-spin components, and the normalized transverse field amplitude, which correspond to the spontaneous magnetization in z- and x-directions, respectively. We have shown that SR state occurs as a result of the interaction between the ordering of the spins in the $z-$ and x-directions and depends on assortativity or disassortativity of the network medium. We have shown that non-trivial topological behavior associated with large fluctuations of network parameters inherent to assortative networks reduces of PM-FM phase transition temperature, while dissasortative networks exhibit high temperature phase transitions. Our findings demonstrate new opportunities to design of quantum materials which may be implemented for current quantum technologies at relatively high temperatures.

1. Introduction

These days, quantum, and especially superconductor materials, represent one of the most promising toolbox for quantum computing hardware [1]. Superconductor circuits are suitable for cavity QED in the strong coupling regime [2] and demonstrate various remarkable features including superradiant phase (SR) transitions [3, 4]. Notably, transverse Ising model is at the heart of current quantum technologies and quantum annealing algorithms proposed for quantum circuits implementation [5, 6]. In particular, it is proven that NP-hard quadratic unconstrained binary optimization (QUBO) problem on a graph may be solved by means of adiabatic quantum computation, which presumes implementation of Ising Hamiltonian with transverse field [7].

In this work we have examined the influence of quantized transverse Ising model non-trivial topology on its critical properties, cf. [8]. In particular, we have studied materials which possess complex (scale-free) network structure cf. [9, 10, 11]. It is important that such networks presume high temperature ferromagnetic (FM)-paramagnetic (PM) ([12, 13]), or superconductor-insulator phase transitions which may appear in high-temperature superconductors [14].

To be more specific, we have examined the scale-free, so-called Barabási-Albert (BA) network model, that considers a preferential link connection. We used the Ising annealed network approach that possesses a power law distribution of degrees [15, 16]. The role of degree exponent γ for scale-free networks and degree correlation (assortativity) is discussed, [17].

In our model, called Dicke–Ising model, the transverse field represents some variable which allows second quantization procedure. Spin systems are established as two-level artificial atoms (TLAA) [18], which interact with the quantized field in the framework of the Dicke model. For our model, quantized field is responsible for superradiant (SR) second order phase transition, cf. [19, 20, 21, 22]. The SR phase transition establishes some certain (non-zero) spontaneous polarization, that occurs in a thermodynamically equilibrium ensemble of TLAA interacting with the quantized field. Notably, the phase transitions problem for the Dicke–Ising model specified for complex networks has not been studied yet at all. The aim of this work is to elucidate the interplay between various order parameters inherent to second order phase transitions in complex quantum materials. We have investigated the role of particular network characteristics, such as node degree, degree exponent for scale-free networks, degree correlation in the SR phase transition that activates the network structure.

2. Model

The model of quantum material considered here is established by a complex network. This network possesses an ensemble of N spin-1/2 systems (TLAA) that randomly occupy N nodes of a network. We represent an assortative, non-assortative and dissortative scale-free network in the form of a graph with nontrivial (specific) topological properties, which assumes a power-law distribution of degrees and exhibits degree correlations.

In particular, some networks can be recognized as assortative if nodes with a high (low) degree are connected to other nodes with a high (low) degree on average. Otherwise, when high-degree nodes are connected to low-degree nodes, the network is disassortative.

The spins placed at the nodes of the graph, are assumed to interact with the classical (local) magnetic field h_i which causes the excitation of a two-level system in the same direction z and a quantized (transverse) field. These two-level systems are characterized by the corresponding Pauli operator σ_i^z . We describe the transverse field by means of annihilation (a) and creation ($a^\dagger$) operators. This field couples with another transverse spin component σ_i^x, $i = 1, ..., N$. The total Hamiltonian of Dicke–Ising model is:

$$\mathrm{H} = -\sum_{ij} J_{ij}\sigma_i^z\sigma_j^z - \frac{1}{2}\sum_i h_i\sigma_i^z + \omega_a a^\dagger a - \frac{1}{2\sqrt{N}}\sum_i \chi_i\sigma_i^x(a + a^\dagger), \quad (1)$$

where χ_i in (1) characterizes the coupling of spin x-component σ_i^x with the quantum transverse field in dipole approximation. Below we restrict the study by a homogeneous problem, when $h_i = h$ and $\chi_i = \chi$ for any i.

The sum is performed over the graph vertices with certain adjacency matrix A_{ij} proportional to J_{ij}, which stores the information about the graph structure: matrix element $A_{ij} = 1$ if two vertices are linked and $A_{ij} = 0$ otherwise. The first two terms in Eq. (1) describe the Ising model, while the last two terms are inherent to the Dicke one. In this paper we put the Planck and Boltzmann constants $\hbar = 1$, $k_B = 1$ for simplicity.

In this work, the average photon number $N_{ph} \equiv \langle a^\dagger a\rangle$ accumulated in the transverse field is a variable which relates to the order parameter of SR phase transition. We recast parameter J_{ij} that indicates the coupling between the nodes in Eq. (1) through probability p_{ij} as $J_{ij} = Jp_{ij}$, where J is a constant. Thus, the strength of two spins interaction J_{ij} is a variable parameter and depends on network characteristics. In particular, it is greater for two pairs of nodes with the highest k coefficient.

The properties of the system described in Eq. (1) were determined by means of three order parameters. The first of them, λ, is a normalized mean transverse field amplitude

$\lambda = \frac{|\alpha|}{\sqrt{N}} = \sqrt{\frac{N_{ph}}{N}}$, where coherent state basis $|\alpha\rangle$ is used for the quantized transverse field that contains $N_{ph} = \langle\alpha|a^\dagger a|\alpha\rangle = |\alpha|^2$ photons on average.

We considered mean-field variational (thermodynamic) approach for the system described by Hamiltonian (1). At this limiting case we can recast the first term in Eq. (1) with $\sum_{ij} J_{ij}\sigma_i^z\sigma_j^z$ using Wick's theorem for factorizing σ^z operators' combination $\sigma_i^z\sigma_j^z = \sigma_i^z\langle\sigma_j^z\rangle + \sigma_j^z\langle\sigma_i^z\rangle - \langle\sigma_i^z\rangle\langle\sigma_j^z\rangle$. Thus, we introduced weighted (S_z) and total (S_0) average spin components in z-direction, respectively as

$$S_z = \frac{1}{N\langle k\rangle}\sum_i k_i\langle\sigma_i^z\rangle, \tag{2a}$$

$$S_0 = \frac{1}{N}\sum\langle\sigma_j^z\rangle. \tag{2b}$$

To determine mean-field equations for the order parameters we exploit partition function $Z = Tr(e^{-\beta \mathrm{H}})$, which presumes factorization of total state $|\Psi\rangle$ representing it as $|\Psi\rangle = |\alpha\rangle|s_1...s_N\rangle$; $|s_1...s_N\rangle$ defines the N-spin state. For partition function Z we obtain

$$Z = N\int d\lambda^2 e^{-\beta N\omega_a\lambda^2} e^{-\beta\frac{JNS_z\langle k\rangle}{(\langle k\rangle+A)}(S_z\langle k\rangle+AS_0)}$$
$$\prod_i 2\cosh\left(\frac{\beta}{2}\sqrt{\left(\frac{4J}{(\langle k\rangle+A)}[S_z\langle k\rangle(k_i+0.5A)+0.5AS_0k_i]+h\right)^2+4\chi^2\lambda^2}\right). \tag{3}$$

Evaluating the integral in Eq.(3) by Laplace method after some calculations we get the mean-field equations for collective spin variables S_z, S_0 and normalized transverse field amplitude λ as follow:

$$S_z = \frac{1}{\langle k\rangle}\int_{k_{min}}^{k_{max}} p(k)\frac{\tanh[\frac{\beta}{2}\Gamma]}{\Gamma}\left(\Xi(k)+H\right)kdk, \tag{4a}$$

$$\lambda\Omega_a = \lambda\int_{k_{min}}^{k_{max}} p(k)\frac{\tanh[\frac{\beta}{2}\Gamma]}{\Gamma}dk, \tag{4b}$$

$$S_0 = \int_{k_{min}}^{k_{max}} p(k)\frac{\tanh[\frac{\beta}{2}\Gamma]}{\Gamma}(\Xi(k)+H)dk, \tag{4c}$$

where $\Xi(k)$ is defined as $\Xi(k) \equiv \frac{\Theta}{\langle k\rangle+A}\left(S_z\langle k\rangle(k+\frac{1}{2}A)+\frac{1}{2}AS_0k\right)$ and Γ is defined as $\Gamma \equiv \sqrt{(\Xi(k)+H)^2+4\lambda^2}$; in (4) we also introduced the normalized dimensionless parameters as

$$\Theta = 4J/\chi;\quad H = h/\chi;\quad \Omega_a = 2\omega_a/\chi;\quad \beta\chi \mapsto \beta. \tag{5}$$

In (4) the number of nodes is assumed large enough, $N \gg 1$. In (4) substitution $\frac{1}{N}\sum_i ... \to \int_{k_{min}}^{k_{max}} ...p(k)dk$ was performed, where $p(k)$ is continuous degree distribution with the minimal and maximal values of node degree k, k_{min} and k_{max} respectively.

3. Scale-free networks

In this research, we are interested in an assortative network approach that features a weighted, fully connected graph model. The network is dynamically rebuilding. The two nodes i and j are connected with the probability p_{ij} as:

$$p_{ij} = P(A_{ij} = 1) = \frac{k_i(k_j + A)}{\sum k_i + A}, \tag{6}$$

where k_i is i-th node degree taken from distribution $p(k)$ and, indicating an expected number of the node neighbors, A_{ij} is adjacency matrix element, A is a constant that characterizes network assortativity; thereafter, we will call it as assortativity parameter.

To design a network architecture with assortative/disassortative properties, we used the following preferred embedding algorithm. At each moment of time t, a new node with m links appears, it joins an already existing complete graph with $\langle k \rangle = m$. Moreover, the connection of two or more links related to the node at its appearance with the similar node is neglected.

Combining the continuous approximation for k and the density function, we found a distribution of degree $p(k)$ in the form:

$$p(k) = \frac{1}{\varpi(m + A)} \left(\frac{m + A}{k + A} \right)^{\gamma} \theta(k - m), \tag{7}$$

where $\varpi = \frac{m}{2m+A}$ and $\gamma = 3 + \frac{A}{m}$. m is the number of initial seed links. Therefore, we take the minimum degree of a network node $k_{min} \sim m$.

At $A \neq 0$ Eq.(7) characterizes correlation degree introducing assortativity coefficient as:

$$r = (M^{-1} \sum l_j k_j - [M^{-1}\frac{1}{2} \sum (l_j + k_j)]^2) * (M^{-1}\frac{1}{2} \sum (l_j^2 + k_j^2) - [M^{-1}\frac{1}{2} \sum (l_j + k_j)]^2), \tag{8}$$

where M is the total number of links, and k_j and l_j is the number of connections at the ends of j node.

Fig.1 shows the network architectures with different values of the correlation coefficient r. Fig.2a shows dependence of the correlation degree on A and N. The maximum value of the correlation coefficient $r \approx 0.3$ was achieved in the limit $A, N \gg 1$. The distribution $p(k)$ in (7) with $A = 0$ corresponds to BA network with degree exponent $\gamma|_{A=0} = 3$.

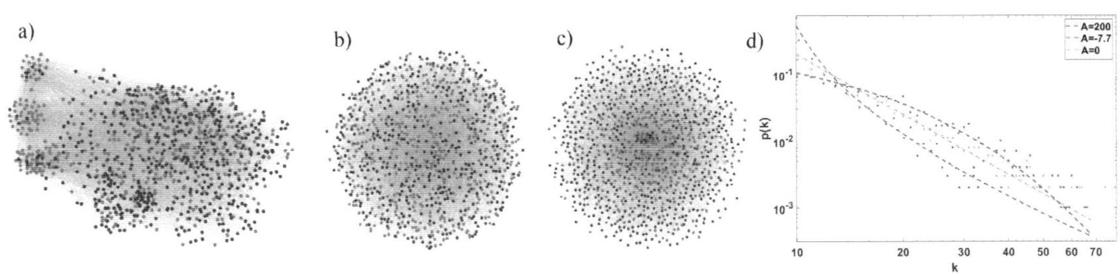

Figure 1. a) Assortative, (b) disassortative, (c) non-assortative networks and (d) the relevant degree distributions $p(k)$ are plotted for structures with assortativity coefficient $r = 0.28$ ($A = 200$), $r = -0.28$ ($A = -7.7$) and $r = 0$ ($A = 0$), respectively. d) The degree distribution $p(k)$ of the network structure for different values of A in a logarithmic scale. Red, blue and green curves represent assortative, disassortative and non-assortative degree distributions respectively. The parameters are: $m = 10$, $N = 1000$.

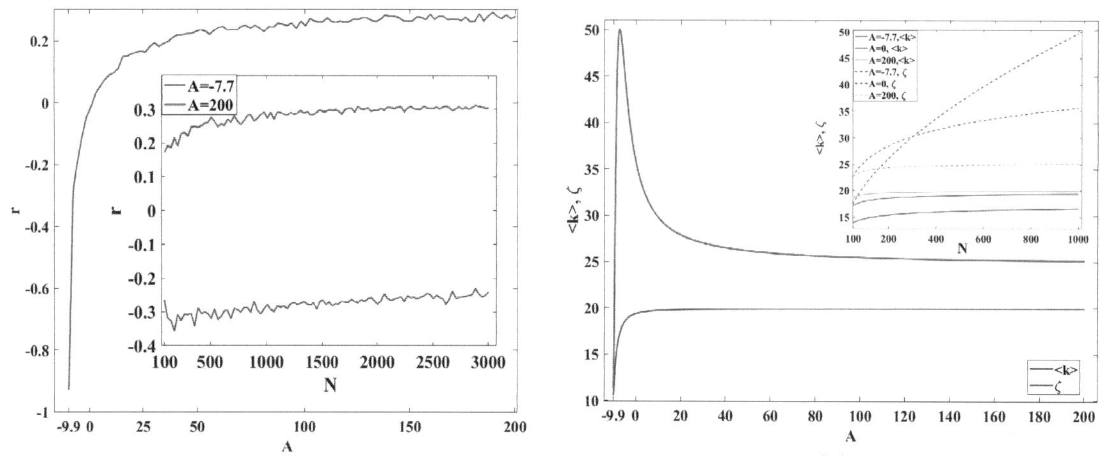

(a) (b)

Figure 2. Dependence of (a) correlation coefficient r and (b) node average degree $\langle k \rangle$ and ζ against assortativity parameter A for $N = 1000$. Inserts show relevant dependence on N for different values of A; $k_{min} = 10$.

From (7) we can define the maximum degree k_{max} of the node using the degree distribution cut-off in finite size scale-free network as:

$$\int_{k_{max}}^{\infty} p(k)dk = \frac{1}{N}. \tag{9}$$

Thus, from (7), (9) we immediately get

$$k_{max} = [k_{min} + A]N^{\frac{1}{\gamma-1}} - A. \tag{10}$$

The network statistical properties are characterized by means of the first ($\langle k \rangle$) and second ($\langle k^2 \rangle$) moments for the degree distribution, which are defined as:

$$\langle k^n \rangle = \int_{k_{min}}^{k_{max}} k^n p(k)dk, \quad n = 1, 2. \tag{11}$$

In particular, using (7) we obtain an analytical expressions for these parameters in the form of

$$\langle k \rangle = 2k_{min} + \frac{A}{N} - \frac{2k_{min} + A}{N^{\frac{\gamma-2}{\gamma-1}}}, \tag{12a}$$

$$\langle k^2 \rangle = \frac{1}{3-\gamma}[\frac{1}{N}(2A^2 + 2Ak_{min}(\gamma-1)N^{\frac{1}{\gamma-1}} + (\gamma-1)((k_{min}+A)N^{\frac{1}{\gamma-1}} - A)^2) - (4Ak_{min} + (\gamma-1)k_{min}^2)]. \tag{12b}$$

Fig. 2b illustrates main features of $\langle k \rangle$ and ζ. From (12a) we see that in the limit of large N, the mean degree of the network is $\langle k \rangle \approx 2k_{min}$. When $A \to -k_{min}$ the mean degree approaches $\langle k \rangle \to k_{min}$, which is inherent to blue line saturation in Fig. 2b. Moreover in $A \to -k_{min}$ limit we obtain $\langle k^2 \rangle = k_{min}^2$ and ζ parameter is

$$\zeta = k_{min}. \tag{13}$$

In the limit $A \to \infty$, the average degree of the network approaches to $\langle k \rangle \to 2k_{min}$ and the second moment $\langle k^2 \rangle \to k_{min} \frac{4A+k_{min}\gamma}{\gamma}$. Thus, we have: $\langle k^2 \rangle = k_{min}^2$ and ζ parameter approaches to

$$\zeta = \frac{4A + k_{min}\gamma}{2\gamma}, \tag{14}$$

which is inherent to red curve behaviour in Fig.2b. In the case $A = 0$ the Eqs. (12), transforms as $\langle k \rangle = 2k_{min}$ and $\langle k^2 \rangle = k_{min}^2 ln(N)$ and ζ parameter increases logarithmically as $\zeta = \frac{k_{min}}{2} ln(N)$.

4. The phase transition

Here we consider phase transition problem for quantum material represented by scale-free network. We examined some important limiting cases which correspond to vanishing classical field $H = 0$.

Fig.3 demonstrates phase diagrams that correspond to numerical solutions of Eqs.(4). The curves in Fig.3a exhibit behavior of the order parameter λ as a function of S_z (solid curves) and S_0 (dashed curves), respectively. Maximal value of λ may be obtained at zero temperature, which is relevant to quantum phase transition paradigm. Dashed green curve with $A = 0$ in Fig.3a corresponds to non-assortative BA examined in [8].

Fig.3a clearly demonstrates a shift between variables S_z and S_0 due to topologically dependent ordering in spin systems. In particular, assortative network with $\gamma > 3$ supports local (node-dependent) interaction between spins due to some specific topological features (hubs and clusters), cf. Fig.1a. This interaction leads to the existence of an effective local field responsible for establishing FM order in the z-direction even without external field H. Thus, SR state occurs as a result of the interaction between the ordering of spins in the $z-$ and x-directions and depends on the topological features of the network in the limit $\lambda \to 0$. For large correlation values of the node degree A (magenta curves in Fig.3a), spin network system becomes extremely sensitive to temperature changes T and exhibits essentially nonlinear features. This curve demonstrates non-trivial behavior associated with large fluctuations of network parameters, such as $\langle k \rangle$ and $\langle k^2 \rangle$. On the other hand, disassortative network is characterized by a more random degree distribution, which implies ordering decreasing in z-direction.

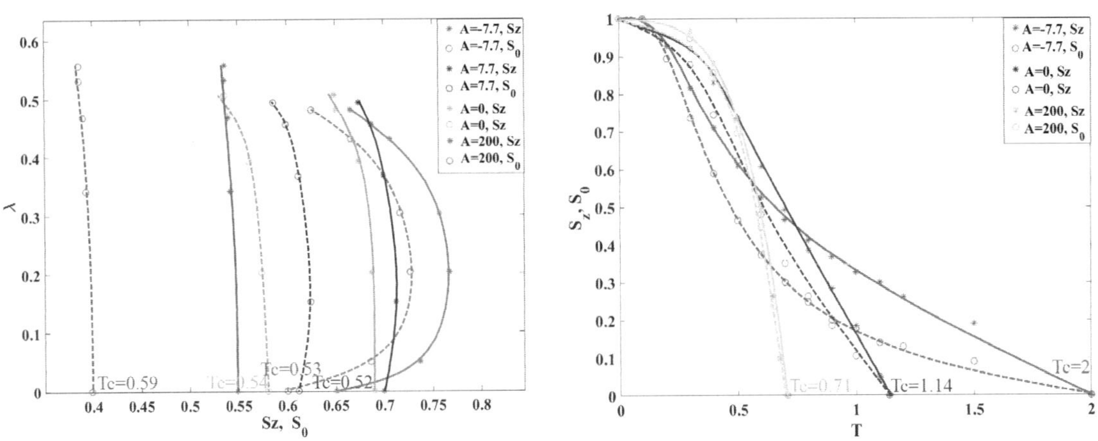

(a) (b)

Figure 3. (a) Phase diagram λ–S_z, S_0 for various A; (b) temperature dependence of S_z and S_0 parameters for $\lambda = 0$. The network parameters for both cases are: $k_{min} = 10$, $N = 1000$, $\Omega_a = 0.8$, $H = 0$, $\Theta = 0.0625$. The maximal degree for the scale-free networks is defined in (10).

Eqs.(4) admits simple analytical treatments in the vicinity of critical point $\lambda_c = 0$.

Fig.3b demonstrates dependence of weighted S_z and total S_0 average spin components in z direction vs. temperature T inherent to different values of assortativity parameter A. In particular, curves in Fig.3b demonstrate second-order phase transition peculiarities. In the absence of a magnetic field $H \to 0$ the S_z and S_0 tend to zero simultaneously (see Fig.3b). Thus, the critical phase transition temperature is:

$$T_c = \frac{\lambda_c}{\tanh^{-1}[\frac{2\lambda_c(\langle k\rangle + A)}{\Theta(\langle k^2\rangle + 0.5A(\langle k\rangle + \zeta))}]}, \tag{15}$$

where $\zeta = \frac{\langle k^2\rangle}{\langle k\rangle}$ determines basic statistical properties of the network, cf. [8].

The critical temperature T_c represented in (15) corresponds to the ferromagnetic-paramagnetic phase transition in network medium. In the limit of vanishing λ_c, i.e. for $\lambda \to 0$ limit, the critical temperature T_c is defined as:

$$T_c = \frac{1}{2}\Theta\frac{\langle k^2\rangle + 0.5A(\langle k\rangle + \zeta)}{\langle k\rangle + A}. \tag{16}$$

Eq.(16) represent critical temperature to phase transition from disorder ($S_0, S_z \simeq 0$, $\lambda \simeq 0$) to some ordering state with non-zero values of the order parameters. The T_c, as it is follows from Fig.3b and (16), decreases for assortative networks, which imply large values of the A, and increases for disassortative ones, respectively. Such a behaviour can be explained by an increase in the number of clusters which are poorly connected to each other, thereby preventing the establishment of a ferromagnetic state with increasing temperature.

5. Conclusion

In this work we considered superradiant phase transition problem which occurs in quantum medium represented by complex network topology. We established Dicke-Ising model which deals with assortative, disassortative and nonassortative scale-free (BA) network structures.

The model presumes the ensemble of spin-1/2 systems in the network nodes that interact with local classical magnetic and weak quantum transverse (photonic) fields. Applying the mean-field approach, we got a set of Eqs. (4), which describe three order parameters S_z, S_0 and λ, respectively. The S_z and S_0 are relevant to the PM-FM phase transition problem and characterize magnetization in z-direction. While λ is normalized transverse field amplitude it describes to the phase transition to the superradiance for the spin network system in x-direction. SR state with $\lambda \neq 0$ may be characterized by a non-zero temperature dependent energy gap for the spin system.

To be more specific, we have inspected the scale-free networks with the power-law degree distribution. Superradiance occurs in a ferromagnetic spin system in the limit of a finite (non-zero) classical magnetic field. The physical features of the system becomes richer and more complex due to classical external field H disappearance. The critical temperature of the disorder-order phase transition for the networks under discussion demonstrated finite-size effects which imply dependence on average node degree and assortativity parameter. Our mean-field approach allowed to consider degree correlations in the the scale-free network structure, which we described with the parameter ζ, see (14); the obtained critical temperature can be quite high but finite which depends on the statistical properties (ζ-parameter) for the scale-free networks, respectively. At low temperatures an FM state with $S_z \simeq 1$ is established; this state is difficult to change using spin interaction with a weak quantized field or even with a moderate external magnetic field. However, the situation changes with increasing the temperature. Simultaneously, decreasing in the effective spin-spin interaction leads to the suppression of the collective spin components S_z, S_0 and creates favorable conditions for PM-FM and/or SR phase transitions. In the scale-free network ordering spin system arises as a result of the interaction between magnetizations in the x- and z-directions. The SR phase transition in this case occurs for some FM states, see Fig.3b. The critical temperature of order-disorder phase transition vanishes for assortative networks due to increasing γ.

Our findings allow new opportunities in quantum materials design which exhibit some features important for quantum technologies at relatively high temperatures.

Acknowledgments

This work was partially supported by the the Ministry of Science and Higher Education of Russian Federation, Goszadanie No. 2019-1339, and financially supported by the Grant of RFBR, No 19-52-52012 MHT.

References

[1] Mohseni M, Read P, Neven H and et al 2017 *Nature* **543** 171
[2] Puri S, Andersen C, Grimsmo A and et al 2017 *Nature Commun.* **8** 15785
[3] Lee C F and Johnson N 2004 *Phys. Rev. Lett.* **93** 083001
[4] Bamba M, Inomata K and Y N 2016 *Phys. Rev. Letts.* **117** 173601
[5] Suzuki J, Inoue J and Chakrabarti B 2012 *Springer* **862**
[6] Zurek W, Dorner U and Zoller P 2005 *Phys. Rev. Letts.* **95** 105701
[7] Choi V 2008 *Quantum Inf Process* **7** 193
[8] Bazhenov A Y, Tsarev D V and Alodjants A P 2021 *Physical Review E* **103** 062309
[9] Dorogovtsev S N 2010 *Oxf. Master Ser. Phys* **20**
[10] Newman M 2018 *Oxford University Press* 727
[11] Barabási A L 2016 *Cambridge University Press* 475
[12] Albert R and Barabási A L 2002 *Nature Reviews Neuroscience* **74** 47
[13] Bianconi G and Barabási A L 2001 *Physical Review Letters* **86** 5632
[14] Bianconi G 2012 *Journal of Statistical Mechanics: Theory and Experiment* **2012** P07021
[15] Bianconi G 2002 *Physics Letters A.* **303** 166–168
[16] Suchecki K and Hołyst J A 2013 *World Scientific Publishing* **3** 167–200
[17] Krishnan J, Torabi R, Di Napoli E, Honerkamp C and Schuppert A 2020 *arXiv:2005.05045*
[18] Buluta I, Ashhab S and Franco N 2011 *Rep. Prog. Phys.* **74** 104401
[19] Hepp K and Lieb E H 1973 *Physical Review A.* **8** 2517
[20] Wang Y K and Hioe F T 1973 *Physical Review A* **7** 831
[21] Emary C and Brandes T 2003 *Physical Review E* **67** 066203
[22] Larson J and Irish E K 2017 *Journal of Physics A: Mathematical and Theoretical* **50** 174002

Journal of Physics: Conference Series **2249** (2022) 012015 doi:10.1088/1742-6596/2249/1/012015

Comparison of multimode GRIN-fiber Raman lasers with FBG and random DFB cavity

Yizhu Chen[1], Chenchen Fan[1], Tianfu Yao[1], Hu Xiao[1], Jiangming Xu[1], Jinyong Leng[1], and Pu Zhou[1], Alexey G. Kuznetsov[2], Ilya N. Nemov[2], Alexey A. Wolf[2,3], Sergey I. Kablukov[2], Evgeny V.Podivilov[2,3], Sergey A. Babin[1,2]

[1]College of Optoelectronic Science and Engineering, NUDT, Changsha, China
[2]Institute of Automation and Electrometry SB RAS, Novosibirsk, Russia
[3]Novosibirsk State University, Novosibirsk, Russia

Corresponding author: zhoupu203@163.com; babin@iae.nsk.su

Abstract. Raman lasing in multimode GRIN fibers is accompanied by sufficient improvement of the output beam quality in comparison with that for pump radiation, which offers opportunities to wavelength-agile fiber lasers of new type. Here we compare power scaling and brightness enhancement capabilities of Raman laser based on multimode GRIN-fiber of 62.5/125 um core/cladding diameters pumped by ~700 W multimode source with beam quality M^2~10, performed in two different cavity configurations: 1) linear cavity based on two fiber Bragg gratings and 2) half-open cavity with one FBG and random distributed feedback via Rayleigh backscattering along the GRIN fiber.

1. Introduction

Raman fiber lasers (RFLs) have such advantages as flexible operating wavelength, low quantum defect, absence of unfavorable effects of amplified spontaneous emission and photo-darkening [1]. In addition, RFL may operate without conventional cavity, via Rayleigh backscattering forming random distributed feedback along a singlemode passive fiber, see [2] and citation therein. Another direction which attracts lots of attention deals with the great potential in achieving brightness enhancement (BE): high-brightness signal laser at low-brightness pumping, in particular, in commercially available multimode GRIN fibers, either in laser [3,4] or in amplifier [5] configurations. In the GRIN-fiber RFL, linear cavity consisting of two bulk mirrors [3] which may be replaced by fiber Bragg gratings (FBGs) inscribed directly in the fiber core near its ends thus providing all-fiber configuration [4], or half-open cavity with one FBG and random distributed feedback (DFB) via Rayleigh backscattering along the GRIN fiber [6,7] are utilized for the multimode Raman laser operation with brightness enhancement. As a pump, either combined multimode LDs [3,4,6] or high power Yb-doped fiber lasers (YDFL) [5,7] are used, with sufficiently higher power in the latter case.

Here we compare two cavity configurations of Raman laser based on multimode GRIN-fiber of 62.5/125 um core/cladding diameters pumped by ~700 W multimode source with beam quality M^2~10 based on combined YDFLs. The first configuration consists of highly- reflective (R~90%) fiber Bragg grating (HR FBG) inscribed in the GRIN fiber by conventional UV technology and output coupler (OC) FBG, inscribed in the central part of GRIN fiber core by point-by-point femtosecond laser technology so that fundamental mode has reflection R~4% whereas higher order modes have 10 dB lower

Journal of Physics: Conference Series 2249 (2022) 012015 doi:10.1088/1742-6596/2249/1/012015

reflection. The second configuration is based on the half open-cavity with the same UV-inscribed HR FBG and random DBF via Rayleigh backscattering along the GRIN fiber instead of OC FBG, with a potential of transverse mode selection via Rayleigh backscattering which is radially dependent in the GRIN fiber with a parabolic refractive index profile. It appears that these two configurations have similar maximum values for the main output RFL characteristics, but significantly differ in the character of their dependence on the pump/signal power.

2. Experiment and results

Experimental scheme of the multimode (MM) RFL in two configurations is shown in Fig.1(a,b). A pump source with a maximum output power of about 700 W is based on combined YDFLs operating at 1070 nm. The output fiber of pump source has a core diameter of 20 μm with numerical aperture (NA) of 0.06 which is identical with all of the seven input fibers of the (6+1) × 1 fiber pump combiner made in house. The free input fibers of the combiner are angle cleaved to prevent unexpected facet feedback and to monitor backward lasing. The output fiber of the combiner is a commercial GRIN fiber with 62.5/125 μm core/cladding diameter. The refractive index of the fiber core has a parabolic transverse profile with numerical aperture NA=0.275. A piece of 70 m identical GRIN fiber is employed as Raman gain medium and the Raman laser cavity was formed by two FBGs written in the GRIN fiber core. The output fiber of the oscillator is spliced to an end cap that coated with anti-reflection film to eliminate Fresnel reflection. Laser emission from the end cap is collimated by a high-NA collimator.

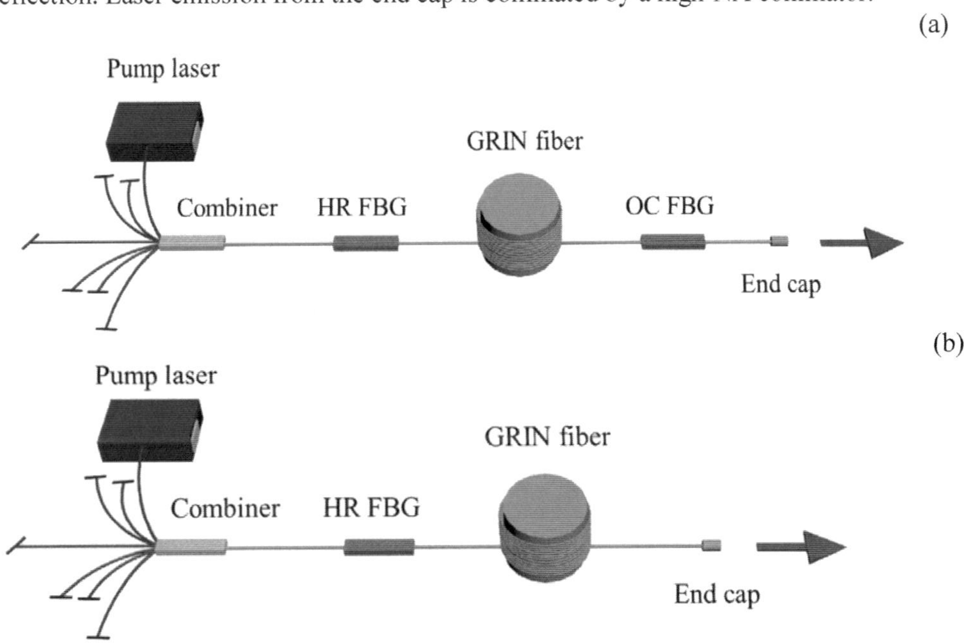

Fig. 1. Multimode RFL in two configurations: (a) linear cavity with two FBGs; (b) half-open cavity with one FBG and random DFB via Rayleigh backscattering.

In the case of MM RFL with linear cavity configuration, Fig.1(a), two FBG reflectors were formed at the both ends of the GRIN fiber. The employed highly reflective (HR) FBG was written by a conventional interferometric technique using a continuous-wave (CW) UV laser. The weakly reflective FBG used as output coupler (OC) was written by a femtosecond point-by-point fabrication technique which has a special transverse profile concentrated in the central area of the GRIN fiber core, see review [4] for more details. The reflectivity of the HR and OC FBGs at their spectral maxima are about 98% and 4%, respectively. The 3 dB bandwidths of their reflection spectra are around 1.1 nm and 0.26 nm, respectively. It should be noted that the reflection coefficients of higher-order modes in the OC FBG spectrum are by ~10 dB lower than the fundamental mode resonance which brings up additional spatial filtering to the Raman beam cleanup effect [8]. At the same time, the broad reflection peak of HR FBG involves resonances of several low-order mode groups spaced by ~0.6 nm in a 62.5-μm core GRIN fiber [4]. In the case of half-open cavity, Fig.1(b), the OC FBG has been removed and random DFB via Rayleigh backscattering along the GRIN fiber is employed in this case instead of output coupler. At the same time, the GRIN fiber length has been increased to 120 m in order to have appropriate Raman threshold value.

Power characteristics of the GRIN-fiber RFL with two different cavity configurations are shown in Fig.2(a,b) including transmitted pump power at 1070 nm, 1st order Stokes power at 1120 nm and total output power. One can see that the total output power is slightly higher in case of random RFL (RRFL) configuration, Fig.2(b), because it does not have an OC FBG inserting additional losses. At the same time, this configuration has ~2 times higher threshold value. However, due to the much higher slope efficiency (fitted as 174%) the Stokes output power at the maximum pump power of 695 W reaches 306 W, that is only slightly lower of the output power in linear FBG cavity amounting to 332 W, Fig. 2(a).

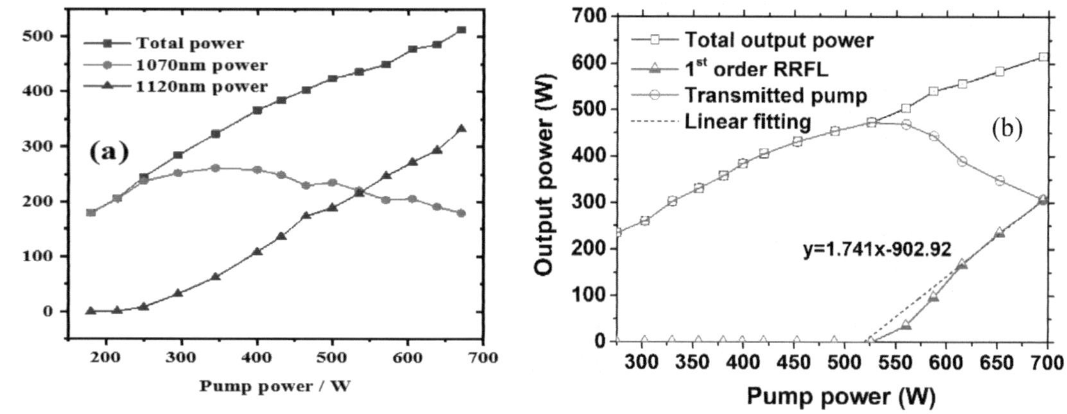

Fig.2. Output power of multimode RFL in two configurations: (a) linear cavity with two FBGs; (b) half-open cavity with one FBG and random DFB via Rayleigh backscattering.

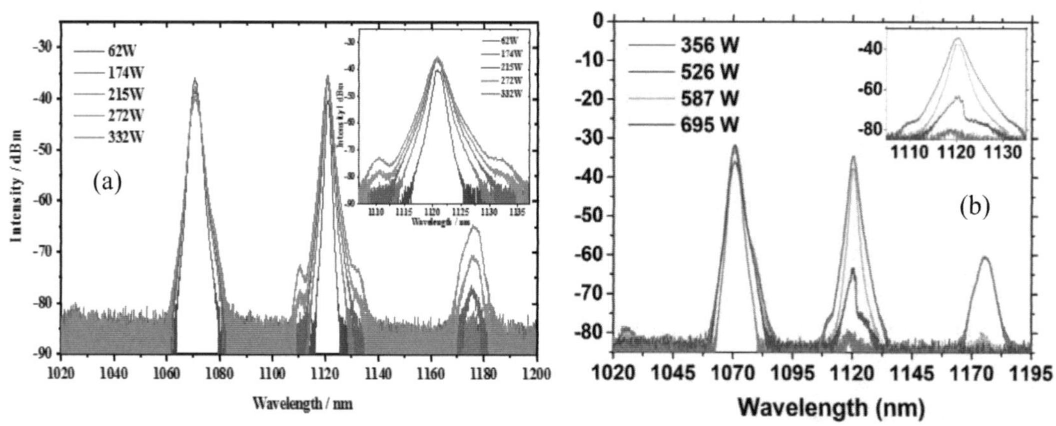

Fig.3. Output spectra of multimode RFL in two configurations: (a) linear cavity with two FBGs, Stokes power is shown as parameter; (b) half-open cavity with one FBG and random DFB via Rayleigh backscattering, pump power is parameter.

The measured Stokes spectra of GRIN-fiber RFL with two different cavity configurations are shown in Fig. 3(a,b). It can be seen that they are similar, the linewdith is slightly different amounting to about 1.8 nm and 2.1 nm at maximum power for RFL and RRFL configurations respectively.

Results of the comparison for output beam quality M^2 and corresponding brightness enhancement factor calculated as $BE=[P/(M^2\lambda)^2]_{Stokes}$ / $[P/(M^2\lambda)^2]_{Pump}$ for two cavity configurations are shown in Fig.4. Though RRFL configuration has sufficiently higher threshold, it has higher slope efficiency and as a result comparable output power (>300W) at maximum pumping. At the same time, it has better beam quality, M^2~2.3, see Fig.4(a) and, as a result, higher value of frightness enhancement factor BE=6.1 in comparison with RFL, see Fig.4 (b).

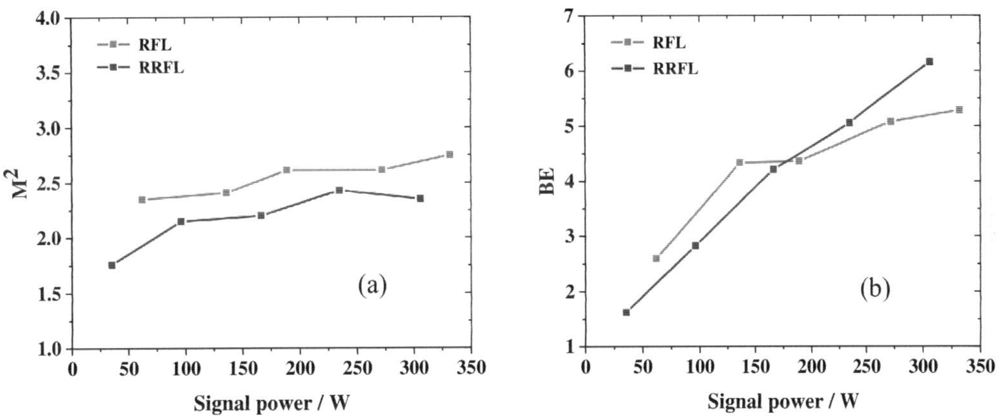

Fig.4. Comparison of the measured beam quality M^2(a) and brightness enhancement factor BE (b) for RFL and RRFL configurations.

3. Conclusion

Thus, we have compared power scaling and brightness enhancement capabilities of multimode GRIN-fiber Raman lasers in two cavity configurations with the same multimode pumping of ~700 W maximum power and M^2~10 beam quality. The RFL with linear cavity consisting of HR and OC FBGs demonstrates lower threshold and slightly higher maximum power of 332 W, whereas the RRFL with half-open cavity consisting of HR FBG and random distributed feedback via Rayleigh backscattering along GRIN fiber demonstrates better beam quality ($M^2\approx2.3$) at maximum power of 306 W and, as a result, higher brightness enhancement factor, which linearly grows with increasing Stokes signal power reaching BE=6.1 at maximum. This means that the Rayleigh backscattering provides better spatial filtering effect than the OC FBG, fabricated by fs pulses in the central area of GRIN fiber. At the same time, OC FBG inserts additional spectral filtering resulting in slightly narrower signal spectrum of RFL - 1.8 nm at 332 W, in comparison with 2.1 nm at 306 W for RRFL. Nevertheless, the proposed approach demonstrates a great potential for power scaling and brightness enhancement of random RFL based on multimode fiber with multimode pumping, in spite of that the earlier demonstrated random RFL based on large-mode-area (LMA) singlemode passive fiber with high-power singlemode pumping approached kW level [9].

Acknowledgements

The work is supported by National Natural Science Foundation of China (61911530134, 11704409, 61605246) and Russian Science Foundation (21-42-00019).

References

[1] V. R. Supradeepa, Y. Feng, and J. W. Nicholson, "Raman fiber lasers," J. Opt. **19** (2), 23001-23026 (2017).
[2] D. V. Churkin, S. Sugavanam, I. D. Vatnik, Z. Wang, E. V. Podivilov, S. A. Babin, Y. J. Rao, and S. K. Turitsyn, "Recent advances in fundamentals and applications of random fiber lasers," Advances in Optics and Photonics **7** (3), 516-569 (2015).
[3] Y. Glick, V. Fromzel, J. Zhang, N. Ter-Gabrielyan, and M. Dubinskii, "High efficiency, 154 W CW, diode-pumped Raman fiber laser with brightness enhancement," Appl.Opt. **56**, B97-B102 (2017).
[4] S. A. Babin, E. A. Zlobina, and S. I. Kablukov, "Multimode Fiber Raman Lasers Directly Pumped by Laser Diodes," IEEE J. of Selected Topics in Quantum Electronics **24** (3), 1-10 (2018).
[5] Y. Chen, T. Yao, H. Xiao, J. Leng, and P. Zhou, "Greater than 2 kW all-passive fiber Raman amplifier with good beam quality," High Power Laser Sci. Eng. **8**, e33 (2020).
[6] E. A. Evmenova, A. G. Kuznetsov, I. N. Nemov, A. A. Wolf, A. V. Dostovalov, S. I. Kablukov, and S. A. Babin, "2nd-order random lasing in a multimode diode-pumped graded-index fiber," Sci. Rep. **8** (1), 17495 (2018).
[7] Y. Chen, C. Fan, T. Yao, H. Xiao, J. Leng, P. Zhou, I. N. Nemov, A. G. Kuznetsov, and S. A. Babin, "Brightness enhancement in random Raman fiber laser based on a graded-index fiber with high-power multimode pumping," Opt. Lett. **46** (5) 1185-1188 (2021).
[8] N. B. Terry, T. G. Alley, and T. H. Russell, "An explanation of SRS beam cleanup in graded index fibers and the absence of SRS beam cleanup in step-index fibers," Opt. Express **15** (26), 17509 (2007).
[9] Hanwei Zhang, Long Huang, Jiaxin Song, Han Wu, Pu Zhou, Xiaolin Wang, Jian Wu, Jiangming Xu, Zinan Wang, Xiaojun Xu, and Yunjiang Rao, "Quasi-kilowatt random fiber laser," Opt. Lett. **44** (11), 2613-261 (2019).

Er-doped fiber laser with regular and random distributed feedback

M.I. Skvortsov[1], A.A. Wolf[1], E.A. Fomiryakov[2], V.N. Treshchikov[2], S.P. Nikitin[2], A.A. Vlasov[1], A.V. Dostovalov[1] and S.A. Babin[1]

[1]Institute of Automation and Electrometry SB RAS, Novosibirsk, Russia
[2]"T8" company group, Moscow, Russia

Corresponding author: qwertymikhails@gmail.com

Abstract. Using a point-by-point fs inscription technology, a Rayleigh reflector with unique characteristics has been designed and fabricated. The reflection coefficient of the structure with a length of 5 cm amounts to 0,03%, which corresponds to the reflection of 6-km SMF-28e+ fiber spool. The designed artificial Rayleigh scattering fiber in combination with distributed feedback (DFB) erbium-doped fiber laser allows to narrow the linewidth. A similar effect was observed in configuration with 25-km SMF-28e+ fiber spool, where it was used as an additional spectral filter. In both cases, an identical narrowing of the linewidth by an order of magnitude was observed at times less than 0,1 msec.

1. Introduction

Distributed feedback (DFB) fiber lasers are known as a unique source of single-frequency lasing for a wide variety of applications from spectroscopy [1] to precision sensing [2], thanks to the high stability, low noise level (signal to noise ratio > 65 dB), and narrow linewidth (<1 kHz). The cavity of distributed feedback fiber lasers is usually based on a periodic structure such as pi-phase-shifted fiber Bragg grating (FBG) which is embedded into an active fiber, so that a stable single-frequency regime is provided [3]. In contrast, random distributed feedback (RDFB) fiber lasers usually utilize Rayleigh scattering on a natural or artificial random refractive index structures. One of the examples of random fiber lasers was realized with RDFB via Rayleigh scattering on the refractive index fluctuations of the medium, and light amplification due to the effect of stimulated Raman scattering (SRS) in a passive optical fiber [4]. Random lasing in active fibers is possible either with the use of FBG with multiple random phase shifts (in Er-doped fiber [5]), or an FBG with both randomly varying phase and amplitude (in Yb-doped fiber [6]). In addition to FBGs, RDFB fiber lasers employ distributed reflectors based on refractive-index modification of arbitrary shape inscribed by means of a localized exposure of fiber to a CO2 laser radiation [7] or by a direct writing technique with the use of femtosecond laser [8]. For example, ring-cavity erbium fiber laser is demonstrated, in which RDFB is provided by eight 1-cm single-mode fiber segments containing randomly-spaced deep refractive index planes [8]. The authors show that randomly spaced planes exhibit a complex interference pattern caused by the presence of Fabry-Perot interference (via core-core modes coupling) and Mach-Zehnder interference (via core-cladding modes coupling). As a result, numerous low-finesse spectral filters provided a single-frequency operation of the laser within a wavelength-locking range. The generation linewidth did not exceed 2.1 kHz and the output power reached 2.9 mW. It is also possible to create an artificial Rayleigh reflector by femtosecond-pulse inscription of random refractive-index structure in hundred-meter long passive fibers and to use it as a distributed laser mirror [9].

The development of point-by-point femtosecond (fs) inscription technology makes it possible the formation of both regular and random refractive index structures of arbitrary shape either in active or passive fibers, thus offering new opportunities for the development of DFB fiber laser schemes. Here we report on the development and characterization of hybrid scheme based on a regular Er-doped DFB fiber laser at 1550 nm combined with random distributed feedback (RDFB) based on Rayleigh reflection. In this scheme a 6-cm pi-shifted FBG inscribed in Er-doped fiber played a role of a regular DFB structure and a 25-km SMF-28e+ fiber spool or a fiber section containing 5-cm fs-pulse-inscribed artificial random structure plays a role of random reflector. The fabricated artificial Rayleigh reflector provides +50 dB/mm back scattered power relative to the SMF-28e+ natural scattering level. The narrowing of the linewidth for configurations with a composite cavity has been demonstrated.

2. Artificial Rayleigh scattering fiber

To create a compact artificial Rayleigh reflector, we applied the femtosecond laser direct writing technique allowing for the distributed modification of refractive index inside the fiber [10]. For inscription of the Rayleigh structure we used a Yb:KGW Light Conversion Pharos 6W laser, which generates pulses of 232 fs duration, 1026 nm wavelength, 10 kHz pulse repetition rate, and ~280 nJ energy. Femtosecond radiation was focused into the core region of the Corning SMF-28e+ fiber using a Mitutoyo 50X Plan Apo NIR HR microobjective (NA=0.65). During irradiation, the fiber was moved along optical axis using a linear stage at a constant speed of ≈ 0.1 mm/s. Under these conditions, the pulse density amounts to 100 pulses/μm, thus resulting in structural changes inside the fiber core region and formation of inhomogeneities enhancing both Rayleigh and Mie scattering.

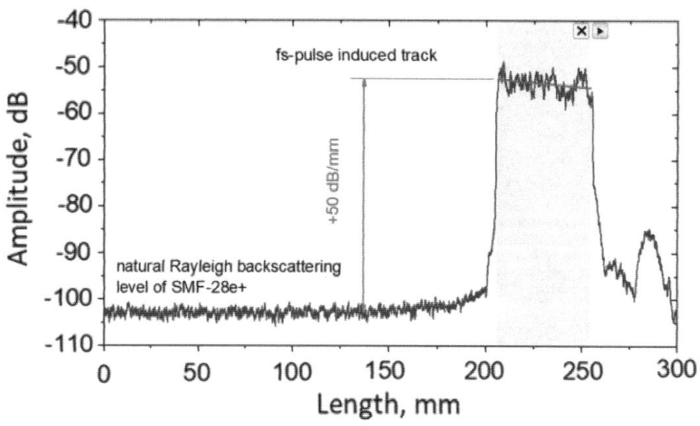

Fig.1. Reflectogram of a 5-cm-long fs-pulse-inscribed Rayleigh reflector.

To monitor the induced backscattering level, we used a high-resolution LUNA OBR4600 backscatter reflectometer. The total length of the fabricated sample was 5 cm, and the averaged induced backscattering level was +50 dB/mm relative to the natural Rayleigh scattering level of the fiber used (Fig. 1), which corresponds to 6-km SMF-28e+ fiber spool. Note that an important feature of the manufactured sample is a low level of induced loss, which in our case did not exceed 2dB.

3. Experimental setup

The experimental scheme of the DFB laser based on the 6 – cm pi-shifted FBG inscribed in Er-doped fiber is presented in Fig. 2. The DFB cavity is pumped with a single-mode laser diode with a wavelength of 980 nm and an output power up to 600 mW through a 980/1550 nm wavelength division multiplexer (WDM). The laser generates at 1550 nm with 0.7 mW output power at maximum pumping. In different configurations of the experimental scheme, the artificial RDFB structure or 25-km SMF-28e+ fiber spool were spliced to the end of the DFB laser resonator. To measure the linewidth, we used a method based on registration of the beating signal [11] of the radiation passing through a Mach-Zehnder interferometer with a delay line of 4.6 km. Signal is analyzed by oscilloscope LecroyWavepro725Zi-A/5-GHz Thorlabs DET08CFC set.

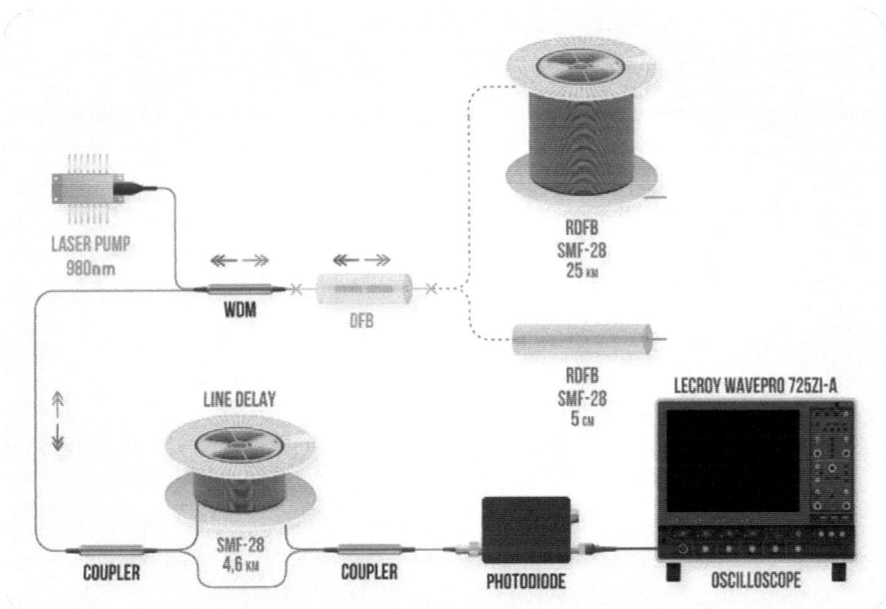

Fig.2. Experimental scheme.

Figure 3 shows the spectral density of the phase noise for different configurations of laser on the short time scale. The spectral density of the phase noise for the 6-cm DFB laser is indicated by the red line, for the DFB laser combined with the 25-km SMF-28e+ fiber spool spliced to the resonator is indicated by the blue line, and the black line shows the value for the DFB laser combined with the 5-cm fs-pulse inscribed artificial Rayleigh reflector. The dotted lines correspond to the Lorentzian profile with different linewidths (from 1 Hz to 1 kHz).

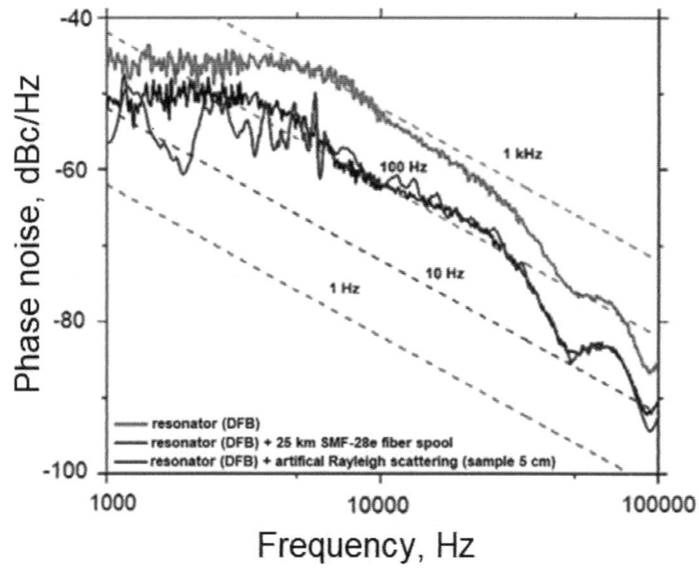

Fig.3. Comparison of the phase noise characteristics of the Er-doped DFB laser with a hybrid DFB+RDFB configuration where the RDFB is provided by either the 5-cm fs-pulse-inscribed artificial Rayleigh reflector or by a 25-km SMF-28e+ fiber spool with naturally present Rayleigh scattering.

A linewidth of the 6-cm DFB laser was measured to be about a 1 kHz at the time interval less than 0,1 ms. A hybrid DFB+RDFB cavity demonstrates a narrowing of linewidth (<100 Hz at time less than 0,1 ms) coupled with unique phase noise characteristics, similar to those obtained when using the same DFB Er-doped fiber laser supplemented with 25-km SMF-28e+fiber spool. Thus, the artificial RDFB structure with a relatively short length serves as an additional spectral filter with characteristics corresponding to the 25-km fiber spool.

4. Conclusions

RDFB structure with short length of 50 mm is fabricated by fs technology in SMF-28e fiber. The reflection coefficient of the structure is 0.03% at 1550 nm, which corresponds to the Rayleigh scattering in a 6 km spool. The inscription technology has been optimized so that the total losses in the structure do not exceed 2 dB. The application of an artificial Rayleigh reflector as an additional spectral filter for a 1550-nm DFB laser leads to a narrowing of the lasing linewidth. In both configurations - namely DFB laser combined with a 25-km SMF-28e+ and DFB laser combined with fs-pulse inscribed Rayleigh reflector - the narrowing of the lasing line by an order of magnitude is demonstrated. In spite of the immediate practical potential for line narrowing, the effect should be studied in more detail, in particular, a theory of such combined DFB+RDFB cavity should be developed, in order to understand physical mechanisms of the line narrowing and lower limits of the linewidth values as a function of the hybrid cavity parameters.

Acknowledgments

The study is supported by Russian Science Foundation (21-72-30024).

References

[1] Persijn S, Harren F and Van Der Veen A 2010 *Appl. Phys. B Lasers Opt.* **100** 383–90

[2] Cranch G A, Flockhart G M H and Kirkendall C K 2008 *IEEE Sens. J.* **8** 1161–72

[3] Skvortsov M I, Wolf A A, Vlasov A A, Proskurina K V, Dostovalov A V, Egorova O N, Galagan B I, Sverchkov S E, Denker B I, Semjonov S L and Babin S A 2020 *Sci. Rep.* **10** 14487

[4] Turitsyn S K, Babin S A, El-Taher A E, Harper P, Churkin D V, Kablukov S I, Ania-Castañón J D, Karalekas V and Podivilov E V 2010 *Nat. Photonics* **4** 231–5

[5] Gagné M and Kashyap R 2009 *Opt. Express* **17** 19067–74

[6] Abdullina S R, Vlasov A A, Lobach I A, Belai O V, Shapiro D A and Babin S A 2016 *Laser Phys. Lett.* **13** 75104

[7] Li Y, Lu P, Bao X and Ou Z 2014 *Opt. Lett.* **39** 2294–7

[8] Li Y, Lu P, Baset F, Ou Z, Song J, Alshehri A, Bhardwaj V R and Bao X 2014 *Appl. Phys. Lett.* **105** 101105

[9] Wang X, Chen D, Li H, She L and Wu Q 2018 *Appl. Opt.* **57** 258–62

[10] Dostovalov A V, Wolf A A, Parygin A V, Zyubin V E and Babin S A 2016 *Opt. Express* **24** 16232–7

[11] Fomiryakov E, Kharasov D, Nikitin S, Nanii O and Treshchikov V 2021 *J. Light. Technol.* **39** 5191–6

29th annual International Laser Physics Workshop (LPHYS'21) IOP Publishing
Journal of Physics: Conference Series **2249** (2022) 012017 doi:10.1088/1742-6596/2249/1/012017

Precision laser diagnostics for LUXE

Rajendra Prasad "for the LUXE collaboration"
FTX, Deutsches Elektronen-Synchrotron DESY, Notkestr. 85, Hamburg, 22607, Germany
E-mail: Rajendra.Prasad@desy.de

Abstract. Strong field QED is an active research frontier. The investigation of fundamental phenomena such as pair creation, photon-photon and photon-electron interactions in the nonlinear QED regime are a formidable challenge both experimentally and theoretically. Several experiments around the world are being planned or in preparation to probe this strong field regime.

LUXE (Laser Und XFEL Experiment) is an experimental platform which envisages the collision of the high quality 16.5 GeV electron beam from the European XFEL accelerator with a 100 TW class high power laser. One of the unique features of LUXE is to measure the key observables such as pair rates (e^+e^-) with unprecedented accuracy in the characterization of both beams together with ample statistics. The state-of-art detector technologies for high energy particle/ photon detection enable percent level precision. The state-of-art high power lasers offer high quality laser beams, however, the residual shot-to-shot fluctuations coupled with the large nonlinearity of the processes under investigation form a particular challenge. An uncertainty of 5% on the absolute laser intensity already leads to a very large (about 40%) uncertainty in the pair rate. Hence it becomes essential to control the laser parameters precisely.

To mitigate this issue a full suite of laser diagnostics is being currently developed at the JETI40 laser in Jena with the aim of tagging the shot intensity to $< 1\%$. In this presentation, details of the laser and the diagnostics suit for the single shot tagging of all the laser parameters will be presented. Moreover, results from an ongoing campaign to properly relay image the beam without significant distortion of the laser beam parameters for post-diagnosis will be discussed.

1. Introduction

The quantum electrodynamics (QED) is the most precise tested theory in modern physics. It has stood the test of experimental precision by one part in billion for the case of electron anomalous moment, for example [1]. The backbone of the QED prediction power is calculations based on perturbative method. However, a limit appears in the vicinity of a super strong electromagnetic field. Above the so-called critical field number of novel phenomena can occur such as vacuum polarisation, pair production from vacuum, to name a few, and a non-perturbative approach has to be employed to correctly interpret these processes. The critical field is described by following expression [2].

$$E_{cr} = \frac{m_e^2 c^3}{e\hbar} \approx 1.3 \times 10^{18}\ \mathrm{V/m}, \tag{1}$$

where e and m_e are elementary charge and rest mass of the electron, c is the speed of light, and $\hbar$ is the reduced Planck's constant.

At present such fields are not directly available in the laboratories. The state-of-art high power lasers now a days can reach focused intensity in excess of $10^{21}\mathrm{W/cm}^2$, which can enable

Journal of Physics: Conference Series **2249** (2022) 012017 doi:10.1088/1742-6596/2249/1/012017

fields of the order of 10^{14}V/m. However, this still falls short by more than three orders of magnitude. In order to reach the critical field, LUXE [3] envisages to collide the high quality electron beam of EuXFEL with high power laser to boost (relativistic effect) the electric field in the rest frame of electron. In Fig. 1 we show two experimental configurations that are planned to explore all the physics processes.

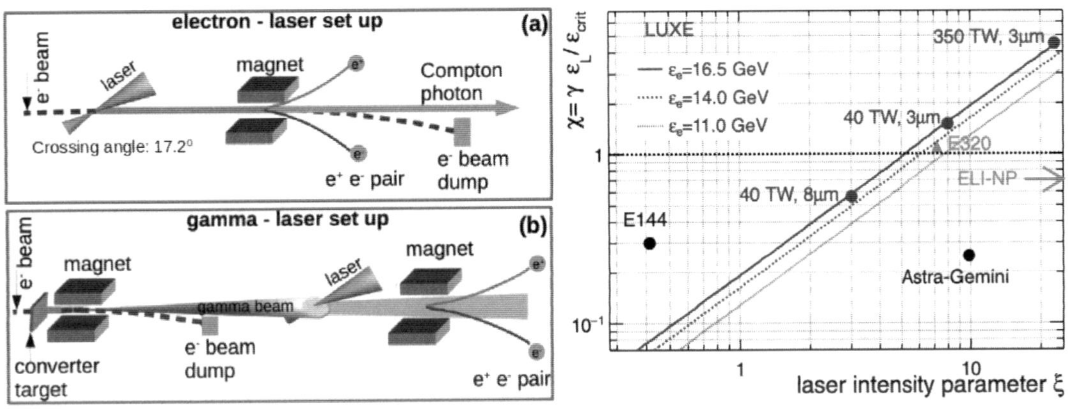

Figure 1. Left panel: two distinct experimental configurations to probe the nonlinear QED regime (a) in electron laser collision "e-laser" mode and (b) in gamma photon and laser collision "g-laser" mode. Right panel: Parameter space covered by LUXE experiment. The normalised laser amplitude ξ is reached up to > 23 leading to accessible χ values > 4. Three blue lines correspond to different electron beam energies: 11.0 GeV (fine dotted line), 14.0 GeV (dotted line) and 16.5 GeV (solid line). Two blue solid points labelled as "40 TW, 8μm" and "40 TW, 3μm" show the laser power and focal spot size for Phase 0, the third one labelled as "350 TW, 3μm" represents the Phase 1, which will be achieved after the upgrade of the laser system.

In Fig. 1(a), the so-called e-laser mode is illustrated. The high energy (up to 16.5 GeV) electron beam collide with a high intensity laser. The main physics process which will be studied here are nonlinear Compton scattering and pair production via nonlinear Breit-Wheeler process [4, 5, 6, 7], which are given by equations 2 and 3.

$$e^- + n\gamma_L \longrightarrow e^-\gamma_c, \tag{2}$$

$$\gamma_c + n\gamma_L \longrightarrow e^+e^-, \tag{3}$$

where γ_L and γ_c represent laser and inverse Compton photons and n is the number of photons involved. The second configuration is shown in Fig. 1(b). Here the high energy photons will be produced using two different mechanisms. In the first case, a converter target will be used to produce high energy photons via bremsstrahlung. The second case will employ an inverse Compton scattering method to produce high energy photons with a narrower energy spectrum. The produced photons in turn then will collide with the high intensity laser pulse. The uniqueness of this geometry is that it will enable a pure photon-photon collision in the nonlinear regime, which has not been experimentally achieved thus far. The main physics process explored here is pair production via nonlinear Breit-Wheeler process:

$$\gamma_B + n\gamma_L \longrightarrow e^+e^-, \tag{4}$$

where γ_B refers the bremsstrahlung photon.

The key parameters in all these studies are the so-called quantum nonlinearity parameter χ and the normalised amplitude of the laser ξ (indicator of laser intensity). They are defined by the following expressions:

$$\xi = \frac{eE}{m_e \omega c} \quad \text{and} \quad \chi = \frac{E^*}{E_{cr}}, \tag{5}$$

where E is the amplitude of the laser electric field, ω is the angular frequency and E^* is the boosted electric field seen by the electron in its rest frame.

The parameter space covered at LUXE is depicted in the right hand panel of Fig. 1. The experiment will be performed in a phased manner. Initially in phase 0, a 40-50 TW laser will be installed. This will enable exploration of the nonlinear regime with $\chi \simeq 1$. In phase 1, an upgrade of laser to 350 TW will expand the explored parameter space to $\chi > 4$. Thus a vast range of scans ranging from $\xi = 0.5 - 23$ will be covered enabling a clear identification of transition from perturbative to non-perturabtive regime, for instance. In addition, the first experiment studying the perturbative transition in the collision of electron beam with laser was performed at SLAC [8, 9] is also marked (black solid points, labelled as "E144"). The parameter range covered in this experiment was mostly in the perturbative regime. We also show the purely laser-based, all-optical schemes that have been employed at Astra-Gemini [10, 11] (black solid point, labelled as "Astra-Gemini") and are planned at, e.g. the Extreme Light Infrastructure (ELI) [12, 13].

2. Requirement of precision laser diagnostics

One of the unique features of LUXE experiment is that the measurements are aimed to be performed with unprecedented accuracy and ample statistics. This poses stringent requirements on both the machines involved (electron and laser). The EuXFEL has very high quality electron beam parameters. However, the residual shot-to-shot fluctuations of laser parameters coupled with the large nonlinearity of the processes under investigation form a particular challenge. First we discuss the important role played by laser intensity on the key physics observable.

High energy Ti:Sapphire lasers using flash-lamp technology for the pump-lasers have typical energy fluctuations of around 2–3% rms. These overall energy fluctuations, while significant, are not the dominant contribution to shot-to-shot intensity fluctuations. The major contributions to such fluctuations are small changes in phase both in real space and in frequency space due to temperature drift, air currents and mechanical vibrations. Small changes in spatial phase result in the spot radius fluctuating at the few % level resulting in up to 10% level intensity fluctuations. Similarly, fluctuations in the spectral phase can lead to 1% rms fluctuations in the pulse duration. In total, the variation in between the highest and lowest intensity shots on a stable laser can reach $\sim$ 15%. Simulations show that an uncertainty of about 5% in intensity can lead up to 40% uncertainty in the pair rate. Hence, it become detrimental to know the laser intensity with very good accuracy.

3. Principle of precision diagnostics

The aim is to diagnose the laser intensity, on a shot-to-shot basis, with percent level accuracy. A dedicated state-of-art diagnostic system capable of measuring the fluctuations will be set up. A schematic of the proposed diagnostic system is shown in Fig. 2. The shots will then be tagged with their precise relative intensity allowing precision relative measurements using various diagnostics. A key factor allowing us to tag the laser parameters on shot-to-shot basis is the fact that during the interaction with electron bunch, the laser beam is not significantly perturbed or absorbed (in contrast to, for example, high-power laser interactions with solid or gas targets). Hence as shown in Fig. 2, the laser beam will be relay imaged to laser area for tagging. One of the key features will be imaging of the laser beam to ensure that the beam intensity profile at the compressor output, the focussing parabola at interaction point and at entry of the returning beam have the same phase and amplitude distribution, thus providing the highest possible fidelity of the diagnostic suite.

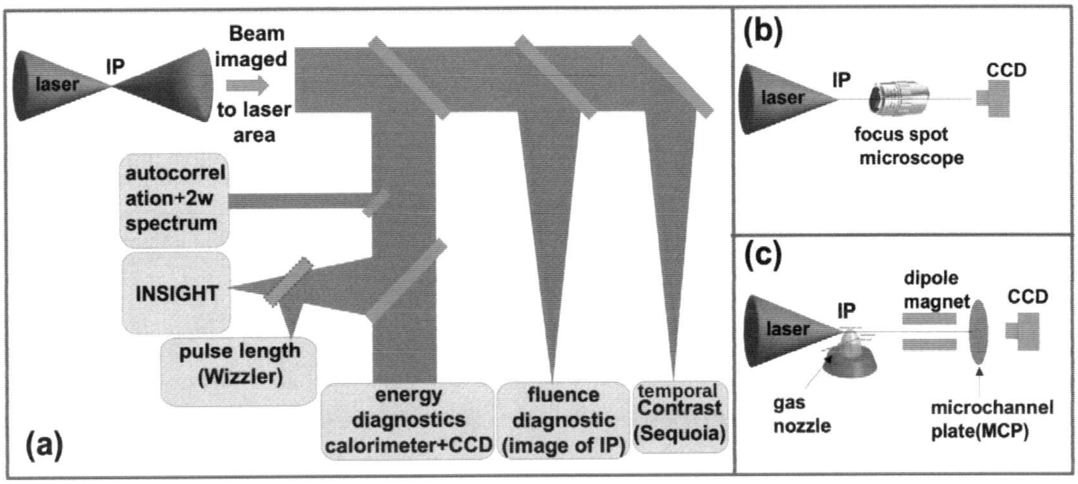

Figure 2. Single shot tagging scheme to enable precision control of the laser parameters. (a) Tagging of primary laser parameters based on the relay imaging of the laser beam to the relevant diagnostics. (b) Focus diagnostic to optimise and measure the focus at the IP. (c) An indirect approach to measure the laser intensity.

The tagged parameters are primarily the laser pulse energy, pulse duration and focal spot size. In addition, we shall also use INSIGHT diagnostics which offers full 3D characterisation of the focus and hence enables determination of the electric-field distribution in the focus at the interaction point (see Fig. 2(a)). We note that this device will be used offline and an average measurement will be taken. The focal spot size at the interaction point will be also checked using high magnification focus diagnostic (see arrangement in Fig. 2(b)). An indirect approach to measure the intensity at the IP is also illustrated in Fig. 2(c). The method is based on ponderomotive scattering of the free electrons [14].

4. Offline diagnostic testing

The offline setup of the diagnostics and testing are already being performed at the JETI40 laser facility of Helmholtz Institute Jena [15]. JETI40 is a Ti:Sa laser system operating at central wavelength of 800 nm. It delivers pulses at 10 Hz repetition rate. The energy in a single pulse can reach in excess of 1 J in a pulse duration of 30 fs (FWHM). The last amplification stage consists of a cryogenic cooled amplifier system. The beam is expanded to 5 cm and enters the grating based compressor. After the compressor, an adaptive optics apparatus enables wavefront correction of the beam and hence a diffraction limited focus spot can be obtained.

As discussed above, the proper imaging of the beam to the diagnostic suite is a very important aspect of the system. The relay imaging of the beam is based on telescope imaging. Based on the calculations with reference parameters, two telescopes lines will be needed. We envisage using an off axis parabola (OAP) in combination with an infinitely corrected microscope objective as first telescope. The second telescope will be based on 4f relay, with no change in the magnification, i.e, 1:1 imaging. The setup is in primary phase of testing.

5. Summary

The LUXE experimental platform presents an exciting opportunity to probe the strong field QED in uncharted territory. The parameter space covered at LUXE enables studies of some novel phenomena proposed by theory such as the intensity-dependent position of the nonlinear Compton edges and the transition from the perturbative to nonperturbative regimes in electron-positron pair production. The test of theoretical predictions relies on accurate measurements of key physics observables. In terms of the particle and photon detection aspects, enabling this goal demands the development of a state-of-art detector suite. On the other hand from the laser side, even more stringent requirements on its parameters demands state-of-art laser diagnostics that can precisely control and monitor the relevant pulse parameters. The single shot tagging method described above aims to tag the shot to shot intensity with $< 1\%$ level of accuracy.

6. Acknowledgments

Author acknowledges the "Bundesministerium für Bildung und Forschung" for the support via the 2018 Helmholtz Innovation Pool. This work has also benefited from computing services provided by the German National Analysis Facility (NAF).

References

[1] D. Hanneke, S. Fogwell Hoogerheide and G. Gabrielse 2011 Cavity control of a single-electron quantum cyclotron: Measuring the electron magnetic moment *Phys. Rev. A* **83** 052122
[2] A. Di Piazza, C. Müller, K. Z. Hatsagortsyan and C. H. Keitel 2012 Extremely high-intensity laser interactions with fundamental quantum systems *Rev. Mod. Phys.* **84** 1177
[3] H. Abramowicz et al 2021 Conceptual Design Report for the LUXE Experiment *Eur. Phys. J. ST* **230** 2445
[4] G. Breit and J.A. Wheeler 1934 Collision of Two Light Quanta *Phys. Rev.* **46** 1087
[5] Howard R. Reiss 1971 Production of Electron Pairs from a Zero-Mass State *Phys. Rev. Lett.* **26** 1072
[6] A. Hartin, A. Ringwald, and N. Tapia 2019 Measuring the boiling point of the vacuum of quantum electrodynamics *Phys. Rev. D* **99** 036008
[7] T.G. Blackburn and M. Marklund 2018 Nonlinear Breit-Wheeler pair creation with bremsstrahlung rays *Plasma Phys. Control. Fusion* **60** 054009
[8] C. Bula *et al.* 1996 Observation of nonlinear effects in Compton scattering *Phys. Rev. Lett.* **76** 3116
[9] D. L. Burke *et al.* 1997 Positron production in multi - photon light by light scattering *Phys. Rev. Lett.* **79** 1626
[10] J. M. Cole *et al.* 2018 Experimental Evidence of Radiation Reaction in the Collision of a High-Intensity Laser Pulse with a Laser-Wakefield Accelerated Electron Beam *Phys. Rev. X* **8** 011020
[11] K. Poder *et al.* 2018 Experimental Signatures of the Quantum Nature of Radiation Reaction in the Field of an Ultraintense Laser *Phys. Rev. X* **8** 31004
[12] S. Weber *et al.* 2017 An installation for high-energy density plasma physics and ultra-high intensity laser-matter interaction at ELI-Beamlines *Matter and Radiation at Extremes* **2** 149
[13] S. Gales *et al.* 2018 The extreme light infrastructure - Nuclear physics (ELI-NP) facility: New horizons in physics with 10 PW ultra-intense lasers and 20 MeV brilliant gamma beams *Reports on Progress in Physics* **81** 94301
[14] FCC Collaboration 2021 Complementary diagnostics of high-intensity femtosecond laser pulses via vacuum acceleration of protons and electrons *Plasma Phys. Control. Fusion* **63** 014002
[15] JETI40 laser system: https://www.hi-jena.de (accessed in 2021)

Spin-Polarized Particle Beams from Laser-Plasma Based Accelerators

L Reichwein[1], A Hützen[2,3], M Büscher[2,3], A Pukhov[1]

[1]Institut für Theoretische Physik I, Heinrich-Heine-Universität Düsseldorf, 40225 Düsseldorf, Germany

[2]Peter Grünberg Institut (PGI-6), Forschungszentrum Jülich, 52425 Jülich, Germany

[3]Institut für Laser- und Plasmaphysik, Heinrich-Heine-Universität Düsseldorf, 40225 Düsseldorf, Germany

E-mail: `lars.reichwein@hhu.de`

Abstract. Current laser-plasma based accelerators are promising options with respect to the acceleration of spin-polarized particle beams. We give an overview over the effects relevant during the acceleration process and more specifically discuss the acceleration of protons via Magnetic Vortex Acceleration (MVA). With the aid of particle-in-cell simulations we show that the length of the density down-ramp at the end of the plasma target affects the final beam quality regarding its collimation. The average spin-polarization of the obtained bunch remains largely robust at about 80% and only decreases for significantly longer ramps.

1. Introduction

Spin-polarized particle beams are of interest for a variety of applications, such as testing the Standard Model of Particle Physics [1], scrutinizing nuclear reactions [2] or even measuring transient magnetic fields in plasmas [3]. With the advances is laser-plasma based accelerator technology, new acceleration schemes have become feasible; an overview with respect to spin-polarized particle beams is given in the review by Büscher *et al* [4].

The most prominent option for electron acceleration in laser-plasma interactions would be laser- (or particle beam-) driven plasma wakefield acceleration [5]. This method allows for high energies over much shorter acceleration distances compared to conventional accelerators. With respect to spin, Wu *et al* have examined both the laser- and the particle beam-driven case and found that the main problem is the injection process, where low energy electrons experience the high azimuthal magnetic fields of the wakefield. In the laser-driven case this could be mitigated to some extent by using a Laguerre-Gaussian beam. Still, both setups necessitate an initially spin-polarized target for a final polarization of up to 80%. Only recently, Nie *et al* proposed a setup similar to Trojan Horse wakefield acceleration that allows for the acceleration of spin-polarized electron beams (up to 30%) from an initially unpolarized target [6].

For protons, many acceleration methods are conceivable. Since pre-polarization still is a requirement for spin-polarized beams, only some of these are experimentally feasible. One promising option is Magnetic Vortex Acceleration (MVA), in which a near-critical density target is irradiated by a high-intensity circularly polarized laser pulse [7, 8].

This leads to the transverse expulsion of particles by the ponderomotive force, leaving behind a low-density channel. A wake induced by the laser can accelerate electrons which leads to the

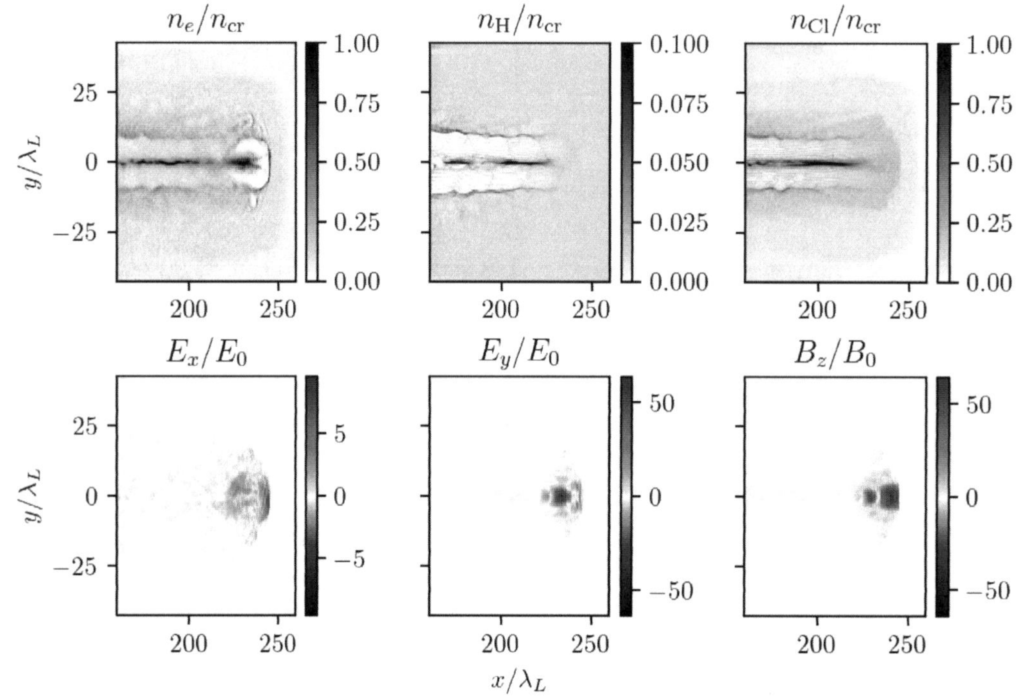

Figure 1. Particle densities (top row, clipped for better visibility) and electromagnetic fields (bottom row) for the MVA process after $t = 160\tau_0$. Notice the wakefield structure in the plots for n_e and E_x as well as the central filament in the plot for n_H. The latter will later yield the accelerated proton beam.

formation of a central filament (cf. Fig. 1). The current flowing through this filament and the (opposing) current in the channel walls create a strong azimuthal magnetic field. Additionally, some protons get accelerated in the filament. The resulting displacement of electrons and protons induces strong electric fields. When exiting the plasma, the protons can get further accelerated by the remaining fields. For spin-polarized proton beams this scheme was first examined by Jin *et al* [9], where energies $\gtrsim$ 100 MeV could be obtained from a laser pulse of $a_0 = eA_0/(m_e c) = 100$ interacting with a near-critical HCl target. Here, m_e denotes the electron mass and c the vacuum speed of light.

In this paper, we will first discuss spin effects relevant for laser-plasma based accelerators. Subsequently, we will present the acceleration of spin-polarized protons beams from a near-critical density target via MVA. There, we will specifically examine the influence of density down-ramps on the final beam quality. A full discussion of the results can be found in [10].

2. Effects involving spin

Generally speaking, we need to consider the influence of three different effects regarding spin polarization. In the following, we will discuss which of these actually become relevant for the regime we consider. A full discussion of spin effects in relativistic laser-plasma interactions is given by Thomas *et al* [11].

Let us consider an ensemble of particles (electrons or protons) in a homogeneous magnetic field. These particles will precess uniformly if no other external influences are present. Once we introduce inhomogeneities into our fields, e.g. laser fields or fields from an existing plasma, different parts of the ensemble experience different fields, leading to variations in particle precession. In turn, an initially polarized ensemble would start to lose its polarization. The change of the spin vector $\mathbf{s}$ in some arbitrary configuration of electric and magnetic fields $\mathbf{E}, \mathbf{B}$ can be described via the Thomas-Bergmann-Michel-Telegdi (T-BMT) equation

$$\frac{\mathrm{d}\mathbf{s}}{\mathrm{d}t} = -\boldsymbol{\Omega} \times \mathbf{s} \,, \tag{1}$$

where

$$\boldsymbol{\Omega} = \frac{qe}{mc}\left[\Omega_B \mathbf{B} - \Omega_v \left(\frac{\mathbf{v}}{c}\cdot\mathbf{B}\right)\frac{\mathbf{v}}{c} - \Omega_E \frac{\mathbf{v}}{c} \times \mathbf{E}\right] \tag{2}$$

is the precession frequency of a particle with charge $q \cdot e$ (e being the elementary charge), mass m and velocity $\mathbf{v}$ [12]. The three prefactors

$$\Omega_B = a + \frac{1}{\gamma}, \qquad \Omega_v = \frac{a\gamma}{\gamma+1}, \qquad \Omega_E = a + \frac{1}{1+\gamma} \tag{3}$$

depend the anomalous magnetic moment $a = \frac{1}{2}(g-2)$ with the Landé g-factor, and the Lorentz factor $\gamma = \sqrt{1 - \mathbf{v}^2/c^2}$. For electrons, the anomalous magnetic moment is $a_e \approx 10^{-3}$, while for protons it is $a_p \approx 1.793$.

Looking at the equations above, we can see that stronger fields lead to stronger spin precession. The prefactors show that, depending on what part of the acceleration process we are in, the precession becomes more (smaller γ) or less (higher γ) pronounced. In the case of electrons the importance of the prefactors was discussed by Wu *et al* [13, 14]. For protons, due to the larger anomalous magnetic moment and the comparatively lower γ that can be reached during acceleration, those prefactors stay relevant for a longer amount of time.

Until now, we have seen how a particle's trajectory through some electromagnetic field configuration can induce a change in spin polarization. But also the change of trajectory due to spin in terms of the Stern-Gerlach force is possible. The force is most famously known from the seminal experiment by Stern and Gerlach [15] as $\mathbf{F}_{\mathrm{SG}} = \nabla(\boldsymbol{\mu} \cdot \mathbf{B})$, but the more general, relativistic formula is

$$\mathbf{F}_{\mathrm{SG}} = \left(\nabla - \frac{\mathrm{d}}{\mathrm{d}t}\nabla_v\right)(\boldsymbol{\Omega} \cdot \mathbf{s}) \,. \tag{4}$$

This force contribution would need to be added to the equations of motion to account for the possible splitting of particle trajectories with respect to their spin. Thomas *et al* have shown that the separation distances due to spin are negligible for both electrons and protons. Therefore, we neglect any contributions by the Stern-Gerlach force in our simulations.

Lastly, radiative effects may play a role in our setup. It is well known that spin-polarization can build up over time in storage rings (typically minutes or hours) due to the Sokolov-Ternov effect [16]. This is due to the fact that there is a different probability for a spin flip from up to down than for the inverse process. The polarization depending on time can be described by

$$P(t) = \frac{P_\uparrow - P_\downarrow}{P_\uparrow + P_\downarrow}\left[1 - e^{-t/\tau_{\mathrm{pol}}}\right] \,, \tag{5}$$

where $P_\uparrow, P_\downarrow$ are the aforementioned probabilities. The characteristic time for polarization build-up τ_{pol} can be approximated both for electrons and protons. Thomas *et al* found a scaling law depending on the particles' kinetic energy T and the maximum field strength $F = \max(|\mathbf{E}|, |\mathbf{B}|)$, namely $\tau_{\text{pol}} \propto T^{-2}F^{-3}$.

While in conventional accelerators these timescales are quite long, laser-plasma based accelerators exhibit larger field strengths meaning that Sokolov-Ternov might eventually become important for future setups. It has further been shown in generalizations of this effect, that electron beams can obtain significant polarization if they collide with intense laser pulses [17]. For our following setup concerning polarized protons from MVA, however, we can safely neglect Sokolov-Ternov, leaving only the T-BMT equation to be considered for our simulations. This has already been implemented in our particle-in-cell (PIC) code VLPL [18].

3. Magnetic Vortex Acceleration

For proton acceleration, MVA is an experimentally feasible method if spin-polarization is of concern. In a preceding work [9], Jin *et al* examined the influence of the laser intensity on the final beam quality with respect to spin polarization. Now, we keep the laser parameters fixed but instead change the plasma parameters.

The plasma target we consider consists of HCl gas with a peak density of $n_{\text{H}} = n_{\text{Cl}} = 0.0122n_{\text{cr}}$ with n_{cr} being the critical density. Such a target can be pre-polarized (see Ref. [14] for a description of the process). In our simulations, we assume a homogeneous initial distribution with $s_y = 1$ (y being transverse to the direction of laser propagation). The interaction volume is built up the following way: After an up-ramp (vacuum to peak density) of $5\lambda_L$ in length, the peak density is kept for $200\lambda_L$. The length of the subsequent down-ramp is varied from $0\lambda_L$ to $100\lambda_L$ to see its influence on the final beam quality. The circularly polarized laser pulse in all of the simulations is fixed to $a_0 = 25$ and wavelength $\lambda_L = 800$ nm. It has a focal radius of $w_0 = 10\lambda_L$ (at $1/e^2$ of the intensity) and a length of $\tau_0 = 10\lambda_L/c$.

The box size in our PIC simulations is $(100 \times 60 \times 60)\lambda_L$, with grid size $h_x = 0.05\lambda_L$ (direction of propagation) and $h_y = h_z = 0.4\lambda_L$. In order to reduce computational effort, the grid size is increasing in the transverse direction. The time step is $\Delta t = h_x/c$, as required by the RIP solver we have used [19]. The simulation box is co-moving with our laser pulse until it has left the plasma.

All of the conducted simulations start alike, as the initial setting for all of them is the same: The formation of the ion channel due to the ponderomotive force of the laser pulse leads to the central proton filament that is pushed out of the plasma (see Fig. 1). Differences start to occur once the down-ramp region is reached: In the case of a sharp border ($L_{\text{ramp}} = 0\lambda_L$), a highly collimated proton bunch (peak density $n_{\text{peak}} \approx 0.209n_{\text{cr}}$) with an average polarization of $\langle s_y \rangle \approx 81\%$ is ejected from the ion channel.

In contrast, longer ramp lengths lead to less collimation. For example, $L_{\text{ramp}} = 50\lambda_L$ yields only $n_{\text{peak}} \approx 0.044n_{\text{cr}}$. However, the degree of polarization stays roughly the same. Only in the case of our longest down-ramp ($L_{\text{ramp}} = 100\lambda_L$), we observe a decrease of polarization to about 63%. This means that spin polarization generally seems quite robust against longer down-ramps and that the main limiting factor for high quality proton beams is the process of beam collimation (Fig. 2).

These results can be explained as an interplay of several effects: the longer ramp lengths also yield an effectively longer interaction between the laser pulse and the target. During this interaction, the particles being accelerated experience the influence of the prevalent electromagnetic fields for a longer time. Further, the laser is being depleted of energy over a longer period, changing the interaction after some time.

This, however, could also be seen in simply a longer target with a sharp boundary instead of a down-ramp. Still, the down-ramp itself affects the final beam quality significantly. When decreasing the density, the electromagnetic fields are able to expand transversely; in turn, the maximum strength of those fields decreases (cf. ref. [7]). This behaviour also induces the

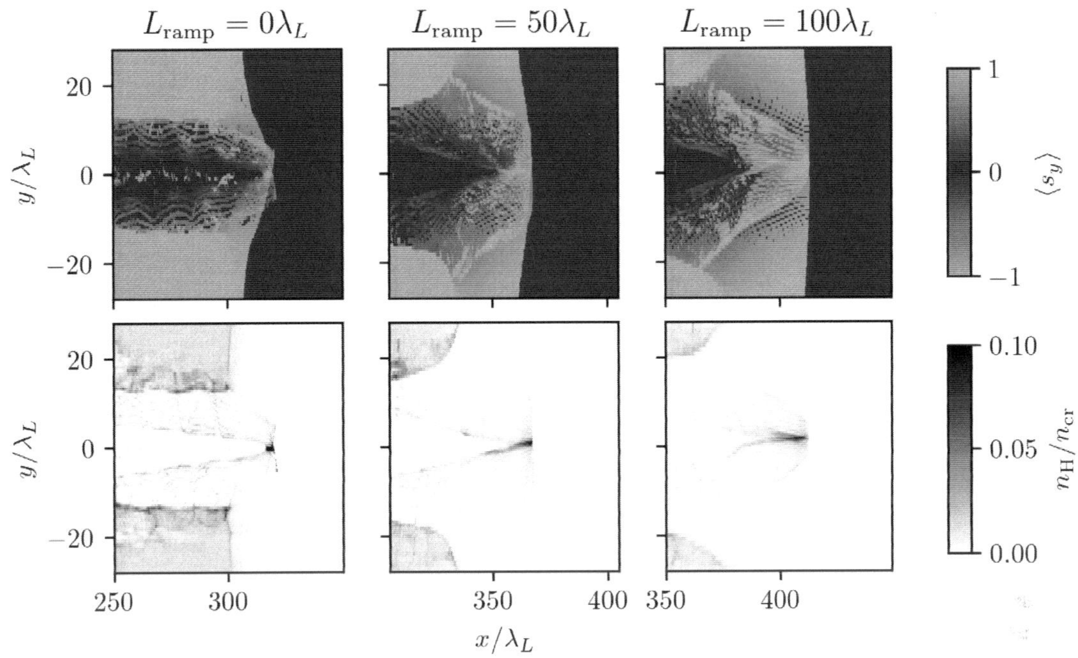

Figure 2. Average spin polarization (top row) and proton density (bottom row, clipped for better visbility) for the different ramp lengths. Higher ramp lengths mainley lead to worse collimation of the protons. Also, significantly long ramps introduce lower-polarization protons into the bunch.

transverse growth of the proton bunch, i.e. the beam gets collimated less in that direction. Longitudinally, lower-polarization protons get introduced into the final beam for longer ramps. This is also due to missing compression of the bunch when leaving the target. As shown in Fig. 3, protons get depolarized in the vicinity of the channel walls where the azimuthal magnetic fields are the strongest. Combined with the transverse spread of the central proton filament this leads to lower polarization. It has to be noted though, that for ramps $\geqslant 75\lambda_L$ it is increasingly hard to properly define the accelerated proton beam and visually isolate it from non-collimated parts (see Fig. 2 for a comparison). Further discussion about the proton bunch and its constituents (not shown here) can be found in Ref. [10].

4. Conclusion

In this paper, we have shown that spin-polarized proton beams can be obtained from a near-critical HCl target via Magnetic Vortex Acceleration. The polarization (ca. 80%) is kept for a variety of down-ramp lengths. Only for very long ramps is the polarization reduced. The main limiting factor is the quickly decreasing collimation of the proton beam: longer ramps allow for more transverse spread of the prevalent electromagnetic fields and therefore the reduction of beam focusing. This behaviour could be explained in terms of the scaling laws for spin polarization that were previously derived. Therefore, as-short-as-possible targets would be ideal, which, however, stands in contrast to experimental feasibility. Alternatively, for realistic targets, different laser and plasma parameters could be used, e.g. using a different species target altogether. Another option would be to introduce a foil in front of the target which could shield some part of the laser fields (similar to Radiation Pressure Acceleration [20]).

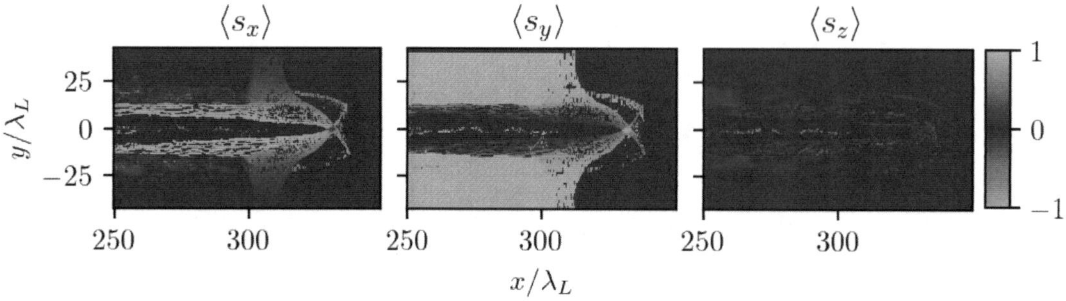

Figure 3. Spin distribution after $t = 160\tau_0$ for the simulation without down-ramp. Towards the channel wall, protons gain s_x contributions due to the azimuthal magnetic field. The final proton beam just leaving the plasma (see $\langle s_y \rangle$ at $x \approx 325\lambda_L$) maintains a high degree of polarization.

Acknowledgments
This work has been supported by the DFG (project PU 213/9-1). The authors gratefully acknowledge the Gauss Centre for Supercomputing e.V. (www.gauss-centre.eu) for funding this project by providing computing time on the GCS Supercomputer JUWELS at Jülich Supercomputing Centre (JSC). The work of A H and M B has been carried out in the framework of the *Ju*SPARC (Jülich Short-Pulsed Particle and Radiation Center) [21] and has been supported by the ATHENA (Accelerator Technology Helmholtz Infrastructure) consortium.

References

[1] Androic D *et al* 2018 *Nature* **557** 207
[2] Glashausser C 1979 *Annu. Rev. Nucl. Part. Sci.* **29** 33
[3] Gong Z, Hatsagortsyan K Z and Keitel H 2021 *Preprint* arXiv:2103.12164
[4] Büscher M et al 2020 *High Power Laser Sci* **8** e36
[5] Pukhov A and Meyer-ter Vehn J 2002 *Appl. Phys. B* **74** 355
[6] Nie Z *et al* 2021 *Phys. Rev. Lett.* **126** 054801
[7] Nakamura T, Bulanov S V, Esirkepov T Z and Kando M 2010 *Phys. Rev. Lett.* **105** 135002
[8] Park J *et al* 2019 *Phys. Plasmas* **26** 103108
[9] Jin L, Wen M, Zhang X M, Hützen A, Büscher M and Shen B F 2020 *Phys. Rev. E* **102** 011201(R)
[10] Reichwein L, Hützen A, Büscher M and Pukhov A 2021 *Plasma Phys. Control. Fusion* **63** 085011
[11] Thomas J *et al* 2020 *Phys. Rev. Accel. Beams* **23** 064401
[12] Mane S R, Shatunov Y M and Yokoya K 2005 *Rep. Prog. Phys.* **68** 1997
[13] Wu Y *et al* 2019 *Phys. Rev. E* **100** 043202
[14] Wu Y *et al* 2019 *New J Phys.* **21** 073052
[15] Gerlach W and Stern O 1922 *Zeitschrift für Physik* **9** 349-352
[16] Sokolov A A and Ternov I M 1971 *Sov. Phys. J.* **24** 47
[17] Geng X S *et al* 2020 *New J. Phys.* **22** 013007
[18] Pukhov A 1999 *J. Plasma Phys.* **61** 425-33
[19] Pukhov A 2020 *J. Comput. Phys.* **418** 109622
[20] Macchi A *et al* 2017 *JINST* **12** C04016
[21] JuSPARC – The Jülich Short-Pulsed Particle and Radiation Center 2020 *JLSRF* **6** A138

Journal of Physics: Conference Series **2249** (2022) 012019 doi:10.1088/1742-6596/2249/1/012019

Plane Wave Backgrounds in the Worldline Formalism

James P. Edwards [1] and Christian Schubert [1]

[1]Instituto de Física y Matemáticas, Universidad Michoacana de San Nicolás de Hidalgo.

E-mail: jpedwards@cantab.net, christianschubert137@gmail.com

Abstract. Plane-wave backgrounds play a special role in strong-field QED as non-trivial field configuration simple enough to be treated analytically whilst still leading to rich physical consequences. In vacuum and in constant backgrounds, the first quantised, string-inspired "Worldline Approach" to field theoryoffers substantial simplifications and calculational efficiency. We present a new, general approach to incorporating plane wave backgrounds into the Worldline Formalism extending initial work by Ilderton and Torgrimsson. The method uses resummation techniques to take the background into account non-perturbatively and yields "Master Formulae" for scattering amplitudes in the background that may offer an alternative tool to studying QED in plane waves as has been achieved in the constant field case.

1. Introduction

The behaviour of quantum fields in the presence of electromagnetic backgrounds – such as those produced at high intensity laser experiments – can be very different to that of fields in vacuum. In particular, additional interactions with photons from the background make possible various new phenomena, some of which form the basis for current and future experimental searches. It is important that theoretical calculations take into account these effects with realistic models of such external fields, especially considering the high precision of experiments planned to come online in the years ahead. Moreover, the mathematical structure of the theory in a background field is of theoretical interest, especially its strong field asymptotics in the context of the famous Ritus-Narozhny conjecture [1, 2] regarding the perturbative expansion parameter of dominant loop contributions that first arose for constant-crossed backgrounds.

In this contribution we shall consider one-loop N-photon amplitudes in three cases: quantum electrodynamics (QED) in vacuum, in constant backgrounds and in an arbitrary *plane wave* background, using the *Worldline Formalism* of quantum field theory (see [3–6]). This alternative approach to field theory, appreciated for its computational advantages and efficient representation of such scattering amplitudes encoded in "Master Formulae," has only recently been adapted to a general treatment in plane wave fields in [7] building on previous work focussing on the on-shell case in [8] which treated a specific $N = 2$-photon process.

These backgrounds (described in more detail below) are of both phenomenological importance as models of the fields produced by the intense lasers mentioned above, and of intrinsic mathematical interest as fields for which the Dirac equation can be solved exactly [9, 10] (as explained later, this allows field theory processes to be studied non-perturbatively with respect to the background field, albeit in perturbation theory in the fine structure constant, $\alpha \equiv \frac{e^2}{4\pi}$ with e the electric charge, according to the Furry picture [11, 12]). As such they are very special examples that permit analytic treatment without losing most of the essential features of QED in more general backgrounds that cannot be solved exactly.

On the other hand, despite great efforts towards studying various processes in these backgrounds, calculations in the standard approach often turn out to be rather laborious except for the simplest cases (see, for example, [13–28]). This work extending the Worldline Formalism to plane wave backgrounds is motivated by its success in providing compact "Master Formulae" for scattering amplitudes in vacuum and constant backgrounds, that have led to useful insight and efficient analysis of field theory processes and their gauge structure in those cases; here we describe these results and the derivation of a Master Formula for one-loop N-photon amplitudes in a plane wave background recently presented for the first time in [7].

We begin in section 2 with an introductory description of the worldline approach to quantum field theory, illustrating its application in vacuum, before recapitulating the worldline description of QED in a constant background in section 3. In section 4 we show how a plane wave background can be treated in the Worldline Formalism before concluding. Contrary to the majority of the works cited, we present our results in Minkowski space throughout (recalling that functional integration is more conveniently defined after Wick rotation to Euclidean space).

2. Photon amplitudes in vacuum

To set the scene, we consider photon amplitudes in QED, starting from the Dirac action coupling the electron field, $\Psi(x)$, to the Maxwell field with gauge potential $\mathrm{A}(x) = A_\mu(x)dx^\mu$,

$$S[\Psi, A] = \int d^D x\big[-\frac{1}{4}\operatorname{tr}(F^2) + \bar{\Psi}(i\not{D} - m)\Psi\big]\,, \tag{1}$$

with covariant derivative $D_\mu := \partial_\mu + ieA_\mu$ and field strength tensor $\mathrm{F}(x) = \mathrm{dA}(x)$. Here we consider A to be a semi-classical field, consisting of a "background field" and a "photon field" so that $\mathrm{A}(x) = \bar{\mathrm{A}}(x) + \mathrm{A}^\gamma(x)$. For general backgrounds, $\bar{\mathrm{A}}(x)$, a full analytic solution to the system described by (1) is out of reach, but amongst the few background fields that allow for an exact classical solution are the constant and plane wave backgrounds we consider in later sections.

2.1. Worldline representation — effective action

For now we consider photon scattering in vacuum, for which we set $\bar{\mathrm{A}}(x) = 0$, which we access via the one-loop effective action for the photon field, $\mathrm{A}^\gamma(x)$. As outlined in [3] and reviewed in [4, 6], this is obtained by integrating out the fermion degrees of freedom according to

$$\begin{aligned}
\mathrm{e}^{i\Gamma[A]} &:= \int \mathscr{D}\bar{\Psi}(x)\mathscr{D}\Psi(x)\exp\big[i\int d^D x\,\bar{\Psi}(x)(i\gamma\cdot D - m)\Psi(x)\big]\\
&= i\mathrm{Det}\big[i\gamma\cdot D - m\big]\\
&= \frac{i}{2}\mathrm{Det}^{\frac{1}{2}}\big[(\gamma\cdot D)^2 + m^2\big]\,.
\end{aligned} \tag{2}$$

A *first quantised* representation of the functional determinant is obtained by first extending a familiar trick to operators, $\mathcal{O}$, namely that $\log\mathrm{Det}[\mathcal{O}] = \mathrm{Tr}\log[\mathcal{O}]$; then employing Schwinger's representation of the logarithm and evaluating the trace in the position basis we arrive at

$$\begin{aligned}
\Gamma[A] &= -\frac{1}{2}\int_0^\infty \frac{dT}{T}\mathrm{e}^{-im^2T}\int d^D x\,\langle x|\mathrm{e}^{-iT(\gamma\cdot D)^2}|x\rangle\\
&= -\frac{1}{2}\int_0^\infty \frac{dT}{T}\mathrm{e}^{-im^2T}\oint_{PBC}\mathscr{D}x(\tau)\oint_{ABC}\mathscr{D}\psi(\tau)\,\mathrm{e}^{iS[x,\psi]}\,,
\end{aligned} \tag{3}$$

where the last line uses the path integral representation of the (diagonal) matrix elements, summing over T-periodic trajectories, $x^\mu(\tau)$, that represent the orbital degrees of freedom of the field and anti-periodic Grassmann fields, $\psi(\tau)$, on these worldlines that produce the spin interaction with the background field. The "worldline action" inherited from the Dirac operator (that has been transformed to second order form [29, 30] in the first line of (3)) is given by

$$S[x,\psi] = \int_0^T d\tau \Big[\frac{\dot{x}^2}{4} + \frac{i}{2}\psi\cdot\dot{\psi} - eA(x)\cdot\dot{x} + ie\psi\cdot F(x)\cdot\psi\Big], \tag{4}$$

describing dynamics of a relativistic spin-1/2 point particle minimally coupled to the background (τ is the particle's proper time and T the proper time of propagation around the loop).

In scalar QED the effective action follows from integrating out the spin-zero Klein-Gordon field and leads to an analogous formula differing only by the absence of the Grassmann fields and overall normalisation (for degrees of freedom and the difference in statistics)

$$\Gamma_{\rm scal}[A] = \int_0^\infty \frac{dT}{T}\mathrm{e}^{-im^2T} \oint_{PBC} \mathscr{D}x(\tau)\, \mathrm{e}^{i\int_0^T d\tau\left[\frac{\dot{x}^2}{4} - eA(x)\cdot\dot{x}\right]}. \tag{5}$$

These proper time integral representations of the effective action are the basic objects of the Worldline Formalism. It is worth pointing out that analogous representation for the field theory propagators in an electromagnetic background were first suggested by Feynman in the appendices of [31, 32] (recently extended to a complete worldline description in vacuum in [33, 34]).

Although less well-known than the standard formalism, the worldline approach has proven to offer various conceptual and computational advantages. These largely stem from the fact that, as in string theory, it combines multiple Feynman diagrams into single, compact integral representations, thereby manifesting symmetry under exchange of external legs. As such it offers superior organisation of gauge information and removal of spurious divergences that would normally cancel between diagrams already at the level of the integrand. Moreover, as in the calculations presented below virtual loop momenta are already integrated out.

2.2. Scattering amplitudes — Master Formulae

To obtain photon amplitudes, we now specialise $\mathrm{A}(x) = \mathrm{A}^\gamma(x)$ to represent asymptotic wavefunctions of N photons with fixed polarisations and momenta by

$$A^\gamma_\mu(x) = \sum_{i=1}^N \varepsilon_{i\mu}\,\mathrm{e}^{ik_i\cdot x} \tag{6}$$

and expand the interactions in (3), (5) to multi-linear order in the ε_i. Doing this represents each photon emission or absorption off the loop by a "vertex operator" familiar from string theory,

$$\begin{aligned} V^\gamma[k,\varepsilon] &:= \int_0^T d\tau \big[\varepsilon\cdot\dot{x}(\tau) + \psi(\tau)\cdot f\cdot\psi(\tau)\big]\mathrm{e}^{ik\cdot x(\tau)}, \\ V^\gamma_{\rm scal}[k,\varepsilon] &:= \int_0^T d\tau\,\varepsilon\cdot\dot{x}(\tau)\,\mathrm{e}^{ik\cdot x(\tau)}, \end{aligned} \tag{7}$$

where in the first line $f_{\mu\nu} := k_\mu\varepsilon_\nu - \varepsilon_\mu k_\nu$ is the (linearised) photon field strength tensor. This procedure leads easily to the following representations of the N-photon amplitudes (first arising from Bern and Kosower's analysis of the infinite tension limit of string theory amplitudes [35,36]):

$$\begin{aligned} -2\Gamma[\{k_i,\varepsilon_i\}] &= (-ie)^N \int_0^\infty \frac{dT}{T}\mathrm{e}^{-im^2T}\oint_{PBC}\mathscr{D}x(\tau)\,\mathrm{e}^{i\int_0^T d\tau\frac{\dot{x}^2}{4}}\oint_{ABC}\mathscr{D}\psi(\tau)\,\mathrm{e}^{i\int_0^T d\tau\frac{i}{2}\psi\cdot\dot{\psi}}\prod_{i=1}^N V^\gamma[k_i,\varepsilon_i] \\ \Gamma_{\rm scal}[\{k_i,\varepsilon_i\}] &= (-ie)^N \int_0^\infty \frac{dT}{T}\mathrm{e}^{-im^2T}\oint_{PBC}\mathscr{D}x(\tau)\,\mathrm{e}^{i\int_0^T d\tau\frac{\dot{x}^2}{4}}\prod_{i=1}^N V^\gamma_{\rm scal}[k_i,\varepsilon_i]. \end{aligned} \tag{8}$$

Note that both integrals are now Gaussian in their corresponding variables.

To compute these path integrals, then, we only require the Green functions of the free kinetic operator and overall normalisation. In the orbital sector, for closed loops this operator has a zero mode, $x_0 = \text{const}$, so we separate this by expanding about the loop centre of mass,

$$x^\mu(\tau) = x_0^\mu + q^\mu(\tau)\,, \quad x_0^\mu = \frac{1}{T}\int_0^T d\tau\, x^\mu(\tau)\,, \quad q^\mu(0) = 0 = q^\mu(T)\,, \quad \int_0^T d\tau\, q^\mu(\tau) = 0\,, \qquad (9)$$

which has led to so-called "string-inspired" (SI) boundary conditions on the deviation $q(\tau)$ which remains to be integrated over: $\oint \mathscr{D}x(\tau) \longrightarrow \int d^D x_0 \int_{\text{SI}} \mathscr{D}q(\tau)$. With this, the appropriate Green functions in the Hilbert space orthogonal to the zero mode are defined by

$$\big\langle q^\mu(\tau_i) q^\nu(\tau_j)\big\rangle_\perp = -G_{Bij}\eta^{\mu\nu}\,, \qquad G_{Bij} \equiv G_B(\tau_i,\tau_j) = |\tau_i - \tau_j| - \frac{(\tau_i-\tau_j)^2}{T}\,, \qquad (10)$$

$$\big\langle \psi^\mu(\tau_i)\psi^\nu(\tau_j)\big\rangle = \frac{1}{2}G_{Fij}\eta^{\mu\nu}\,, \qquad G_{Fij} \equiv G_F(\tau_i,\tau_j) = \sigma(\tau_i - \tau_j)\,. \qquad (11)$$

In the scalar case, we can borrow another trick from string theory by exponentiating the prefactor in the vertex operator, writing $V^{\gamma}_{\text{scal}}[k,\varepsilon] = \int_0^T d\tau\, \mathrm{e}^{ik\cdot x(\tau) + \varepsilon\cdot\dot{x}(\tau)}\big|_{\text{lin }\varepsilon}$ which allows the path integral to be computed by completing the square. With the free particle path integral normalisation $\int_{\text{SI}} \mathscr{D}q(\tau)\, \mathrm{e}^{i\int_0^T d\tau\, \frac{\dot{q}^2}{4}} = (4\pi i T)^{-\frac{D}{2}}$ we arrive at the Bern-Kosower Master formula

$$\Gamma_{\text{scal}}[\{k_i,\varepsilon_i\}] = (-ie)^N (2\pi)^D \delta^D\Big(\sum_{i=1}^N k_i\Big) \int_0^\infty \frac{dT}{T}(4\pi i T)^{-\frac{D}{2}}\mathrm{e}^{-im^2T}$$
$$\prod_{i=1}^N \int_0^T d\tau_i\, \mathrm{e}^{\frac{i}{2}\sum_{i,j=1}^N\left[k_i\cdot k_j G_{Bij} - 2i\varepsilon_i\cdot k_j \dot{G}_{Bij} + \varepsilon_i\cdot\varepsilon_j \ddot{G}_{Bij}\right]}\Big|_{\text{lin }\varepsilon_1\ldots\varepsilon_N}\,, \qquad (12)$$

where the momentum conserving $\delta^D\big(\sum_{i=1}^N k_i\big)$ came from integrating over the loop centre of mass (saying that the scattering photons cannot impart or receive momentum from the loop).

In the spinor case a similar master can be derived by using a *superspace formalism* that linearises the spin part of the interaction vertex. An alternative, often more useful method decomposes contributions to the effective action according to the number of spin and orbital interactions selected from the product of vertex operators in the first line of (8):

$$\Gamma = \sum_{S=0}^N \Gamma_{NS}\,, \qquad \Gamma_{NS} = \sum_{\{i_1\ldots i_S\}} \Gamma_{NS}^{\{i_1\ldots i_S\}}\,, \qquad (13)$$

in which S indicates the number of times the spin part of the vertex is chosen and the sum in the second equality is over partitions of the N indices to assign S photons to this interaction. If we define polynomials P_N according to the expansion of the Bern-Kosower exponent in (12) by

$$\mathrm{e}^{\frac{i}{2}\sum_{i,j=1}^N\left[k_i\cdot k_j G_{Bij} - 2i\varepsilon_i\cdot k_j\dot{G}_{Bij} + \varepsilon_i\cdot\varepsilon_j\ddot{G}_{Bij}\right]}\Big|_{\text{lin }\varepsilon_1\ldots\varepsilon_N} := (-i)^N P_N(\dot{G}_{ij},\ddot{G}_{ij})\,\mathrm{e}^{\frac{i}{2}\sum_{i,j=1}^N k_i\cdot k_j G_{Bij}}\,, \qquad (14)$$

then we can write (12) more explicitly as (we omit the momentum conserving δ-function)

$$\Gamma_{\text{scal}}[\{k_i,\varepsilon_i\}] = (-e)^N \int_0^\infty \frac{dT}{T}(4\pi i T)^{-\frac{D}{2}}\mathrm{e}^{-im^2T}\prod_{i=1}^N\int_0^T d\tau_i\, P_N\, \mathrm{e}^{\frac{i}{2}\sum_{i,j=1}^N k_i\cdot k_j G_{Bij}}\,. \qquad (15)$$

This in turn opens two avenues for the spinor case. Firstly the polynomials can be generalised for use in the spin-orbit decomposition by removing the S photons assigned to the spin interaction:

$$\mathrm{e}^{\frac{i}{2}\sum_{i,j=1}^{N}\left[k_i\cdot k_j G_{Bij}-2i\varepsilon_i\cdot k_j\dot{G}_{Bij}+\varepsilon_i\cdot\varepsilon_j\ddot{G}_{Bij}\right]}\Big|^{\varepsilon_{i_1}=0=\dots=0=\varepsilon_{i_S}}_{\varepsilon_{i_{S+1}}\cdots\varepsilon_{i_N}} := (-i)^N P_{NS}^{\{i_1\dots i_S\}}\,\mathrm{e}^{\frac{i}{2}\sum_{i,j=1}^{N}k_i\cdot k_j G_{Bij}}\,. \quad (16)$$

Then, evaluating the spin-terms with, $W(f_{i_1},\dots f_{i_S}) := \langle\psi_{i_1}\cdot f_{i_1}\cdot\psi_{i_1}\cdots\psi_{i_S}\cdot f_{i_S}\cdot\psi_{i_S}\rangle$ where

$$W(f_{i_1},\dots f_{i_S}) := \sum_{\text{Partitions of S}} (-1)^{cy}G_F(i_1\dots i_{n_1})G_F(i_{n_1+1}\dots i_{n_1+n_2}) \cdots G_F(i_{n_1+\dots+n_{cy-1}+1}\cdots i_{n_1+\dots+n_{cy}})\,, \quad (17)$$

with the sum over inequivalent partitions of the indices between cy bi-cycles defined for either the bosonic or fermionic Green function by $G(i_1\dots i_n) := G_{i_1i_2}G_{i_2i_3}\cdots G_{i_ni_1}Z_n(i_1\dots i_n)$ with $Z_n(i_1\dots i_n) = 2^{-\delta_{n2}}\,\mathrm{tr}\left(\prod_{j=1}^{n} f_{i_j}\right)$, we can write

$$\Gamma_{NS}^{\{i_1\dots i_S\}} = -2(-e)^N\int_0^\infty\frac{dT}{T}(4\pi iT)^{-\frac{D}{2}}\mathrm{e}^{-im^2T}\prod_{i=1}^{N}\int_0^T d\tau_i\,W(f_{i_1},\dots f_{i_S})P_{NS}^{\{i_1\dots i_S\}}\mathrm{e}^{\frac{i}{2}\sum_{i,j=1}^{N}k_i\cdot k_j G_{Bij}}\,. \quad (18)$$

These functions completely describe the N-photon amplitudes in vacuum: this formalism shall be generalised to a constant electromagnetic field and plane wave background in what follows. The second option is to follow Bern and Kosower's integration by parts procedure and apply a "cyclic replacement rule" to convert the scalar amplitudes into their spinor counterparts [4].

3. Constant electromagnetic background

The worldline formalism has had success in constant backgrounds, which of course are relevant for processes where the field can be approximated as slowly varying (e.g. where we may apply the LCFA). Guided by standard techniques, which incorporate the field into the Feynman rules using the exact solution to the Dirac equation, it turns out that little modification is needed to treat the field non-perturbatively in the worldline approach [37–42].

To facilitate the worldline computation in this background it is convenient to use Fock-Schwinger gauge centred at the loop centre of mass whereby (for a constant field)

$$\bar{A}_\mu(x(\tau)) = -\frac{1}{2}\bar{F}_{\mu\nu}\big(x(\tau)-x_0\big)^\nu = -\frac{1}{2}\bar{F}_{\mu\nu}q(\tau)^\nu\,, \quad (19)$$

with which the background dependent part of the worldline action, (4), becomes

$$S_{\bar{A}}[x,\psi|\bar{F}] = \int_0^T d\tau\Big[\frac{1}{4}x\cdot\Big(-\frac{d^2}{d\tau^2}-2e\bar{F}\partial_\tau\Big)\cdot x+\frac{i}{2}\psi\cdot\Big(\frac{d}{d\tau}+2e\bar{F}\Big)\cdot\psi\Big]\,. \quad (20)$$

Now the advantage of Fock-Schwinger gauge is clear: the worldline action remains quadratic in the path integration variables so the field can be accounted for exactly by modifying the kinetic terms as indicated in (20). Indeed, the presence of external photons (i.e. the inclusion of A^γ as in (6)) does not spoil this Gaussian nature so the only modification required for evaluation of the effective actions are the path integral normalisations in the presence of the background:

$$\mathop{\mathrm{Det}}_{\mathrm{SI}}\Big[-\partial_\tau^2-2e\bar{F}\partial_\tau\Big] = (4\pi iT)^{-\frac{D}{2}}\,\mathrm{det}^{-\frac{1}{2}}\Big[\frac{\sinh\mathcal{Z}}{\mathcal{Z}}\Big]\,,\quad \mathop{\mathrm{Det}}_{\mathrm{ABC}}\Big[\partial_\tau+2e\bar{F}\Big] = 2^{\frac{D}{2}}\,\mathrm{det}^{-\frac{1}{2}}\Big[\cosh\mathcal{Z}\Big]\,, \quad (21)$$

where $\mathcal{Z} := e\bar{F}T$, and the appropriate Green functions (inverses of the kinetic operators)

$$G_{Bij}\eta^{\mu\nu}\to\mathcal{G}_{Bij}^{\mu\nu} = \frac{iT}{2\mathcal{Z}^2}\left(\frac{\mathcal{Z}}{\sinh\mathcal{Z}}e^{-\mathcal{Z}\dot{G}_{Bij}}+\mathcal{Z}\dot{G}_{Bij}-1\right)\,,\qquad G_{Fij}\eta^{\mu\nu}\to\mathcal{G}_{Fij}^{\mu\nu} = G_{Fij}\frac{e^{-\mathcal{Z}\dot{G}_{Bij}}}{\cosh\mathcal{Z}}\,, \quad (22)$$

and their derivatives. This treats the external field non-perturbatively in the path integral.

3.1. Euler-Heisenberg Lagrangians

We illustrate this formalism with the $N = 0$, bare path integral to recover compact versions of the (unrenormalised) one-loop Euler-Heisenberg (EH) and Weisskopf (WK) Lagrangians:

$$\mathcal{L}^{(1)}_{\rm EH} = -2\int_0^\infty \frac{dT}{T}(4\pi i T)^{-\frac{D}{2}}\,{\rm e}^{-im^2T}\,{\rm det}^{-\frac{1}{2}}\Big[\frac{\tanh\mathcal{Z}}{\mathcal{Z}}\Big]\,, \tag{23}$$

$$\mathcal{L}^{(1)}_{\rm WK} = \int_0^\infty \frac{dT}{T}(4\pi i T)^{-\frac{D}{2}}\,{\rm e}^{-im^2T}\,{\rm det}^{-\frac{1}{2}}\Big[\frac{\sinh\mathcal{Z}}{\mathcal{Z}}\Big]\,. \tag{24}$$

One may write the determinants in terms of the two standard secular invariants of the Maxwell field to show equivalence with the Schwinger proper-time representation these Lagrangians. For worldline calculations of these Lagrangians at higher loop order see, for example, [37,41,43–46].

3.2. N-photon amplitudes

The function forms of the master formula (12) and the spin-orbit decomposition representation of the one-loop N-photon amplitudes (18) remain valid in a constant background up to the replacements of normalisation and Green functions. For instance, for scalar QED we find [37,40]

$$\Gamma_{\rm scal}[k_1,\varepsilon_1;\ldots;k_N,\varepsilon_N] = (-ie)^N(2\pi)^D\delta^D\Big(\sum_{i=1}^N k_i\Big)\int_0^\infty \frac{dT}{T}(4\pi i T)^{-\frac{D}{2}}\,{\rm det}^{-\frac{1}{2}}\Big[\frac{\sinh\mathcal{Z}}{\mathcal{Z}}\Big]{\rm e}^{-im^2T}$$
$$\prod_{i=1}^N\int_0^T d\tau_i\,{\rm e}^{\frac{i}{2}\sum_{i,j=1}^N\big[k_i\cdot\mathcal{G}_{Bij}\cdot k_j - 2i\varepsilon_i\cdot\dot{\mathcal{G}}_{Bij}\cdot k_j + \varepsilon_i\cdot\ddot{\mathcal{G}}_{Bij}\cdot\varepsilon_j\big]}\Big|_{{\rm lin}\;\varepsilon_1\ldots\varepsilon_N}\,. \tag{25}$$

Expanding the Bern-Kosower exponent to multi-linear order will define polynomials $\mathcal{P}_N(\dot{\mathcal{G}},\ddot{\mathcal{G}})$ in the background; the extension to the spin-orbit decomposition in spinor QED should be clear.

Amongst the various applications of these formulae we mention worldline calculations of vacuum polarisation tensors [47,48], to effects in a constant magnetic field [49,50], and more recently to rederiving and extending the lately discovered one-particle-reducible contributions to the EHL starting at two-loop order in [51–54] which gave a streamlined proof that such contributions can be generated by differentiating lower order, irreducible diagrams.

4. Worldline approach in plane wave backgrounds

In this section we follow [7] by incorporating plane wave backgrounds into the worldline framework. As indicated in the introduction we are motivated by models of intense laser environments and the efficiencies often gained with first quantised techniques. This work also extends [8] and clarifies important aspects of their analysis of helicity flip on the worldline. Indeed, since their work used a semi-classical analysis, which is usually exact only for Gaussian path integrals, it had been unclear how to expose this structure in a more general framework.

4.1. Plane wave fields

Our general setup is to define the plane wave background by a vector potential depending on a null vector, $n^2 = 0$, so that $e\bar{\rm A}(x) = {\rm a}(n\cdot x)$ with ${\rm a}(-\infty) = 0$. The null vector defines light-cone coordinates $x^+ := n\cdot x$ and $x^-, {\bf x}^\perp$ which imply the inner product

$$k\cdot x = -k^+x^- - k^-x^+ + {\bf x}^\perp\cdot{\bf x}^\perp\,. \tag{26}$$

We will assume light-front gauge, $n\cdot a = 0$, and later on it will be convenient to specify the gauge of external photons to satisfy $n\cdot\varepsilon_i = 0$ (valid for arbitrary momenta).

The worldline representation of the one-loop N-photon amplitudes in this background is

$$\Gamma[k_1,\varepsilon_1;\ldots;k_N,\varepsilon_N] = -\frac{1}{2}(-ie)^N \int_0^\infty \frac{dT}{T}\mathrm{e}^{-im^2T} \oint_{PBC} \mathscr{D}x(\tau)\mathrm{e}^{i\int_0^T d\tau\,\left[\frac{\dot{x}^2}{4}-\dot{x}^\mu a_\mu(x^+)\right]} \oint_{ABC} \mathscr{D}\psi(\tau)\,\mathrm{e}^{i\int_0^T d\tau\,\left[\frac{i}{2}\psi\cdot\dot{\psi}+i\psi^\mu f_{\mu\nu}(x^+)\psi^\mu\right]} \prod_{i=1}^N V^\gamma[k_i,\varepsilon_i]\,, \tag{27}$$

where $f_{\mu\nu}(x^+) = n_\mu a'_\nu(x^+) - a'_\mu(x^+)n_\nu$. Inspecting the x^+ dependence of the plane wave background potential it is clear that the integrals over x^- and $\mathbf{x}^\perp$ go through as before – in particular integrating over x_0^+ and $\mathbf{x}_0^\perp$ gives momentum conservation in these directions.

The integral over x^+ seems to be more challenging due to the awkward $q(\tau)$ dependence of $a_\mu(x^+) \equiv a_\mu(x_0^+ + q^+(\tau))$, but it turns out we can achieve a resummation that reveals the true Gaussian nature of the path integral. To see this, for the orbital coupling we expand

$$\mathrm{e}^{-i\int_0^T d\tau\,\dot{q}\cdot a(x_0^+ + q^+(\tau))} = 1 - i\int_0^T d\tau\,\dot{q}\cdot a(x_0^+ + q^+(\tau)) + \frac{(-i)^2}{2!}\int_0^T d\tau\,\dot{q}\cdot a(x_0^+ + q^+)\int_0^T d\tau'\,\dot{q}'\cdot a(x_0^+ + q'^+) + \cdots\,, \tag{28}$$

which will enter in the expectation value ($q_i \equiv q(\tau_i)$ and $\psi_i \equiv \psi(\tau_i)$ etc)

$$\left\langle \mathrm{e}^{-i\int_0^T d\tau\,\dot{q}\cdot a(x_0^+ + q^+(\tau))} \prod_{i=1}^N V^\gamma[k_i,\varepsilon_i] \right\rangle = \left\langle \mathrm{e}^{-i\int_0^T d\tau\,\dot{q}\cdot a(x_0^+ + q^+(\tau))} \prod_{i=1}^N \left[\varepsilon_i\cdot\dot{q}_i + \psi_i\cdot f_i\cdot\psi_i\right]\mathrm{e}^{ik_i\cdot(x_{0i}+q_i)} \right\rangle. \tag{29}$$

Considering the possible Wick contractions of the field q, we see that for a non-vanishing result

- We cannot contract a $q_m^+ = n\cdot q_m$ with another such factor because $n^2 = 0$.
- We cannot contract a $q_m^+ = n\cdot q_m$ with a $\dot{q}_l\cdot a$ because $n\cdot a = 0$.

This leaves the only remaining option, that each $n\cdot q_m$ must be contracted with the external photon vertex operators. As shown in [7] this implies a replacement in the argument of a_μ

$$q_m^+ \longrightarrow n\cdot\sum_{i=1}^N\left[-iG_{mi}k_i + \dot{G}_{mi}\varepsilon_i\right] \tag{30}$$

and the resummations that remove the path integration variable from the interaction exponential

$$a_\mu(x^+) \longrightarrow a_\mu(\tau) \equiv a_\mu\Big(x_0^+ + \sum_{i=1}^N\left[-iG_{mi}k_i^+ + \dot{G}_{mi}\varepsilon_i^+\right]\Big)\,,\quad \mathrm{e}^{-i\int_0^T d\tau\,\dot{q}\cdot a(x_0^+ + q^+(\tau))} \longrightarrow \mathrm{e}^{-i\int_0^T d\tau\dot{q}\cdot a(\tau)}\,, \tag{31}$$

and render the integral in Gaussian form. This explains why the semi-classical solution in [8] obtained by expanding about the trajectory satisfying the classical equation of motion was exact.

4.2. Master formula

Having removed the path integration variable from the argument of a we can compute the N-photon amplitudes with the simple procedure of completing the square in the exponent of

$$\int_{\mathrm{SI}} \mathscr{D}q(\tau)\,\mathrm{e}^{i\int_0^T d\tau\,\left[\frac{\dot{q}^2}{4} - \dot{q}\cdot a(\tau) + \sum_{i=1}^N\left(k_i\cdot q_i - i\varepsilon_i\cdot\dot{q}_i\right)\right]}\,. \tag{32}$$

The resulting effective action is most conveniently expressed via *worldline averages* defined by[1]

$$\langle\langle f\rangle\rangle := \frac{1}{T}\int_0^T d\tau f(\tau)\,, \tag{33}$$

in terms of which the analysis of [7] yields the Master Formula for scalar QED

$$\begin{aligned}\Gamma_{\rm scal}[k_1,\varepsilon_1;\ldots;k_N,\varepsilon_N] &= (-ie)^N(2\pi)^3\delta\Big(\sum_{i=1}^N k_i^1\Big)\delta\Big(\sum_{i=1}^N k_i^2\Big)\delta\Big(\sum_{i=1}^N k_i^+\Big)\int_{-\infty}^{\infty}dx_0^+ \mathrm{e}^{-ix_0^+\sum_{i=1}^N k_i^-}\\ &\times\int_0^\infty\frac{dT}{T}(4\pi iT)^{-\frac{D}{2}}\prod_{i=1}^N\int_0^T d\tau_i \mathrm{e}^{\frac{i}{2}\sum_{i,j=1}^N\left[G_{ij}k_i\cdot k_j-2i\dot G_{ij}\varepsilon_i\cdot k_j+\ddot G_{ij}\varepsilon_i\cdot\varepsilon_j\right]}\\ &\times\,\mathrm{e}^{-i\left(m^2+\langle\langle a^2\rangle\rangle-\langle\langle a\rangle\rangle^2\right)T+2i\sum_{i=1}^N k_i\cdot\left(I(\tau_i)-\langle\langle I\rangle\rangle\right)+2\sum_{i=1}^N\left(a(\tau_i)-\langle\langle a\rangle\rangle\right)\cdot\varepsilon_i}\Big|_{\varepsilon_1\cdots\varepsilon_N}\end{aligned} \tag{34}$$

where we also introduced the periodic integral of the background potential,

$$I_\mu(\tau)\equiv\int_0^\tau d\tau'\big(a_\mu(\tau')-\langle\langle a_\mu\rangle\rangle\big)\,. \tag{35}$$

We can be more explicit by imposing the further gauge condition $n\cdot\varepsilon_i=0$ which eliminates the polarisations from the argument of $a(\tau)$ in (31) and facilitates the projection onto the multi-linear sector, realised with the help of new polynomials $\mathfrak{P}_N$ defined now by

$$(-i)^N\mathfrak{P}_N\equiv \mathrm{e}^{\frac{i}{2}\sum_{i,j=1}^N\left(-2i\dot G_{ij}\varepsilon_i\cdot k_j+\ddot G_{ij}\varepsilon_i\cdot\varepsilon_j\right)+2\sum_{i=1}^N\left(a(\tau_i)-\langle\langle a\rangle\rangle\right)\cdot\varepsilon_i}\Big|_{\varepsilon_1\cdots\varepsilon_N}\,. \tag{36}$$

With this we can write

$$\begin{aligned}\Gamma_{\rm scal}[k_1,\varepsilon_1;\ldots;k_N,\varepsilon_N] &= (-e)^N(2\pi)^3\delta\Big(\sum_{i=1}^N k_i^1\Big)\delta\Big(\sum_{i=1}^N k_i^2\Big)\delta\Big(\sum_{i=1}^N k_i^+\Big)\int_{-\infty}^{\infty}dx_0^+ \mathrm{e}^{-ix_0^+\sum_{i=1}^N k_i^-} \qquad (37)\\ \times\int_0^\infty\frac{dT}{T}(4\pi iT)^{-\frac{D}{2}}&\mathrm{e}^{-i\left(m^2+\langle\langle a^2\rangle\rangle-\langle\langle a\rangle\rangle^2\right)T}\prod_{i=1}^N\int_0^T d\tau_i\,\mathfrak{P}_N\,\mathrm{e}^{\frac{i}{2}\sum_{i,j=1}^N G_{ij}k_i\cdot k_j+2i\sum_{i=1}^N k_i\cdot\left(I(\tau_i)-\langle\langle I\rangle\rangle\right)}\,.\end{aligned}$$

Note that the proper time exponent has been completed to the effective mass in the background.

4.2.1. Spinor QED For the spinor case, the spin-orbit decomposition still goes through based on the polynomials $\mathfrak{P}_N$ and a resummation of the spin interaction with the background:

$$f_{\mu\nu}(x_0^++q^+)\longrightarrow f_{\mu\nu}(\tau)\equiv n_\mu a'_\nu(\tau)-a'_\mu(\tau)n_\nu\,. \tag{38}$$

Thus Wick contractions for the spin interaction only need the Green function for the operator $\mathcal{O}_{\mu\nu}:=\frac{\eta_{\mu\nu}}{2}\partial_\tau+ia'_\mu(\tau)n_\nu-in_\mu a'_\nu(\tau)$, inverted in [7] by $\langle\psi^\mu(\tau)\psi^\nu(\tau')\rangle=\frac{1}{2}\mathfrak{G}_F^{\mu\nu}(\tau,\tau')$ with

$$\mathfrak{G}_F^{\mu\nu}(\tau,\tau')=\Big\{\delta^{\mu\nu}+2in^\mu\mathcal{J}^\nu(\tau,\tau')+2i\mathcal{J}^\mu(\tau',\tau)n^\nu+2\Big[\mathcal{J}^2(\tau,\tau')-\frac{T^2}{4}\langle\langle a'\rangle\rangle^2\Big]n^\mu n^\nu\Big\}G_F(\tau,\tau')\,, \tag{39}$$

[1] We direct the interested reader to [8] to see how to convert these to *spacetime* averages.

in terms of further periodic functions of the background field

$$J_\mu(\tau) \equiv \int_0^T d\tau' \Big(a'_\mu(\tau') - \langle\langle a'_\mu \rangle\rangle\Big) , \qquad \mathcal{J}_\mu(\tau, \tau') \equiv J_\mu(\tau) - J_\mu(\tau') - \frac{T}{2}\dot{G}(\tau, \tau')\langle\langle a'_\mu \rangle\rangle . \tag{40}$$

Then, generalising (17) to $\mathfrak{W}(f_{i_1} \ldots, f_{i_S}) := \big\langle \psi_{i_1} \cdot f_{i_1} \cdot \psi_{i_1} \cdots \psi_{i_S} \cdot f_{i_S} \cdot \psi_{i_S} \big\rangle$ with

$$\mathfrak{W}(f_{i_1} \ldots, f_{i_S}) = \sum_{\text{partitions}} (-1)^{cy} \mathfrak{G}_F(i_1 i_2 \ldots i_{n_1}) \mathfrak{G}_F(i_{n_1+1} \ldots i_{n_1+n_2}) \cdots \mathfrak{G}_F(i_{n_1+\ldots+n_{cy-1}+1} \cdots i_{n_1+\ldots+n_{cy}}) , \tag{41}$$

with $\mathfrak{G}_F(i_1 i_2 \ldots i_n) := 2^{-\delta_{n2}} \mathrm{tr}(f_{i_1} \cdot \mathfrak{G}_{F i_1 i_2} \cdot f_{i_2} \cdot \mathfrak{G}_{F i_2 i_3} \cdots f_{i_n} \cdot \mathfrak{G}_{F i_n i_1})$, and likewise (16) to

$$(-i)^{N-S} \mathfrak{P}_{NS}^{\{i_1 i_2 \ldots i_S\}} \equiv \mathrm{e}^{\frac{i}{2}\sum_{i,j=1}^N \left(-2i\dot{G}_{ij}\varepsilon_i \cdot k_j + \ddot{G}_{ij}\varepsilon_i \cdot \varepsilon_j\right) + 2\sum_{i=1}^N \left(a(\tau_i) - \langle\langle a \rangle\rangle\right) \cdot \varepsilon_i} \Big|^{\varepsilon_{i_1} = 0 = \cdots = 0 = \varepsilon_{i_S}}_{\varepsilon_{i_{S+1}} \cdots \varepsilon_{i_N}} , \tag{42}$$

we may write explicitly the elements of the spin-obit decomposition in the background

$$\Gamma_{NS}^{\{i_1 i_2 \ldots i_S\}} = -2(-e)^N (2\pi)^3 \delta\Big(\sum_{i=1}^N k_i^1\Big) \delta\Big(\sum_{i=1}^N k_i^2\Big) \delta\Big(\sum_{i=1}^N k_i^+\Big) \int_{-\infty}^{\infty} dx_0^+ \mathrm{e}^{-ix_0^+ \sum_{i=1}^N k_i^-} \int_0^\infty \frac{dT}{T} (4\pi i T)^{-\frac{D}{2}}$$
$$\times\, \mathrm{e}^{-i\left(m^2 + \langle\langle a^2 \rangle\rangle - \langle\langle a \rangle\rangle^2\right)T} \prod_{i=1}^N \int_0^T d\tau_i\, \mathfrak{W}(f_{i_1}, \ldots, f_{i_S}) \mathfrak{P}_{NS}^{\{i_1 i_2 \ldots i_S\}} \mathrm{e}^{\frac{i}{2}\sum_{i,j=1}^N G_{ij} k_i \cdot k_j + 2i \sum_{i=1}^N k_i \cdot \left(I(\tau_i) - \langle\langle I \rangle\rangle\right)} . \tag{43}$$

In [7] the Master Formulae for both scalar and spinor QED are expanded to order $N = 2$ photons, first working completely off-shell, before specialising to the on-shell case and finally confirming the amplitude for helicity flip studied using the semi-classical approach in [8].

5. Conclusion

This contribution has reviewed one-loop photon scattering amplitudes in the Worldline Formalism of QFT in vacuum and in a constant and later plane wave backgrounds. We presented Master Formulae for N-photon amplitudes expected to streamline calculations, especially for higher N, since they combine various Feynman diagrams and the propagators in the loop into compact integral representations that avoid fixing a specific ordering or integrate explicitly over Volkov functions. It also has the advantage of being a unifying description for the scalar and spinor theories which appear more similar in this formalism than in the standard approach.

In ongoing work we are analysing the application of the Master Formulae for the $N = 3$ (photon-splitting) and $N = 4$ (light-by-light scattering) amplitudes in the plane wave background. The formalism is also being extended to the open-line case to describe the scalar and spinor propagators in the plane wave background, building upon [33, 34, 42].

Acknowledgments

The authors thank Anton Ilderton and Antonino di Piazza for helpful correspondence. JPE acknowledges funding through a CIC–U.M.S.N.H project.

References

[1] Ritus V 1970 *Sov. Phys. JETP*
[2] Narozhnyi N B 1980 *Phys. Rev. D* **21** 1176–1183
[3] Strassler M J 1992 *Nucl. Phys. B* **385** 145–184

[4] Schubert C 2001 *Phys. Rept.* **355** 73–234 (*Preprint* hep-th/0101036)
[5] Corradini O, Schubert C, Edwards J P and Ahmadiniaz N 2015 Spinning Particles in Quantum Mechanics and Quantum Field Theory (*Preprint* 1512.08694)
[6] Edwards J P and Schubert C 2019 Quantum mechanical path integrals in the first quantised approach to quantum field theory (*Preprint* 1912.10004)
[7] Edwards J P and Schubert C 2021 *Phys. Lett. B* **822** 136696 (*Preprint* 2105.08173)
[8] Ilderton A and Torgrimsson G 2016 *Phys. Rev. D* **93**(8) 085006
[9] Dittrich W and Gies H 2000 *Probing the Quantum Vacuum. Perturbative Effective Action Approach in Quantum Electrodynamics and Its Application* vol 166 (Springer Tracts Mod. Phys.)
[10] Fradkin E, Gitman D and Shvartsman S 1991 *Quantum Electrodynamics with Unstable Vacuum* (Springer)
[11] Furry W H 1951 *Phys. Rev.* **81** 115–124
[12] Seipt D 2017 Volkov States and Non-linear Compton Scattering in Short and Intense Laser Pulses *Quantum Field Theory at the Limits: from Strong Fields to Heavy Quarks* pp 24–43 (*Preprint* 1701.03692)
[13] Schwinger J S 1951 *Phys. Rev.* **82** 664–679
[14] Nikishov A I and Ritus V I 1964 *Sov. Phys. JETP* **19** 529–541
[15] Nikishov A I and Ritus V I 1964 *Sov. Phys. JETP* **19** 1191–1199
[16] Baier R and Breitenlohner P 1967 *Nuovo Cim. B* **47** 117–120
[17] Batalin I A and Fradkin E S 1970 *Teor. Mat. Fiz.* **5** 190–218
[18] Adler S L 1971 *Annals Phys.* **67** 599–647
[19] Neville R A and Rohrlich F 1971 *Phys. Rev. D* **3** 1692–1707
[20] Tsai W y 1974 *Phys. Rev. D* **10** 1342
[21] Baier V N, Katkov V M and Strakhovenko V M 1974 *Zh. Eksp. Teor. Fiz.* **67** 453
[22] Becker W and Mitter H 1975 *J. Phys. A* **8** 1638–1657
[23] De Raad L L, Dass N D H and Milton K A 1974 *Phys. Rev. D* **9** 1041–1053
[24] Urrutia L F 1978 *Phys. Rev. D* **17** 1977
[25] Morozov D A, Narozhnyi N B and Ritus V I 1981 *Sov. Phys. JETP* **53** 1103
[26] Di Piazza A, Milstein A I and Keitel C H 2007 *Phys. Rev. A* **76** 032103 (*Preprint* 0704.0695)
[27] Bragin S and Di Piazza A 2020 *Phys. Rev. D* **102** 116012 (*Preprint* 2003.02231)
[28] Di Piazza A and Lopez-Lopez M A 2020 *Phys. Rev. D* **102** 076018 (*Preprint* 2006.15370)
[29] Feynman R P and Gell-Mann M 1958 *Phys. Rev.* **109**(1) 193–198
[30] Hostler L C 1985 *J. Math. Phys.* **26** 1348–1354
[31] Feynman R P 1950 *Phys. Rev.* **80** 440–457
[32] Feynman R P 1951 *Phys. Rev.* **84** 108–128
[33] Ahmadiniaz N, Banda Guzmán V M, Bastianelli F, Corradini O, Edwards J P and Schubert C 2020 *JHEP* **08** 049 (*Preprint* 2004.01391)
[34] Ahmadiniaz N, Guzman V M B, Bastianelli F, Corradini O, Edwards J P and Schubert C 2021 (*Preprint* 2107.00199)
[35] Bern Z and Kosower D A 1991 *Nucl. Phys. B* **362** 389–448
[36] Bern Z and Kosower D A 1992 *Nucl. Phys. B* **379** 451–561
[37] Reuter M, Schmidt M G and Schubert C 1997 *Annals Phys.* **259** 313–365 (*Preprint* hep-th/9610191)
[38] Schmidt M G and Schubert C 1993 *Phys. Lett. B* **318** 438–446 (*Preprint* hep-th/9309055)
[39] Gusynin V P and Shovkovy I A 1999 *J. Math. Phys.* **40** 5406–5439 (*Preprint* hep-th/9804143)
[40] Shaisultanov R 1996 *Phys. Lett. B* **378** 354–356 (*Preprint* hep-th/9512142)
[41] Kors B and Schmidt M G 1999 *Eur. Phys. J. C* **6** 175–182 (*Preprint* hep-th/9803144)
[42] Ahmad A, Ahmadiniaz N, Corradini O, Kim S P and Schubert C 2017 *Nucl. Phys. B* **919** 9–24
[43] Fliegner D, Reuter M, Schmidt M G and Schubert C 1997 *Theor. Math. Phys.* **113** 1442–1451
[44] Dunne G V and Schubert C 2002 *JHEP* **08** 053 (*Preprint* hep-th/0205004)
[45] Dunne G V and Schubert C 2002 *JHEP* **06** 042 (*Preprint* hep-th/0205005)
[46] Huet I, McKeon D G C and Schubert C 2010 *JHEP* **12** 036 (*Preprint* 1010.5315)
[47] Dittrich W and Shaisultanov R 2000 *Phys. Rev. D* **62** 045024 (*Preprint* hep-th/0001171)
[48] Schubert C 2000 *Nucl. Phys. B* **585** 407–428 (*Preprint* hep-ph/0001288)
[49] Adler S L and Schubert C 1996 *Phys. Rev. Lett.* **77** 1695–1698 (*Preprint* hep-th/9605035)
[50] McKeon D and Sherry T 1994 *Modern Physics Letters A* **09** 2167–2178
[51] Edwards J P and Schubert C 2017 *Nucl. Phys. B* **923** 339–349 (*Preprint* 1704.00482)
[52] Ahmadiniaz N, Bastianelli F, Corradini O, Edwards J P and Schubert C 2017 *Nucl. Phys. B* **924** 377–386 (*Preprint* 1704.05040)
[53] Ahmadiniaz N, Edwards J P and Ilderton A 2019 *JHEP* **05** 038 (*Preprint* 1901.09416)
[54] Ahmadiniaz N, Bastianelli F, Corradini O, Edwards J P and Schubert C 2020 Reducible contributions to the propagator and EHL in background fields *19th Lomonosov Conference* (*Preprint* 2003.02508)

Journal of Physics: Conference Series **2249** (2022) 012020 doi:10.1088/1742-6596/2249/1/012020

Generalized Gelfand-Dikii equation for fermionic Schwinger pair production

N Ahmadiniaz[1], **S P Kim**[2,3] **and C Schubert**[4,5]

[1]Helmholtz-Zentrum Dresden-Rossendorf, Bautzner Landstraße 400, 01328 Dresden, Germany
[2]Department of Physics, Kunsan National University, Kunsan 54150, Korea
[3]Center for Relativistic Laser Science, Institute for Basic Science, Gwangju 61005, Korea
[4]Instituto de Física y Matemáticas Universidad Michoacana de San Nicolás de Hidalgo Edificio C-3, Apdo. Postal 2-82 C.P. 58040, Morelia, Michoacán, Mexico
[5]Centro Internacional de Ciencias A.C. Campus UNAM-UAEM Cuernavaca Mor. Mexico C.P. 62100

E-mail: n.ahmadiniaz@hzdr.de, sangkim@kunsan.ac.kr, christianschubert137@gmail.com

Abstract. Schwinger pair creation in a purely time-dependent electric field can be reduced to an effective quantum mechanical problem using a variety of formalisms. Here we develop an approach based on the Gelfand-Dikii equation for scalar QED, and on a generalization of that equation for spinor QED. We discuss a number of solvable special cases from this point of view. In previous work, two of the authors had shown for the scalar case how to use the well-known solitonic solutions of the KdV equation to construct Pöschl-Teller like electric fields that do not pair create at some fixed but arbitrary momentum. Here, we present numerical evidence that this construction can be adapted to the fermionic case by a mere change of parameters.

1. Introduction

The case of a purely time-dependent field is, despite of the idealized character of such a field, a popular testing ground for approaches to Schwinger pair creation in strong-field QED. This is because it can effectively be reduced to a quantum-mechanical time evolution problem, and thus is relatively amenable to exact analytical treatments. Moreover, this time evolution can be formulated in many equivalent ways (see, e.g., [1]). Here we focus on an approach based on the Gelfand-Dikii equation, which offers an interesting link to the Korteweg-de-Vries ('KdV') equation. In previous work by two of the authors [2] this connection was used for the construction of electric fields in scalar QED that are related to the well-known solitonic solutions of that equation, and have the property that they do *not* pair create at some fixed reference momentum. Here we extend this approach to the spinor QED case. We find an evolution equation describing the pair creation of spin half particles that can be considered as a natural generalization of the Gelfand-Dikii equation, and we present numerical evidence that also the concept of solitonic electric fields possesses some extension to the spinor QED case.

2. Basic set-up

Let us consider a purely time-dependent electric field $\mathbf{E}(t)$ pointing into a fixed direction, and its vector potential in temporal gauge $A_0 = 0, \dot{\mathbf{A}}(t) = -\mathbf{E}(t)$. We further assume that the field is localized in time, $E(\infty) = E(-\infty) = 0$, with a finite integral $\int_{-\infty}^{\infty} dt\, E(t) = A(-\infty) - A(\infty)$.

3. Review of the scalar QED case

3.1. The mode equation

The Hamiltonian of a charged field with mass m and charge q in a purely time-dependent electric field can be decomposed into an infinite number of harmonic oscillators with time-dependent frequencies,

$$H(t) = \sum_{\mathbf{k}} \pi^*_{\mathbf{k}}\pi_{\mathbf{k}} + \omega^2_{\mathbf{k}}(t)\phi^*_{\mathbf{k}}\phi_{\mathbf{k}}, \tag{1}$$

where

$$\omega^2_{\mathbf{k}}(t) = \Pi^2 + \mu^2, \quad \Pi = k_{\|} - qA_{\|}, \quad \mu^2 = m^2 + \mathbf{k}^2_{\perp} \tag{2}$$

and $k_{\|}, k_{\perp}$ denote the components of the canonical three-momentum along the field resp. perpendicular to it. Since we assume that the field integral is finite, we can define the limits

$$\omega^{\pm}_{\mathbf{k}} \equiv \lim_{t\to\pm\infty} \omega_{\mathbf{k}}(t)\,. \tag{3}$$

The simplicity of the purely time-dependent electric field case is due to the fact that the individual modes $\phi_{\mathbf{k}}$ obey a time-dependent oscillator equation (" mode equation")

$$\ddot{\phi}_{\mathbf{k}}(t) + \omega^2_{\mathbf{k}}(t)\phi_{\mathbf{k}}(t) = 0\,. \tag{4}$$

Canonical quantization of the scalar field leads to the Wronskian constraint

$$\mathrm{Wr}[\phi_{\mathbf{k}}, \phi^*_{\mathbf{k}}] \equiv \phi_{\mathbf{k}}\dot{\phi}^*_{\mathbf{k}} - \dot{\phi}_{\mathbf{k}}\phi^*_{\mathbf{k}} = i\,. \tag{5}$$

Using Bogolyubov theory, from a solution of the mode equation with index $\mathbf{k}$ one can get the density of created pairs $\mathcal{N}_{\mathbf{k}}(t)$

$$\mathcal{N}_{\mathbf{k}}(t) = \frac{|\dot{\phi}_{\mathbf{k}}(t)|^2 + \omega^2_{\mathbf{k}}(t)|\phi_{\mathbf{k}}(t)|^2}{2\omega_{\mathbf{k}}(t)} - \frac{1}{2}\,. \tag{6}$$

Usually one assumes that initially there are no particles present, $\lim_{t\to-\infty}\mathcal{N}_{\mathbf{k}}(t) = 0$. The relevant solution of the mode equation will then obey (up to an irrelevant phase factor)

$$\phi_{\mathbf{k}}(t) \overset{t\to-\infty}{\longrightarrow} \frac{1}{\sqrt{2\omega_i}}\,\mathrm{e}^{-i\omega_i t}\,. \tag{7}$$

No direct physical meaning can be ascribed to $\mathcal{N}_{\mathbf{k}}(t)$ at intermediate times, since its definition is inherently ambiguous on account of its dependence on the choice of an instantaneous adiabatic basis [1].

3.2. The quantum Vlasov equation

Alternatively, one can use the Quantum Vlasov Equation [3, 4, 5, 6]. This is an evolution equation at fixed $\mathbf{k}$ for the density of pairs $\mathcal{N}_{\mathbf{k}}(t)$:

$$\dot{\mathcal{N}}_{\mathbf{k}}(t) = \frac{\dot{\omega}_{\mathbf{k}}(t)}{2\omega_{\mathbf{k}}(t)}\int_{t_0}^{t} dt' \frac{\dot{\omega}_{\mathbf{k}}(t')}{\omega_{\mathbf{k}}(t')}(1 + 2\mathcal{N}_{\mathbf{k}}(t'))\cos\Big[2\int_{t'}^{t} dt''\omega_{\mathbf{k}}(t'')\Big]\,. \tag{8}$$

Here t_0 is the initial time, usually $-\infty$, and

$$\omega^2_{\mathbf{k}}(t) = (k_{\|} - qA_{\|}(t))^2 + \mathbf{k}^2_{\perp} + m^2\,. \tag{9}$$

$\mathcal{N}_{\mathbf{k}}(t)$ is zero at $t = -\infty$, and for $t \to \infty$ turns into the density of created pairs with fixed momentum $\mathbf{k}$.

3.3. The time-like Sauter field

As an example, let us consider the time-like Sauter field, defined by

$$E(t) = -\dot{A} = -E_0 \operatorname{sech}^2(t/\tau) \tag{10}$$

which we can realize by

$$A^\mu = (0, 0, 0, E_0\tau(1 + \tanh(t/\tau))) \,. \tag{11}$$

The exact solution of the mode equation for this field, obeying the initial conditions, is given by

$$\phi_{\mathbf{k}}(t) = \frac{1}{\sqrt{2\omega_i\, e^{\pi\omega_i\tau}}} z^{-\frac{i}{2}\omega_i\tau}(1-z)^{\frac{1}{2}+i\lambda^{sc}}\, {}_2F_1(\alpha^{sc}, \beta^{sc}; \gamma; z) \tag{12}$$

where $z \equiv -\,e^{\frac{2t}{\tau}}$, and

$$\begin{aligned}
\lambda^{sc} &\equiv \sqrt{(eE_0\tau^2)^2 - \frac{1}{4}}\,, \\
\alpha^{sc} &\equiv \frac{1}{2} - \frac{i}{2}\Big[\tau(\omega_i - \omega_f) - 2\lambda^{sc}\Big]\,, \\
\beta^{sc} &\equiv \frac{1}{2} - \frac{i}{2}\Big[\tau(\omega_i + \omega_f) - 2\lambda^{sc}\Big]\,, \\
\gamma &\equiv 1 - i\tau\omega_i\,, \\
\omega_i &= \omega_0 \equiv \sqrt{\mathbf{k}^2 + m^2}\,, \\
\omega_f &= \sqrt{(k_\parallel - 2qE_0\tau)^2 + \mu^2}\,.
\end{aligned} \tag{13}$$

Using the solution of the mode equation in (6), we plot $\mathcal{N}_{\mathbf{k}}(t)$ for $E_0 = 1$, $\tau = 3$, and $k_\parallel = \omega_0 = 1$ (Fig. 1)

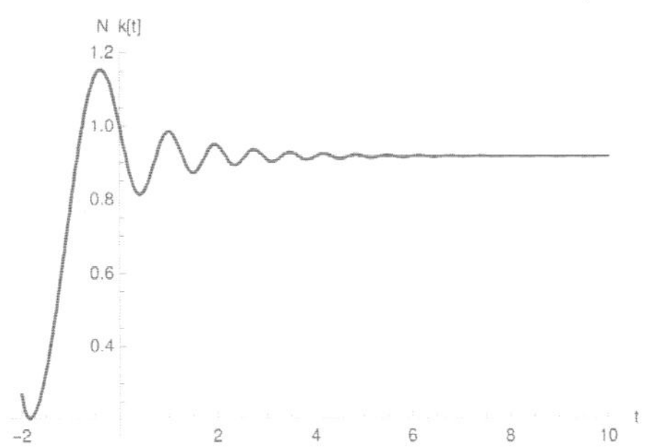

Figure 1. Plot of the pair-creation rate $\mathcal{N}_{\mathbf{k}}(t)$ for the Sauter field with $E_0 = 1$, $\tau = 3$, $k_\parallel = \omega_0 = 1$.

Only its asymptotic value for $t \to \infty$ is physically meaningful.

3.4. The Gelfand-Dikii equation

To obtain $\mathcal{N}_{\mathbf{k}}(t)$, it is sufficient to know $|\phi_{\mathbf{k}}(t)|$. This suggests that it might be more economical to work with $G_{\mathbf{k}}(t) \equiv |\phi_{\mathbf{k}}(t)|^2$, which evolves by a linear third-order differential equation,

$$\dddot{G}_{\mathbf{k}}(t) + 4\omega_{\mathbf{k}}^2(t)\dot{G}_{\mathbf{k}}(t) + 2\big(\omega_{\mathbf{k}}^2(t)\big)^{\cdot} G_{\mathbf{k}}(t) = 0 \tag{14}$$

with the initial condition

$$\lim_{t\to-\infty} G_{\mathbf{k}}(t) = \frac{1}{2\omega_i}\,.$$

This is actually the Gelfand-Dikii equation [7]. In terms of $G_{\mathbf{k}}$, $\mathcal{N}_{\mathbf{k}}(t)$ is given by

$$\mathcal{N}_{\mathbf{k}}(t) = \frac{\ddot{G}_{\mathbf{k}}}{4\omega(t)} + \omega(t)G_{\mathbf{k}} - \frac{1}{2}\,. \tag{15}$$

For example, for the Sauter field one has

$$G_{\mathbf{k}} \;=\; \frac{1-z}{2\omega_0\,\mathrm{e}^{\pi\omega_0\tau}}\Big|{}_2F_1(\alpha^{\mathrm{sc}},\beta^{\mathrm{sc}};\gamma;z)\Big|^2\,. \tag{16}$$

3.5. Solitonic electric fields

The Gelfand-Dikii equation relates to the resolvent of the mode equation, and through it to the (generalized) KdV equation [7]. In [2, 8] the following infinite family of vector potentials was constructed, related to the solitonic solutions of the KdV equation:

$$qA_{\|(p)}(t) \equiv \tilde{k}_\| - \sqrt{\tilde{k}_\|^2 + \frac{p(p+1)\omega_0^2}{\cosh^2(\omega_0 t)}}, \quad p = 1, 2, \ldots \tag{17}$$

where $\tilde{k}_\|$ is a fixed but arbitrary reference momentum. The corresponding squared energies at $k_\| = \tilde{k}_\|$ are of Pöschl-Teller type,

$$\omega^2_{(p)\tilde{\mathbf{k}}}(t) = \omega_0^2\Big(1 + \frac{p(p+1)}{\cosh^2(\omega_0 t)}\Big), \quad p = 1, 2, \ldots$$

All these "solitonic" electric fields do not produce pairs for $k_\| = \tilde{k}_\|$ (although they will generally do so for other values of $k_\|$). For these solitonic fields, the mode equation can be solved exactly:

$$\phi_{(p)}(t) = \frac{e^{-i\omega_0 t}}{\sqrt{2\omega_0}}\Big(1 + e^{2\omega_0 t}\Big)^{p+1}{}_2F_1\Big[1+p; -i+p+1; 1-i; -e^{2\omega_0 t}\Big]\,. \tag{18}$$

Let us show the resulting $\mathcal{N}_{\tilde{\mathbf{k}}}(t)$ for the simplest case $p = 1$:

$$\mathcal{N}_{\tilde{\mathbf{k}}}(t) \;=\; \frac{4 + \mathrm{sech}^4(\omega_0 \mathrm{t})(1 + 2\cosh(2\omega_0 \mathrm{t}))}{8\sqrt{1 + 2\mathrm{sech}^2(\omega_0 \mathrm{t})}} - \frac{1}{2}\,. \tag{19}$$

In Fig. 2 we plot this function for $\omega_0 = 1.1$.

Since $\mathcal{N}_{\tilde{\mathbf{k}}}(t)$ vanishes for $t \to \infty$ there is no pair creation at that particular momentum $\mathbf{k} = \tilde{\mathbf{k}}$. The external field excites the vacuum, but no particles materialize.

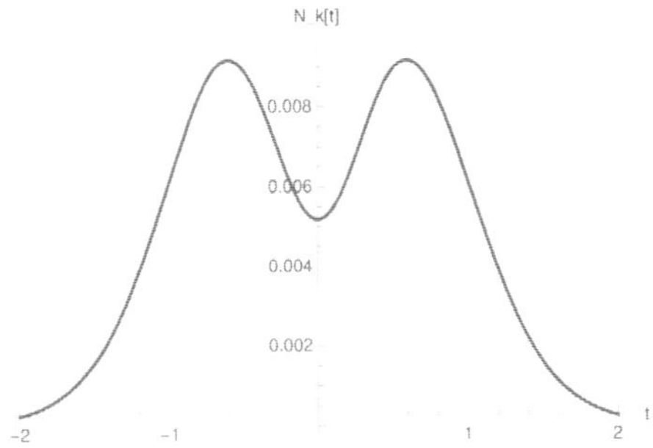

Figure 2. $\mathcal{N}_{\tilde{\mathbf{k}}}(t)$ for the $p = 1$ soliton (with $\omega_0 = 1.1$).

4. Generalization to spinor QED

4.1. The mode equation

For spin $\frac{1}{2}$ particles the mode equation becomes [9]

$$\ddot{\psi}_{\mathbf{k}}^{(\pm)}(t) + \left[\omega_{\mathbf{k}}^2(t) \mp iq\dot{A}_{\parallel}(t)\right]\psi_{\mathbf{k}}^{(\pm)}(t) = 0\,, \tag{20}$$

where the $(\pm)$ denotes the spin. In QFT, this comes from the spin-diagonalized square of the Dirac operator. In [10] it was shown that the constraint equation (5) generalizes to

$$\omega_{\mathbf{k}}^2|\psi_{\mathbf{k}}^{(\pm)}|^2 + |\dot{\psi}_{\mathbf{k}}^{(\pm)}|^2 \mp i\Pi(\dot{\psi}_{\mathbf{k}}^{(\pm)}\psi_{\mathbf{k}}^{(\pm)*} - \psi_{\mathbf{k}}^{(\pm)}\dot{\psi}_{\mathbf{k}}^{(\pm)*}) = \mu^2 \tag{21}$$

where $\Pi = k_{\parallel} - qA_{\parallel}(t)$ and that $\mathcal{N}_{\mathbf{k}}(t)$ with the usual initial condition (7) can be written in terms of the mode solution as

$$\mathcal{N}_{\mathbf{k}}(t) \;=\; \frac{1}{2\omega_{\mathbf{k}}(\omega_{\mathbf{k}} - \Pi)}\left[\omega_{\mathbf{k}}^2|\psi_{\mathbf{k}}^{(\pm)}|^2 + |\dot{\psi}^{(\pm)}|^2 \mp i\omega_{\mathbf{k}}(\dot{\psi}_{\mathbf{k}}^{(\pm)}\psi_{\mathbf{k}}^{(\pm)*} - \psi_{\mathbf{k}}^{(\pm)}\dot{\psi}^{(\pm)*})\right]. \tag{22}$$

Note that $\mathcal{N}_{\mathbf{k}}(t)$ does not depend on the spin orientiation. Combining the last two equations, we can also write

$$\mathcal{N}_{\mathbf{k}}(t) \;=\; \frac{\mu^2}{2\Pi(\omega_{\mathbf{k}} - \Pi)} - \frac{\omega_{\mathbf{k}}^2|\psi_{\mathbf{k}}^{(\pm)}|^2 + |\dot{\psi}_{\mathbf{k}}^{(\pm)}|^2}{2\omega_{\mathbf{k}}\Pi}\,. \tag{23}$$

4.2. Fermionic Quantum Vlasov Equation

There is also a fermionic generalization of the Quantum Vlasov Equation [5]

$$\dot{\mathcal{N}}_{\mathbf{k}}(t) \;=\; \frac{\mu q\dot{A}_{\parallel}(t)}{2\omega_{\mathbf{k}}^2(t)}\int_{-\infty}^{t} dt'\,\frac{\mu q\dot{A}_{\parallel}(t')}{\omega_{\mathbf{k}}^2(t')}\left(1 - 2\mathcal{N}_{\mathbf{k}}(t')\right)\cos\left[2\int_{t'}^{t} dt''\omega_{\mathbf{k}}(t'')\right]. \tag{24}$$

4.3. Fermionic Gelfand-Dikii equation

Defining $G_{\mathbf{k}}^{\pm} \equiv |\psi_{\mathbf{k}}^{\pm}|^2$ like in the scalar case and combining (20), (21), one arrives at the following generalization of the Gelfand-Dikii equation (14)

$$\dddot{G}_{\mathbf{k}}^{\pm} - \frac{E}{\Pi}\ddot{G}_{\mathbf{k}}^{\pm} + 4\omega_{\mathbf{k}}^2\dot{G}_{\mathbf{k}} + \left(2(\omega_{\mathbf{k}}^2)^{\cdot} - 4\frac{E\omega_{\mathbf{k}}^2}{\Pi}\right)G_{\mathbf{k}}^{\pm} = -2\mu^2\frac{E}{\Pi}\,. \tag{25}$$

Differentiating the mode equation one can eliminate $|\dot{\psi}^{(\pm)}|^2$ and rewrite $\mathcal{N}_{\mathbf{k}}$ in terms of $G_{\mathbf{k}}^{\pm}$:

$$\mathcal{N}_{\mathbf{k}} = \frac{1}{2} + \frac{\omega_{\mathbf{k}}}{\Pi}\left(\frac{1}{2} - G_{\mathbf{k}}^{\pm}\right) - \frac{\ddot{G}_{\mathbf{k}}^{\pm}}{4\omega_{\mathbf{k}}\Pi}\,. \tag{26}$$

4.4. Fermions in the solitonic field

A numerical analysis using the Quantum Vlasov Equation (24) suggests that the above solitonic fields (17) are non-pair creating also in the fermionic case, although for different values of the parameter p:

$$p = \frac{1}{2}\left(\pm\sqrt{m^2+m+1}-1\right), \quad m = 2, 4, 6, \ldots \tag{27}$$

As an example, let us show the result of such a numerical evaluation for the simplest cases $m = 2, p = \frac{1}{2}\left(\pm\sqrt{7}-1\right)$ (Fig. 3) and $m = 4, p = \frac{1}{2}\left(\pm\sqrt{21}-1\right)$ (Fig. 4).

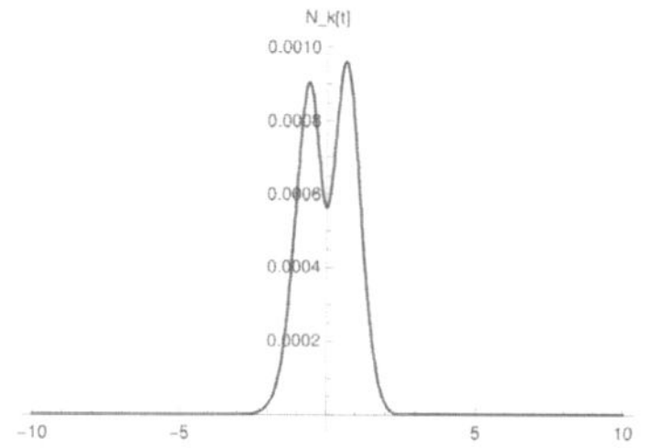

Figure 3. Fermionic $\mathcal{N}_{\tilde{\mathbf{k}}}(t)$ for the $p = \frac{1}{2}\left(\pm\sqrt{7}-1\right)$ soliton.

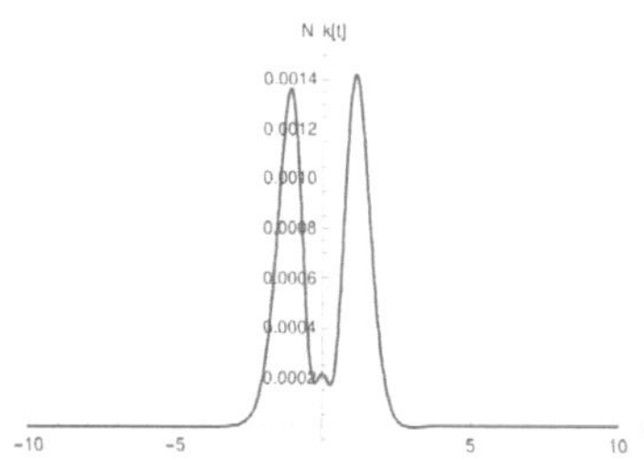

Figure 4. Fermionic $\mathcal{N}_{\tilde{\mathbf{k}}}(t)$ for the $p = \frac{1}{2}\left(\pm\sqrt{21}-1\right)$ soliton.

This should be compared to the case $p = 1$, which would not pair create in the scalar case but clearly does so here (Fig. 5):

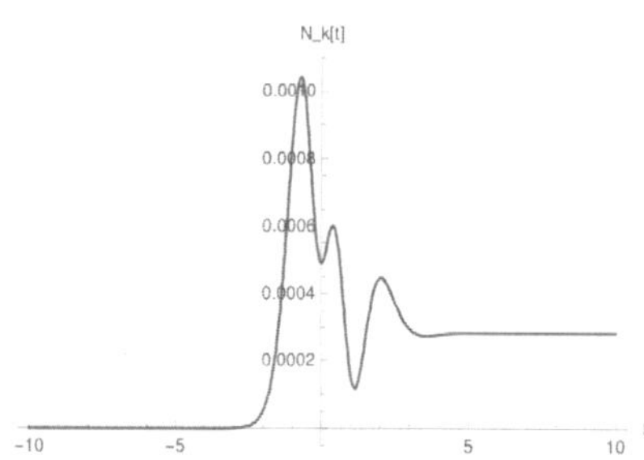

Figure 5. Fermionic $\mathcal{N}_{\tilde{\mathbf{k}}}(t)$ for the $p = 1$ soliton.

5. Summary and Outlook

To summarize,

- We have presented an evolution equation for fermions in a purely time-dependent electric field that can be seen as a fermionic generalization of the Gelfand-Dikii equation.
- Solitonic electric fields allow one to tune the pair creation rate to disappear at some prescribed longitudinal momentum for both scalars and spinors, although not for the same field.
- Thus they provide an example of spin selection in the pair creation process, and also a step towards "designer fields", that is, the construction of fields with a prescribed pair-creation spectrum [11].
- It remains to find an analytic derivation of the "magic numbers" (27).

A more extensive study of fermionic pair creation in solitonic fields is in progress [12].

Acknowledgments

The work of SPK was supported by the Institute for Basic Science (IBS) under IBS-R012-D1. CS thanks the CIC of UMSNH for financial support.

References

[1] R Dabrowski and G V Dunne, Time dependence of adiabatic particle number, Phys. Rev. D 94, 065005 (2016), arXiv:1606.00902 [hep-th].

[2] S P Kim and C Schubert, Non-adiabatic Quantum Vlasov Equation for Schwinger Pair Production, Phys .Rev. D **84** 125028 (2011), arXiv:1110.0900 [hep-th].

[3] Y Kluger, J M Eisenberg, B Svetitsky, F Cooper and E Mottola, Pair production in a strong electric field, Phys. Rev. Lett. **67** (1991) 2427.

[4] Y Kluger, J M Eisenberg, B Svetitsky, F Cooper and E Mottola, Fermion pair production in a strong electric field, Phys. Rev. D **45** (1992) 4659.

[5] S Schmidt, D Blaschke, G Röpke, S A Smolyansky, A V Prozorkevich and V D Toneev, A quantum kinetic equation for particle production in the Schwinger mechanism, Int. J. Mod. Phys. E **7** (1998) 709, arXiv: hep-ph/9809227.

[6] F Hebenstreit, R Alkofer and H Gies, Pair production beyond the Schwinger formula in time-dependent electric fields, Phys. Rev. D **78**, 061701 (2008), arXiv:0807.2785.

[7] I M Gelfand and A Dikii, Asymptotic Behavior of the Resolvent of Sturm-Liouville Equations and the Algebra of the Korteweg-De Vries Equations, Russ. Math. Surveys **30**, 5 (1975).

[8] A Huet, S P Kim and C Schubert, Vlasov equation for Schwinger pair production in a time-dependent electric field, Phys. Rev. D **90** (2014) 125033, arXiv:1411.3074 [hep-th].

[9] A B Balantekin, J E Seger and S H Fricke, Dynamical effects in pair production by electric fields, Int. J. Mod. Phys. A **6** (1991) 695.

[10] A M Fedotov, E G Gelfer, K Yu Korolev and S A Smolyansky, Kinetic equation approach to pair production by a time-dependent electric field, *Phys. Rev. D*, **83**, 025011 (2011), arXiv:1008.2098 [hep-ph].

[11] F Hebenstreit, The inverse problem for Schwinger pair production, Phys. Lett. B **753** (2016) 336, arXiv:1509.08693 [hep-ph].

[12] N Ahmadiniaz, A M Fedotov, E G Gelfer, S P Kim and C Schubert, Generalized Gelfand-Dikii equation and solitonic electric fields for fermionic Schwinger pair production, in preparation.

Axion Electrodynamics in Strong Magnetic Backgrounds

S Villalba-Chávez[1], A E Shabad[2,3], C Müller[1]

1 Institut für Theoretische Physik I, Heinrich-Heine-Universität Düsseldorf, Universitätsstr. 1, 40225 Düsseldorf, Germany
2 Lebedev Physical Institute, 53, Leninsky prospect, 119991, Moscow, Russia
3 Tomsk State University, 36, Lenin Prospekt, 634050, Tomsk, Russia

E-mail: villalba@uni-duesseldorf.de

Abstract. The overcritical regime of axion-electrodynamics (AED) is investigated. For magnetic fields larger than the characteristic scale linked to AED, quantum vacuum fluctuations due to axion-like fields can dominate over those associated with the electron-positron fields. This hypothetical regime of the dominance of AED over QED is predicted to induce strong birefringence and screening effects. We show that, if the magnetic field lines are curved, extraordinary photons could be canalized along the field direction. It is also shown that the running QED coupling depends on the magnetic field strength, and for certain energy regimes, it could be screened almost to zero, making the QED building blocks very weakly interacting between each other. The impact of this phenomenon on the radiation mechanism of pulsars is discussed.

1. Introduction

The Peccei-Quinn mechanism is among the most appealing solutions to the so-called strong CP problem. However, it predicts the existence of a pseudoscalar Nambu-Goldstone boson which has not been observed so far: the axion [1–3]. Constraints on the invisible axion [4–9]– or more general – axion-like particles are deduced from their potential astro-cosmological consequences which are not reflected accordingly by the current observational data of our universe [10–13]. A basic assumption underlying this line of argument is that the interplay between axion-like particles and the well established Standard Model sector – photons in first place – is extremely feeble. As a consequence, axion-like particles produced copiously in the core of stars via the Primakoff effect might escape from there almost freely, constituting a leak of energy that accelerates the cooling of the star and, thus, shortens its lifetime. Therefore, the number of red giants in the helium-burning phase in globular clusters should diminish considerably. That this fact does not take place – at least not significantly – constraints the axion-diphoton coupling g to lie below $g \lesssim 10^{-10}$ GeV^{-1} for axion masses m below the keV scale [14–16].

Precisely on the surface of stellar objects identified as neutron stars [17–20] and magnetars, magnetic fields as large as $B \sim \mathcal{O}(10^{13} - 10^{15})$ G are predicted to exist. As these strengths are bigger than the characteristic QED scale $B_0 = 4.42 \times 10^{13}$ G [21–23], such astrophysical scenarios are propitious for the realization of a variety of yet unobserved quantum processes, which are central in our current understanding of the nature and origin of the pulsar radiation. Notable among them are: the photon capture effect by the magnetic field owing to the resonant

Journal of Physics: Conference Series **2249** (2022) 012021 doi:10.1088/1742-6596/2249/1/012021

behavior [24] of the vacuum polarization in QED [25,26] and the photon splitting effect [27–29]. Clearly, the strong-field environments provided by these compact objects can also be favorable for an axion-like particles phenomenology – as they are in QED – contrary to what is predicted relying on the weak coupling assumption. This occurs, because the aforementioned field strengths could compensate for the weakness of the coupling and significantly stimulate quantum vacuum fluctuations of axion-like fields.

Theoretically, the problem linked to field strengths overpassing the corresponding electric $E_c = m/g$ and magnetic $B_c = m/g$ scales of AED have not yet attracted much attention, save in [30, 31], where their respective relevance has been analyzed in connection with phenomenologies predicted to take place in black holes and neutron stars. While an electric background $E \gtrsim E_c$ induces an axionic instability, for magnetic fields $B \gtrsim B_c$ the theory is stable and – as in QED – the screening and refractive properties induced by quantum vacuum fluctuations of axion-like fields are predicted to be more pronounced.

2. Refraction properties in a constant magnetic field

When the Lagrangian density of AED is embedded in the QED action and a magnetic field background is considered, quantum vacuum fluctuations of the axion-like field introduce a contribution to the polarization tensor, as shown in Fig. 1.[1] For magnetic fields larger than[2] the characteristic scale $B_c = m/g$ linked to AED, the phenomenology of the quantum vacuum could be ruled by virtual ALPs instead of fluctuations of the electron-positron field. This seems to be particularly accessible because the vacuum polarization tensor $\Pi^{\mu\nu}(x,\tilde{x})$ in AED exhibits a quadratic growth in the external field while the corresponding QED polarization tensor grows linearly [33, 34], provided $B \gg B_0$ with $B_0 = m_e^2/e \approx 4.42 \times 10^{13}$ G referring to the characteristic scale in QED. The consequence of this conjecture of axion dominance is explored in this contribution. The starting point of our analysis is the equation of motion for a small-amplitude electromagnetic wave $a_\mu(x)$ with $\partial a = 0$, modified by $\Pi^{\mu\nu}(x,\tilde{x})$

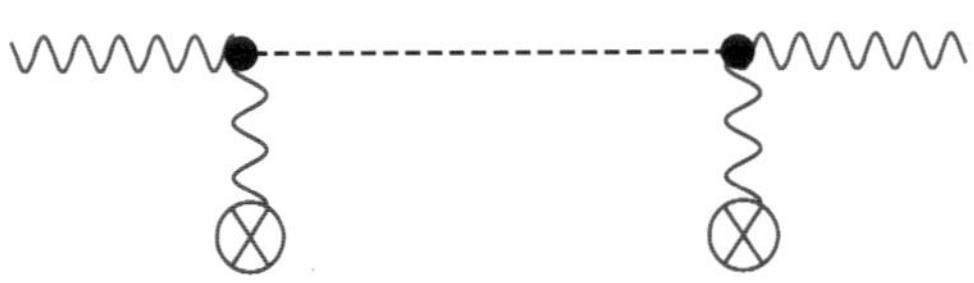

Figure 1. Diagrammatic representation of the magnetic-field dependent part of the vacuum polarization tensor mediated by quantum vacuum fluctuation of an axion-like field.

$$\Box a^\mu(x) + i\int d^4\tilde{x}\, \Pi^\mu_{\ \nu}(x,\tilde{x}) a^\nu(\tilde{x}) = 0. \tag{1}$$

Its solutions in the Fourier space are characterized by a massive mode and two massless states. Among the massless excitations, there is an ordinary mode with dispersion relation $q_{0,o} = |\boldsymbol{q}|$ and an extraordinary mode, the dispersion law of which reads [see Fig. 2]

$$q_{0,e}^2 = \boldsymbol{q}^2 + \frac{1}{2}m_*^2\left[1 - \sqrt{1 + \frac{4\mathfrak{b}^2}{1+\mathfrak{b}^2}\frac{q_\perp^2}{m_*^2}}\right], \tag{2}$$

where $m_* = m\sqrt{1+\mathfrak{b}^2}$ is the ALP mass dressed by the magnetic field parameter $\mathfrak{b} = B/B_c$. Here $q_\perp$ ($q_\parallel$) is the momentum perpendicular (parallel) to the field direction and $|\boldsymbol{q}|^2 = q_\parallel^2 + q_\perp^2$.

[1] Also a one-loop contribution is involved [32]. However, in a strong field $B \gg B_c$ it can be ignored.

[2] We use units with the Planck constant $\hbar$, the speed of light c and the vacuum permittivity ϵ_0 taken to be unity [$\hbar = c = \epsilon_0 = 1$].

In contrast to the ordinary mode, the dispersion relation of the extraordinary one depends on the magnetic field strength which reveals the axion-modified birefringent feature of the vacuum. When calculating the components of the group velocity $v_{\perp,\parallel} = \partial q_0/\partial q_{\perp,\parallel}$ linked to the extraordinary branch, we find $[q_\perp \ll m_*]$

$$v_{\parallel,e} \approx \frac{q_\parallel}{q_{0,e}},$$
$$v_{\perp,e} \approx \frac{1}{1+\mathfrak{b}^2}\frac{q_\perp}{q_{0,e}}\left[1+\frac{2\mathfrak{b}^4}{1+\mathfrak{b}^2}\frac{q_\perp^2}{m_*^2}\right]. \tag{3}$$

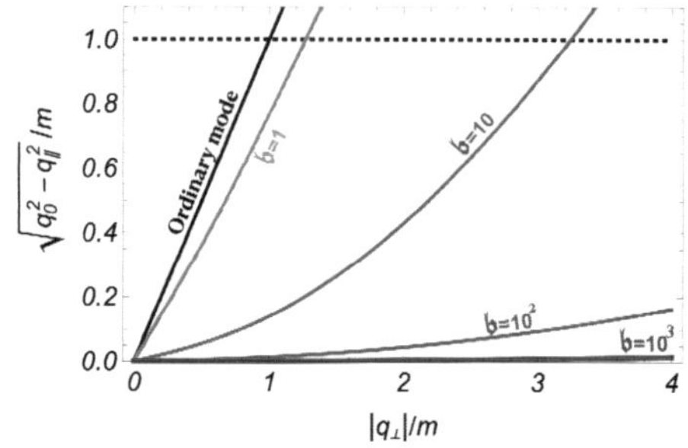

Figure 2. Dispersions curves for the ordinary (in black) and the extraordinary mode (gray, cyan, green and red) [see Eq. (2)].

As a consequence, the angle between the direction of propagation of the electromagnetic energy and the external magnetic field $\theta = \arctan(v_\perp/v_\parallel)$ does not coincide with the one between the photon momentum $\boldsymbol{q}$ and $\boldsymbol{B}$, i.e. $\vartheta = \arctan(q_\perp/q_\parallel)$. Furthermore, we observe that it tends to vanish when $q_\perp \ll m_* \approx m\mathfrak{b}$ the faster, the stronger the field is, since one has $v_{\perp,e} \to 0$ and $v_{\parallel,e} \to 1$. This implies that the group velocity of extraordinary excitations tends to be parallel to the magnetic field for hard, as well as for soft photons with $q_\perp \ll m_*$. Beyond this limit, i.e. for $q_\perp \gg m_*$, the components of the group velocity approach $v_{\perp,e} \approx q_\perp/q_{0,e}$, $v_{\parallel,e} = q_\parallel/q_{0,e}$, and the angle θ between the direction of propagation of the electromagnetic energy and $\boldsymbol{B}$ does not deviate substantially from the one formed by the momentum of the small-amplitude wave and the external field $[\theta \approx \vartheta]$.

3. Running QED coupling and modified Coulomb potential

The problem of determining how the quantum vacuum fluctuation of a pseudoscalar axion field modifies the Coulomb potential $a_C(\boldsymbol{x}) = q/(4\pi|\boldsymbol{x}|)$ reduces, to a large extent, to determine the explicit expression for the photon Green function including the vacuum polarization effect. This can be reached by inverting the operator which defines the equation of motion [see Eq. (1)]. As a consequence, the following expression for the axion-modified Coulomb potential was found [for details we refer the reader to Ref. [31]]:

$$a_0(\boldsymbol{x}) = q\int \frac{d^3q}{(2\pi)^3}\frac{e^{-i\boldsymbol{q}\cdot\boldsymbol{x}}}{\boldsymbol{q}^2}\frac{1}{1+\frac{g^2B^2q_\parallel^2}{\boldsymbol{q}^2+m^2}}, \tag{4}$$

where q is the charge of a point-like source.

Let us consider the electrostatic energy $\mathcal{U}(\boldsymbol{x}) = -ea_0(\boldsymbol{x})$ of an electron. The insertion of Eq. (4) into this formula with $q \to e$ determines how quantum vacuum fluctuations of axion-like fields in the presence of a strong magnetic background $[\mathfrak{b} \gg 1]$ modify the running QED coupling

$$\alpha_{\text{EM}}(q_\parallel^2, q_\perp^2) \approx \frac{\alpha}{1+\frac{g^2B^2q_\parallel^2}{\boldsymbol{q}^2(\boldsymbol{q}^2+m^2)}} \tag{5}$$

with $\alpha = e^2/(4\pi) \approx 137^{-1}$ defined at $\mathfrak{b} \to 0$ and $|\boldsymbol{q}| \to 0$ [21, 22]. Notably, if the dominance of AED over QED takes place, $\alpha_{\text{EM}}(\boldsymbol{q})$ is free of the characteristic Landau pole arising in QED that rises the coupling constant to an infinite value for large enough momentum. Furthermore, its dependence on the magnetic field leads to an anisotropic behavior along and transverse to $\boldsymbol{B}$. This dependence decreases the value of the running QED coupling as the magnetic field B grows while keeping the momentum fixed. Indeed, for $q_\perp^2 \ll q_\parallel^2 \ll m^2$, one finds

Journal of Physics: Conference Series **2249** (2022) 012021 doi:10.1088/1742-6596/2249/1/012021

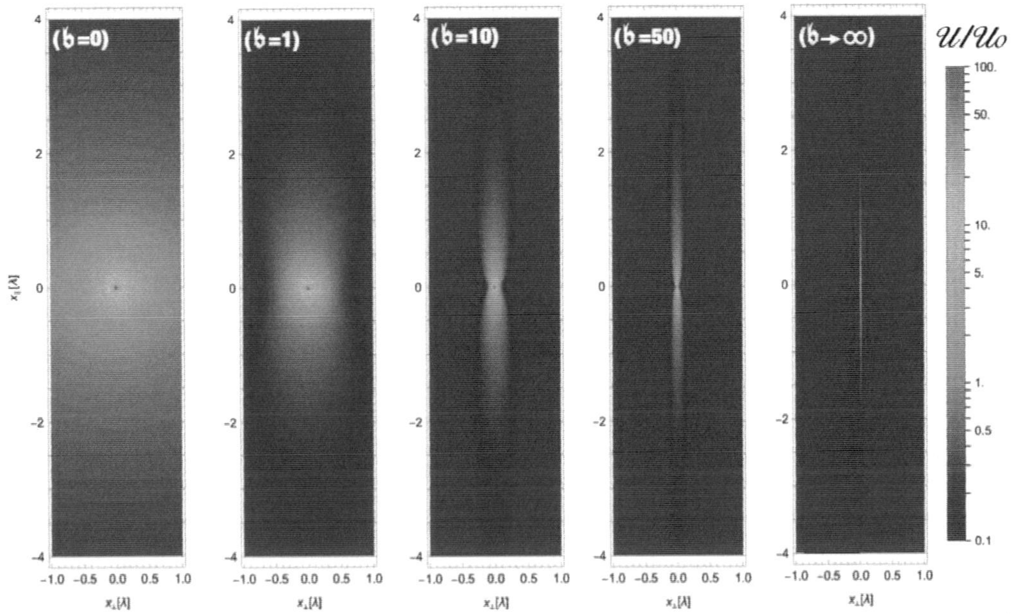

Figure 3. Electrostatic energy of an electron $\mathcal{U}(\boldsymbol{x}) = -ea_0(\boldsymbol{x})$ due to the anisotropic Coulomb potential $a_0(\boldsymbol{x})$ of a point-like static charge surrounded by quantum vacuum fluctuations of an axion-like field over a constant magnetic-like background. Here the energy is shown in units of the electron energy at the Compton wavelength of an axion $\mathcal{U}_0 = -eq/(4\pi\lambda)$ with $\lambda = 2\pi m^{-1}$.

that $\alpha_{\rm EM}(q_\parallel^2, q_\perp^2) \sim \alpha/\mathfrak{b}^2 \ll 1$ tends to be screened to zero. This phenomenon makes QED an almost "trivial" theory in the aforementioned regime, meaning that the corresponding charged matter sector and the small-amplitude electromagnetic waves turn out to be very weakly interacting between each other. This situation becomes particularly dramatic if the axion mass turns out to be larger than the first pair creation threshold, i.e. $m > 2m_e$, where m_e refers to the electron mass. Indeed, under such a restriction, radiation channels such as the recombination of pairs, cyclotron radiation and the photon splitting effect along with the main pair production mechanisms in highly-magnetized pulsars would be suppressed. This sort of quantum triviality [35–37] persists in domains other than the one discussed before [31]. Conversely, for $q_\perp^2/m^2 \ll 1 \ll \mathfrak{b}^2 \ll q_\parallel^2/m^2$ or $1 \ll \mathfrak{b}^2 \ll q_\perp^2/q_\parallel^2 \ll m^2/q_\parallel^2$ the screening effects are negligibly small and the running QED coupling approximates to its standard value $\alpha_{\rm EM}(q_\parallel^2, q_\perp^2) \sim \alpha$.

The described properties of the running electromagnetic coupling have direct implications on the Coulomb potential [see Eq. (4)]. For strong magnetic fields [$\mathfrak{b} \gg 1$] and at distances from the source much smaller than the Compton wavelength of the axion $2\pi m^{-1} \gg |\boldsymbol{x}_{\perp,\parallel}|$, the Coulomb potential of a point-like charge turns out to be of Yukawa-type in the direction perpendicular to the magnetic field while along $\boldsymbol{B}$ it follows approximately half of the Coulomb law:

$$a_0(\boldsymbol{x}_\perp, 0) \approx \frac{q}{4\pi|\boldsymbol{x}_\perp|}e^{-\frac{1}{2}m_*|\boldsymbol{x}_\perp|}, \quad a_0(0, \boldsymbol{x}_\parallel) \approx \frac{1}{2}\frac{q}{4\pi|\boldsymbol{x}_\parallel|} \tag{6}$$

Figure 3 exhibits the evolution of the electrostatic energy of an electron $\mathcal{U}(\boldsymbol{x}) = -ea_0(\boldsymbol{x})$ as the magnetic field grows gradually. This evaluation was carried out from Eq. (4). While the panel most to the left has been obtained by considering the standard Coulomb potential $a_C(\boldsymbol{x}) = q/(4\pi|\boldsymbol{x}|)$, the remaining ones were generated by taking the field parameter $\mathfrak{b} = 1$,

$\mathfrak{b} = 10$, $\mathfrak{b} = 50$ and $\mathfrak{b} \to \infty$, respectively. These results show that, as the field strength increases, the static field of an electric charge is anisotropic and tends to be squeezed towards the axis parallel to $\boldsymbol{B}$. This behavior is a clear manifestation of a dimensional reduction linked to the corresponding gauge sector,[3] and it still appears at distances for which $|\boldsymbol{x}_{\perp,\parallel}| \gg m_*^{-1}$, where the Coulomb's law acquires the form

$$a_0(\boldsymbol{x}) \approx \frac{q}{4\pi\sqrt{|\boldsymbol{x}_\perp|^2(1+\mathfrak{b}^2)+|\boldsymbol{x}_\parallel|^2}}. \tag{7}$$

Unless $|\boldsymbol{x}_\perp| = 0$, this formula tends to vanish as the external field grows and $|\boldsymbol{x}_\perp| \gg \mathfrak{b}^{-1}|\boldsymbol{x}_\parallel|$. Conversely, for $\mathfrak{b} \gg 1$ and $|\boldsymbol{x}_\parallel| \gg \mathfrak{b}|\boldsymbol{x}_\perp|$, the Coulomb potential tends to be one-dimensional $a_0(\boldsymbol{x}) \approx q/(4\pi|\boldsymbol{x}_\parallel|)$. We remark that the expression above resembles closely the longe-range behavior found within a pure QED context, when the external magnetic field exceeds the characteristic scale associated with this framework $B_0 = m_e^2/e \approx 4.42 \times 10^{13}$ G [38–40].

4. Captures of γ quanta: implications

Let us suppose a magnetic field with its lines of force curved as it is adopted in models describing the magnetosphere of pulsars. The fields of these highly magnetized objects rotate with certain angular velocity Ω. However, as long as both its curvature radius and its rotation period $2\pi\Omega^{-1}$ exceed considerably the photon wavelength and the Compton wavelength of ALPs, the analysis developed above can still be applied. Indeed, whenever the previous requirements are fulfilled a geometrical optical and an adiabatic approximation can be used simultaneously to investigate the propagation of those photons subject to the aforementioned limitations. To accomplish them, we shall refer hereafter to the so-called curvature γ-radiation ($q_0 > 2m_e$) which is emitted tangentially to the magnetic field lines. At the moment of its emission this electromagnetic wave decomposes into ordinary and extraordinary modes and both of them are localized at the origin of the diagram in Fig. 2. As the wave propagates, the momentum components $q_\perp$ and $q_\parallel$ of theses modes change, while their energies q_0 are kept constant approximately. Thereby the dispersion relation of the ordinary and extraordinary photons evolve from the moment of their emission following the corresponding curves displayed in Fig. 2.

Since the curvature radiation appears well below the first pair creation threshold, the main process taking place in this regime is the photon splitting: $\gamma + B \to \gamma' + \gamma'' + B$. This inelastic scattering becomes feasible as long as the photon wave vector $\boldsymbol{q}$ and the magnetic field $\boldsymbol{B}$ form a nontrivial angle [$\vartheta \neq 0$]. Under such a circumstance the ordinary mode is expected to decay into two extraordinary photons before reaching the first pair creation threshold, provided $B \gtrsim 0.2\ B_0$ [26]. Mainly because the electromagnetic sector linked to this type of degree of freedom is unaffected by ALPs, and therefore, it is not subject to the aforementioned dominance. As a result of the decay of the ordinary mode, the curvature radiation acquires a high degree of polarization in the plane spanned by $\boldsymbol{q}$ and $\boldsymbol{B}$ before the extraordinary modes reach the first pair creation threshold.

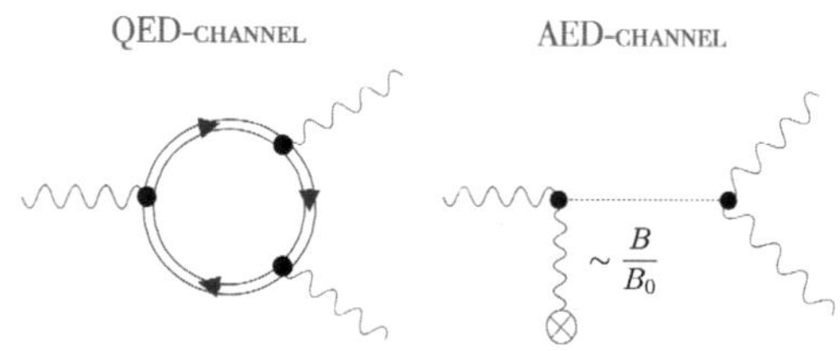

Figure 4. Photon splitting in QED+AED. A double-line in the QED diagram represents the electron-positron propagator including the interaction with the external field. The dashed line in the AED-channel denotes the free axion propagator.

[3] Only virtual modes with the polarization of the extraordinary photons carry the electrostatic interaction.

Journal of Physics: Conference Series **2249** (2022) 012021 doi:10.1088/1742-6596/2249/1/012021

The axion-diphoton interaction opens a new channel for the splitting of a photon [see Fig. 4]. Observe that, owing to the parity invariance, only ordinary γ photons can decay via the AED channel, and that the result of the reaction involves photons with different nature. We remark that, in the equation of motion, the term linked to this photon splitting turns out to be subleading as compared to the term involving the axion-mediated polarization tensor. Mainly because it depends linearly on the external field.

Now, for $q_\perp \ll m_*$ the dispersion law of an extraordinary mode tends to stick to the horizontal axis as $\mathfrak{b}$ grows. This is precisely the region in which the group velocity across the magnetic field disappears [$v_{\perp,e} \to 0$]. Hence, this γ quantum would be canalized along the magnetic line of force immediately after its emission. It is worth remarking that, if this photon capture occurs, a single sufficiently hard – belonging to the γ-range – photon might have no chance to produce an electron-positron pair before it has propagated far enough to leave the region of the strong dipole field of the pulsar where, however, the one-photon pair creation is strongly suppressed. This is guaranteed if the distance traversed by the extraordinary photon is such that the local magnetic field satisfies the relation [$b = B/B_0$]

$$b \ll 1 \ll \mathfrak{b} \tag{8}$$

and, up to this point the lowest pair creation threshold is not reached i.e. $q_{0.e}^2 - q_\parallel^2 < 4m_e^2$. While the former relation introduces the condition $B_c \ll B_0$ along with $\mathfrak{b} \gg 1$, the latter provides a characteristic value for $q_\perp$ [31]:

$$(q_{\perp,\mathrm{thr}})^2 = 4m_e^2 + \frac{1}{2}m^2\left(\sqrt{1+16\mathfrak{b}^2\frac{m_e^2}{m^2}}-1\right) > 4m_e^2. \tag{9}$$

We remark that the inequality above is valid for any value of $\mathfrak{b}$ and m. Observe that, for a dipole magnetic field $B(r) = B_*\left(r_*/r\right)^3$ with B_* the field at the star surface and r_* the star's radius, the condition (8) translates into

$$\mathfrak{b}_*^{1/3} \gg \frac{r}{r_*} \gg b_*^{1/3},$$

where the respective field parameters at the pulsar surface are $\mathfrak{b}_* = B_*/B_c$ and $b_* = B_*/B_0$. We note, that the emission of curvature γ quanta takes place in a vicinity of the pulsar surface [$r \gtrsim r_*$] where the local field can exceed the characteristic scale of QED [$b_* \gtrsim 1$]. It is worth remarking that, for pulsars holding $b_*^{1/3} < 1$, the inequality in the right hand side is bounded by unity. Hence, instead of the relation above we should consider

$$\mathfrak{b}_*^{1/3} \gg \frac{r}{r_*} > \max\{1, b_*^{1/3}\}. \tag{10}$$

For ALPs satisfying the relations $m\mathfrak{b} < 2m_e$ and $\mathfrak{b} \gg 1$ at the moment when $b \ll 1$[4] the characteristic value of $q_\perp$ approximates to $q_{\perp,\mathrm{thr}} \approx 2m_e[1 + m\mathfrak{b}/(2m_e)]$. In a dipole field, the condition $m\mathfrak{b} < 2m_e$ introduces an additional restriction in Eq. (10):

$$\mathfrak{b}_*^{1/3} \gg \frac{r}{r_*} > \max\left\{1, b_*^{1/3}, \left(\frac{m\mathfrak{b}_*}{2m_e}\right)^{1/3}\right\}. \tag{11}$$

This means that the fraction of extraordinary photons with $q_\perp \ll 2m_e$, having traversed a radial distance r which satisfies (11) will escape from the pulsar. This effect could explain the observed hard γ−rays emission $\gtrsim 10^3$ GeV from neutron stars, and might even provide new bounds on the parameter space of ALPs by requiring that the energy loss due to this mechanism does not exceed the corresponding luminosity of the pulsar.

[4] These two conditions imply that locally $m/(2m_e) < \mathfrak{b}^{-1} \ll 1$. Hence, we are basically referring to $m \ll 1$ MeV.

Journal of Physics: Conference Series **2249** (2022) 012021 doi:10.1088/1742-6596/2249/1/012021

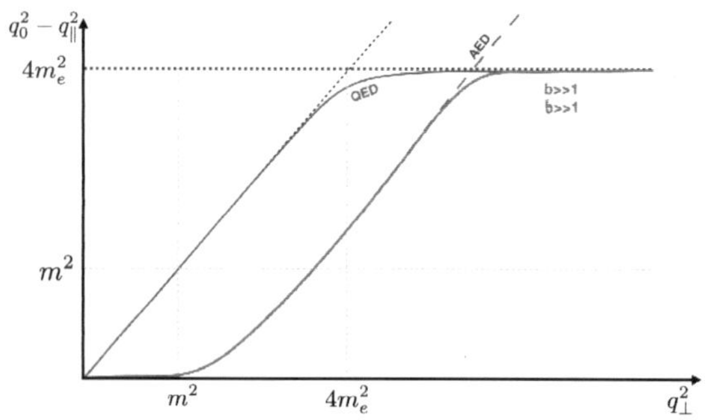

Figure 5. Sketch of the dispersion relation for an extraordinary mode in an inhomogeneous magnetic field (red). The curve describes the scenario in which the γ quantum reaches $q_\perp > 2m_e$ while $\mathfrak{b}$, $b \gg 1$. The green curve corresponds to the pure QED case, whereas the dashed blue line is linked to pure AED. As $q_0^2 - q_\parallel^2 \to 4m_e^2$, the combined scenario of QED+AED leads to a dispersion law which tends to be parallel to the dotted line that demarcates the first pair creation threshold. Once the field is such that $b < 1$ but $\mathfrak{b} \gg 1$, both the green and red curves are expected to cross the threshold.

On the other hand, the fraction of extraordinary photons with $q_\perp > 2m_e[1 + m\mathfrak{b}/(2m_e)]$ that has traversed a distance r according to (11), could decay into electron-positron pairs because their dispersion curves could cross the first threshold. Clearly, until fulfilling the aforementioned condition, these quanta could undergo a magnetic field strength characterized by $b \gg 1$. If during the time interval in which the previous condition applies both $q_\perp > 2m_e[1 + m\mathfrak{b}/(2m_e)]$ and $q_{0,e}^2 - q_\parallel^2 \approx 4m_e^2$ are met, the squared-root singularity of the QED polarization tensor becomes relevant [41]. Under such circumstances, QED restores the control of the phenomenology for a while. This means that an extraordinary photon which has been canalized initially due to the strong axion-induced refraction, i.e. having $q_\perp \ll m\mathfrak{b} < 2m_e$,could be recaptured later on owing to the strong birefringent property of the vacuum driven by the quantum vacuum fluctuations of the electron-positron field. As the field decreases gradually, the capture due to the aforementioned singularity is expected to end, and eventually, the dispersion laws of these photons will cross the first pair creation threshold.

It is worth remarking that, under such a circumstance, the capture undergone by this part of the wave would induce an effective growing of the mean free path of the curvature radiation as compared to the case in which the vacuum polarization effects are considered perturbatively. This fact constitutes a striking effect that puts under scrutiny pulsar models in which the curvature radiation is supposed to propagate straightforwardly and to supply the electron-positron plasma [42–44]. In these models the plasma serves to limit a required polar gap the height of which – relative to the star surface – cannot exceed considerably the mean free path that a curvature $\gamma-$quantum traverses before creating an electron-positron pair [43]. We emphasize that the extension of the polar gap due to the axion-mediated capture of the extraordinary mode would modify the so far accepted expression for the pulsar luminosity which is proportional to the square of the gap height .

5. Discussion and outlook

Magnetic fields stronger than the characteristic scale of AED can induce nontrivial modifications of the QED phenomenology. In this contribution we have explored the scenario in which these modifications prevail hypothetically over the QED. It has been shown that, in super strong magnetic fields [$\mathfrak{b} \gg 1$], electromagnetic processes characterized by perpendicular momentum transfers much smaller than the axion mass tend to be suppressed due to the dimensional reduction that the gauge sector responsible for the electrostatic interaction undergoes. This effect represents a manifestation of the anisotropy of the running electromagnetic coupling that the axion-mediated vacuum polarization causes.

In the second part we evaluated some implications on the γ emission mechanism of pulsars. This preliminary analysis indicates that, in a strong field regime – characterized by the condition $B \gg B_c$ with $B_c \ll B_0$ – ALPs could enable the escape of certain part of the curvature radiation. The escaping of these photons could explain the observed hard $\gamma-$rays emission from pulsars. Hence, establishing how the gap height and the corresponding pulsar luminosity are modified by the axion-diphoton interaction would open an enticing window to probe the existence of axion-like particles. This will be the subject of a forthcoming investigation.

5.1. *Acknowledgments*

Villalba-Chávez S and Müller C gratefully acknowledge the funding by the German Research Foundation (DFG) under Grants No. 258838303 (MU 3149/2-1) and No. 388720772 (MU 3149/5-1). Shabad A E acknowledges the support of the Russian Foundation for Basic Research under the project No. 20-02-00193.

References

[1] Peccei R D and Quinn H R 1977 *CP Conservation in the Presence of Pseudoparticles*, Phys. Rev. Lett. **38** 1440
[2] Wilczek F 1978 *Problem of Strong P and T Invariance in the Presence of Instantons*, Phys. Rev. Lett. **40** 279
[3] Weinberg S 1978 *A New Light Boson?* Phys. Rev. Lett. 40 223
[4] Donnelly T W, Freedman S J, Lytel R S, Peccei R D and Schwartz M 1978 *Do Axions Exist?*, Phys. Rev. D **18** 1607
[5] Zehnder A, Freedman S J, Lytel R S, Peccei R D and Schwartz M 1981 *Axion Search in a Monochromatic Transition: A New Lower Limit for the Axion Mass*, Phys. Lett. **104B** 494
[6] Dine M, Fischler W and Srednicki M 1981 *A simple solution to the strong CP problem with a harmless axion*, Phys. Lett. B **104** 199
[7] Zhitnitskii A R 1980 *On possible suppression of the axion-hadron interactions*, Yad. Fiz **31**, 497; [1980 Sov. J. Nucl. Phys. **31** 260 (translation)].
[8] Kim J E 1979 *Weak-Interaction Singlet and Strong CP Invariance*, Phys. Rev. Lett. **43** 103
[9] Shifman M A, Vainshtein A I and Zakharov V I 1980 *Can confinement ensure natural CP invariance of strong interactions?*, Nucl. Phys. B **166** 493
[10] Jaeckel J and Ringwald A 2010 *The Low-Energy Frontier of Particle Physics* Ann. Rev. Nucl. Part. Sci. **60** 405.
[11] Ringwald A 2012 *Exploring the Role of Axions and Other WISPs in the Dark Universe* Phys. Dark Univ. **1** 116
[12] Alekhin S et al. 2016 *A facility to Search for Hidden Particles at the CERN SPS: the SHiP physics case* Rept. Prog. Phys. **79** 124201
[13] Irastorza I G and Redondo J 2018 *New experimental approaches in the search for axion-like particles* Prog. Part. Nucl. Phys. **102** 89
[14] Raffelt G. G. 1986 *Astrophysical Axion Bounds Diminished By Screening Effects* Phys. Rev. D **33** 897
[15] Raffelt G. G. 1999 *Particle physics from stars* Ann. Rev. Nucl. Part. Sci. **49** 163
[16] Raffelt G. G. 2008 *Astrophysical axion bounds* Lect. Notes Phys. **741** 51
[17] Bloom J S et al. 1998 *Expected characteristics of the subclass of Supernova Gamma-ray Bursts (S-GRBs)* Astrophys. J. **506**, L105
[18] Kouveliotou C et al. 1998 *An X-ray pulsar with a superstrong magnetic field in the soft gamma-ray* Nature **393** 235
[19] Ibrahim A I, Swank J H and Parke W 2003 *New evidence for proton cyclotron resonance in a magnetar strength field from SGR 1806-20* Astrophys. J. Lett. **584** L17
[20] Manchester R M, Hobbs G B, Teoh A and Hobbs M 2005 *The Australia Telescope National Facility Pulsar Catalogue* Astron. J. **129** 1993
[21] Berestetsky V B, Lifshits E M and Pitayevsky L P 1989 *Quantum Electrodynamics* Pergamon Pr. Oxford, New York
[22] Schwartz M D 2014 *Quantum Field Theory and the Standard Model* Cambridge: Univ. Pr.
[23] Dittrich W and Gies H 2000 *Probing the Quantum Vacuum. Perturbative Effective Action Approach in Quantum Electrodynamics and its Application* Springer, New York, Tracts Mod. Phys., **166**
[24] Shabad A E 1972 *Cyclotronic Resonance in the Vacuum Polarization* Lett. Nuovo Cimento **3** 457
[25] Shabad A E and Usov V V 1982 *Gamma-quanta capture by magnetic field and pair creation suppression in pulsars* Nature (London) **295** 215
[26] Shabad A E and Usov V V 1984 *Propagation of γ-radiation in strong magnetic fields of pulsars* Astrophys. Space Sci. **102** 327
[27] Adler S, Bahcall J N, Callan C G and Rosenbluth M N 1970 *Photon Splitting in a Strong Magnetic Field* Phys. Rev. Lett. **25** 1061
[28] Adler S 1971 *Photon splitting and photon dispersion in a strong magnetic field* Ann. Phys. **67** 599

[29] Adler S L and Schubert C 1996 *Photon splitting in a strong magnetic field: Recalculation and comparison with previous calculations* Phys. Rev. Lett. **77** 1965
[30] Bošković M, Brito R, Cardoso V, Ikeda T and Witek H 2019 *Axionic instabilities and new black hole solutions* Phys. Rev. D **99** 035006
[31] Villalba-Chávez S, Shabad A E and Müller C 2021 *Magnetic dominance of axion electrodynamics: photon capture effect and anisotropy of Coulomb potential* Eur. Phys. Jour. C **81**, 331
[32] Villalba-Chávez S, Golub A and Müller C 2018 *Axion-modified photon propagator, Coulomb potential and Lamb-shift* Phys. Rev. D **98**, 115008
[33] Melrose D B and Stoneham R J 1976 *Vacuum polarization and photon propagation in a magnetic field* Nuovo Cimento A **32** 435
[34] Heyl J S and Hernquist L 1997 *Birefringence and dichroism of the QED vacuum* J. Phys. A **30** 6485
[35] Gies H and Jaeckel J 2004 *Renormalization Flow of QED* Phys. Rev. Lett. **93** 110405
[36] Landau L D, Abrikosov A A and Khalatnikov I M 1954 Dokl. Akad. Nauk SSSR **95** 497
[37] Ferrer E J and Sanchez A 2019 *Magnetic field effect in the fine-structure constant and electron dynamical mass* Phys. Rev. D 100 096006
[38] Shabad A E and Usov V V 2007 *Modified Coulomb Law in a Strongly Magnetized Vacuum* Phys, Rev. Lett. **98** 180403
[39] Shabad A E and Usov V V 2008 *Electric field of a point-like charge in a strong magnetic field and ground state of a hydrogen-like atom* Phys, Rev. D **77** 025001
[40] Adorno T C, Gitman D M and Shabad A E, 2016 *Coulomb field in a constant electromagnetic background* Phys, Rev. D **93**, 125031
[41] Shabad A E 1975 *Photon Dispersion in a Strong Magnetic Field* Annals Phys. **90**, 166
[42] Sturrock P 1971 *A Model of Pulsars* Astrophys. J. **164** 529
[43] Ruderman M A and Sutherland P G 1975 *Theory of pulsars: polar gaps, sparks, and coherent microwave radiation* Astrophys. J. **196** 5
[44] Arons J 1981 *Origin of Cosmic Rays* IAU Symp. **164**; Proc. Workshop Plasma Astrophysics, Varenna, Italy, 273

From theory to precision modelling of strong-field QED in the transition regime

Alexander J. Macleod, on behalf of the LUXE collaboration
Institute of Physics of the ASCR, ELI-Beamlines, Na Slovance 2, 18221 Prague, Czechia
E-mail: alexander.macleod@eli-beams.eu

Abstract. The combination of energetic electron beams, delivered from conventional accelerators at a high repetition rate, and ultraintense lasers, makes it possible to perform precision measurements of strong-field QED. The LUXE collaboration aims to perform precision measurements of nonlinear Compton scattering and Breit-Wheeler pair creation in the transition from the perturbative to nonperturbative regimes. Here we present an overview of recent developments in the modelling of strong-field QED processes, which are needed to reach the required precision of a few percent for intensity parameters $0.1 < \xi < 10$. We discuss how to go from plane-wave QED results to numerical simulations and present predicted signals and error estimates.

1. Introduction

Precision tests of QED, such as the measurement of the electron anomalous magnetic moment, have demonstrated extraordinary levels of agreement between theory and experiment. This relies on the ability to perform calculations *perturbatively* in the coupling[1], $\alpha = e^2/4\pi \simeq 1/137$. Whether this agreement persists in the *non-perturbative* regime of QED is an open question.

At high energies and background field strengths we enter the strong-field regime of QED, which requires a non-perturbative treatment of the interaction with the background field. This regime is characterised by two key parameters becoming larger than unity. Firstly, the intensity parameter, $\xi = eE_0 ƛ_C/\omega_0$, which is a dimensionless measure of the work done on an electron by a laser of peak field strength E_0 and frequency ω_0 over a (reduced) Compton wavelength $ƛ_C$ [1, 2]. This is often called the "classical nonlinearity parameter", and describes the effective coupling to the background. Secondly, the "quantum nonlinearity parameter" of a probe particle with momentum p_μ, $\chi = e\sqrt{-(p \cdot F)^2}/m^3$, with $F_{\mu\nu}$ the background field tensor. This gives a measure of the importance of quantum effects and recoil in the interaction. In a plane wave background $\chi = \eta\xi$, where $\eta = k \cdot p/m^2$ is often called the "lightfront energy" of a probe particle with momentum p_μ, where k_μ is the laser wavevector.

The intensity parameter, ξ, roughly corresponds to the number of background photons absorbed over a Compton wavelength, and so determines the onset of nonlinear effects. When $\xi \ll 1$, processes are well approximated using a perturbative treatment. However, when $\xi \gtrsim 1$ a fully *non-perturbative* treatment of the interaction with the background is required. This poses several challenges for theory, particularly in developing analytical models which are useful for the analysis of laser-matter interactions.

[1] We use units where $\hbar = c = \epsilon_0 = 1$ throughout.

Journal of Physics: Conference Series **2249** (2022) 012022 doi:10.1088/1742-6596/2249/1/012022

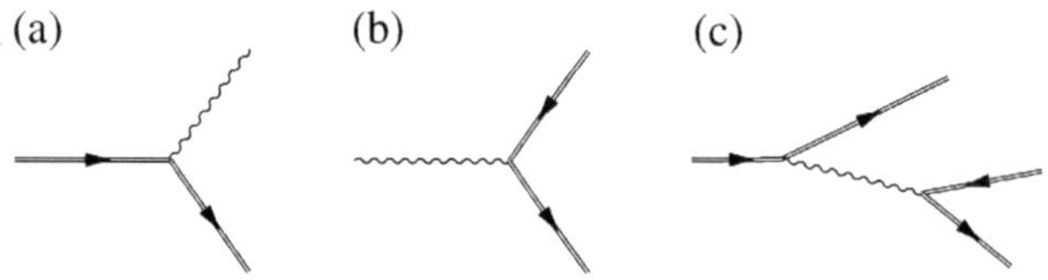

Figure 1. The three key processes which will be measured at LUXE: (a) Nonlinear Compton scattering (b) Breit-Wheeler pair production (c) trident pair production.

One such model, valid when $\xi \gg 1$ *and* $\xi \gg \chi^{1/3}$, is the locally constant field approximation (LCFA) [1]. It forms the basis of many simulation frameworks used to analyse QED effects in laser-matter interactions (e.g. [4, 5, 6]). Local *rates*, which encode quantum effects in these simulations, depend only on χ in the LCFA. However, limitations of the LCFA [7, 8, 9, 10, 11] in the *transition regime* from perturbative to non-perturbative physics, where $\xi \sim \mathcal{O}(1)$, cause large errors when compared with exact QED results. This problem is of immediate concern; a new wave of experiments exploring strong-field QED effects when $\xi \sim \mathcal{O}(1)$ are on the horizon.

The Light Und XFEL Experiment (LUXE) [3] will perform high-precision measurements of strong-field QED processes in the transition regime. This will be achieved using a 16.5 GeV electron beam from the European XFEL and 40 TW JETI laser (Phase-0), with a subsequent upgrade to 350 TW (Phase-1), to measure nonlinear Compton scattering [12, 13] and trident pair production [14] in electron seeded interactions, and Breit-Wheeler pair production [15, 16, 17] in photon seeded interactions (see fig. 1). Schematics for both the electron and photon seeded interactions are given in fig. 2. The parameters accessible to LUXE are shown in fig. 3

To enable accurate modelling of the experimental outcomes of LUXE, and analysis of the experimental data, a robust simulation framework which works in the transition regime is required. The *locally-monochromatic approximation* (LMA) has been shown to agree with exact QED results [23, 24, 25] and can be used to derive local rates that take into account interference effects at the scale of the laser wavelength. These rates have been implemented into the simulation framework PTARMIGAN, which describes strong-field QED effects through combining classical trajectories with probability rates that treat the background 'locally' as a monochromatic plane wave.

The outline of this paper is as follows. In section 2 we discuss how strong-field effects are modelled in the transition regime using the LMA, in particular outlining the steps from theory to the simulation framework PTARMIGAN. The calculated experimental signatures of

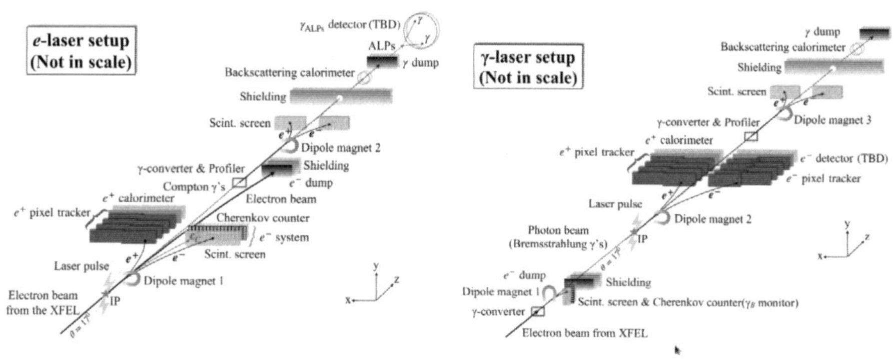

Figure 2. Schematic experimental layouts for: (left) e-laser and (right) γ-laser setup [3].

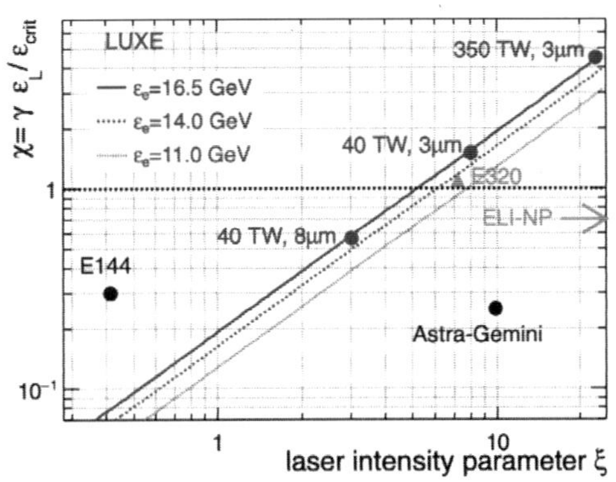

Figure 3. Strong-field QED χ vs ξ parameter space. Isolines represent three different electron beam energies ε_e. Blue dots show parameters accessible to LUXE in each phase [3]. Also shown are parameters for: E-144 [18], E-320 [19] Astra-Gemini [20, 21] and ELI-NP [22].

the transition regime for LUXE are presented in section 3. Finally, we conclude and summarise in section 4.

2. From theory to simulation in the transition regime

2.1. Plane waves and interference

We begin with the Dirac equation for an electron, mass m and charge e, coupled to a gauge potential describing the background field, A_μ,

$$(i\partial\!\!\!/ - eA\!\!\!/ - m)\psi(x) = 0\,. \tag{1}$$

Here, $b\!\!\!/ \equiv b_\mu\gamma^\mu$, with γ^μ the Dirac matrices. To treat the background field non-perturbatively, we seek field configurations for which eq. (1) can be solved *exactly*. This gives *dressed* wavefunctions, $\psi(x)$, where the interaction with the background is included to all orders, and allows one to calculate scattering processes in the *Furry picture* [26].

Exact solutions to the Dirac equation for focused laser pulses are not known. However, if the initial Lorentz factor of the probe particle $\gamma \gg \xi^2$, the probability for a strong-field QED process in a focused background, $\mathsf{P}_{\rm focus}$, can be expressed as a sum over the probability for the process to occur in a plane wave, $\mathsf{P}_{\rm pw}$, weighted by the electron areal probe density, $\rho(x^\perp)$ [27],

$$\mathsf{P}_{\rm focus} \approx \int \mathrm{d}^2x^\perp \rho(x^\perp)\mathsf{P}_{\rm pw}\left[\xi(x^\perp), \eta(x^\perp)\right], \tag{2}$$

where $x^\perp$ represent the directions transverse to the laser propagation direction and the intensity parameter, ξ, and lightfront energy, η, are defined *locally*. At LUXE, the electron beam energy of $\varepsilon_e \leq 16.5$ GeV at the proposed collision angle of $\vartheta = 17.2°$ corresponds to $\gamma \simeq 10^4$, and so to leading order eq. (2) is a good approximation. This motivates us to use the plane wave approximation to calculate the probabilities, and subsequently *rates*, which will be used to simulate the experimental outcomes of LUXE.

Working in a plane wave background, it is convenient to use lightfront variables. A plane wave propagating in the z-direction with light-like ($k^2 = 0$) wavevector $k_\mu = \delta^-_\mu k_- = \omega(1,0,0,1)$ has a phase $\varphi \equiv k\cdot x = \omega x^-$, where $x^\pm = t \pm z$. Here, x^- acts as a "time" coordinate, with x^+ the *longitudinal* direction, and the remaining coordinates $x^\perp = (x,y)$ being *transverse* to the propagation direction. Solving eq. (1) in the gauge potential $eA_\mu(\varphi) = a_\mu(\varphi)$, one arrives at the *Volkov solutions*,

$$\psi_p(x) = \left(1 + \frac{\not{k}\not{a}(\varphi)}{2k \cdot p}\right) u_p \, e^{-iS_{\rm cl}(x)} \,, \tag{3}$$

for an electron with initial momentum p_μ and constant spinor u_p. The exponent, $S_{\rm cl}(x)$, is the *classical action* for an electron in a plane wave field with instantaneous momentum, π_μ,

$$S_{\rm cl}(x) = p \cdot x + \int^{\varphi} {\rm d}t \, \frac{2p \cdot a(t) - a^2(t)}{2k \cdot p} \,, \quad \pi_\mu(\varphi) = p_\mu - a_\mu(\varphi) + \frac{2p \cdot a(\varphi) - a^2(\varphi)}{2k \cdot p} k_\mu \,. \tag{4}$$

On a plane wave background only three components of energy and momentum are conserved, and the S-matrix for a given process can always be expressed as,

$$S_{\rm fi} = (2\pi)^3 \delta^3_{+,\perp}(P_{\rm in} - P_{\rm out}) \mathcal{M} \,, \tag{5}$$

where $\mathcal{M}$ is the invariant part of the amplitude, $P_{\rm in}$ and $P_{\rm out}$ are the sums of the incoming and outgoing momenta, respectively, and $\delta^3_{+,\perp}(q) \equiv \delta(q_+)\delta(q_x)\delta(q_y)$. The non-conservation of all components of the momentum gives rise to the possibility of processes which are kinematically forbidden in vacuum, such as the target processes of LUXE (see fig. 1).

Consider a first-order processes, such as non-linear Compton scattering or Breit-Wheeler pair production (see fig. 1(a) & (b)). The probability takes the form $\mathsf{P}_{\rm pw} \sim \int {\rm d}\Omega_{\rm LIPS} \int {\rm d}\varphi {\rm d}\varphi' |\mathcal{L}_{\rm I}|^2$, where $\mathcal{L}_{\rm I}$ is the interaction Lagrangian and ${\rm d}\Omega_{\rm LIPS}$ is the Lorentz invariant phase space for the process. This integral is non-local; interference between different spacetime points in the field are of crucial importance. This non-locality makes the probability unsuitable for numerical simulations, which require *local rates*. Defining the average $\phi = (\varphi + \varphi')/2$ and interference $\theta = \varphi - \varphi'$ phases, one method of obtaining a local description of the probability is to expand the probability as a series in $\theta \ll 1$. This is the LCFA: interference effects are then completely neglected which leads to large errors in the transition regime [7, 8, 9, 10, 11]. The experimental outcomes of LUXE are analysed using a different approach.

2.2. Locally-monochromatic approximation

Instead of performing an expansion directly in the interference phase, θ, one can include interference effects at the scale of the laser wavelength using the *locally-monochromatic approximation* (LMA). A typical plane wave pulse will have both slow and fast oscillations. Consider a circularly polarised plane wave pulse,

$$a_\mu(\varphi) = m\xi_0 f\left(\frac{\varphi}{\Phi}\right) \left[\alpha_\mu \cos\left(b(\varphi)\right) + \beta_\mu \sin\left(b(\varphi)\right)\right] , \tag{6}$$

where α_μ, β_μ are spacelike vectors satisfying $\alpha^2 = \beta^2 = -1$ and $\alpha \cdot \beta = \alpha \cdot k = \beta \cdot k = 0$, $b(\varphi)$ accounts for chirp effects and ξ_0 is the intensity parameter. The pulse envelope $f(\varphi/\Phi)$ is slowly-varying over the pulse phase duration, Φ, while the carrier frequency terms in square brackets are rapidly oscillating. In the LMA, the fast oscillations are included *exactly*, and only the leading order contribution from a derivative expansion of the pulse envelope is taken into account. This is essentially a combination of two approximations, the *slowly-varying envelope approximation* [28, 29, 30] and a local expansion akin to that used in the LCFA [1, 7, 9, 10, 12].

This has both advantages and disadvantages over the LCFA. Interference effects are included, the low-energy limit is exact [23], and physics in the transition regime is more accurately described. However, it is only suitable for fields which can be well approximated by plane waves, and so cannot be used in environments with significant plasma production. There is an explicit trade-off between versatility and accuracy in the transition regime. The experimental parameters for LUXE [3] make the LMA the ideal candidate for modelling strong-field QED processes, and as such it is the central approximation used in numerical codes.

Consider nonlinear Compton scattering in the circularly polarised background, eq. (6). An electron with initial momentum p_μ, corresponding to a lightfront energy $\eta_0 = k \cdot p/m^2$, collides with the background (wavevector k_μ) and emits a photon of momentum l_μ, which we use to define the lightfront fraction of the emitted photon, $u = k \cdot l/k \cdot p$. The probability can be expressed as, $\mathsf{P}_{\mathrm{LMA}} = \int \mathrm{d}\tau W_{\mathrm{LMA}}(\tau)$, where the *local rate*, W_{LMA}, takes the form,

$$W_{\mathrm{LMA}}(\tau) = \sum_{n=1}^{\infty} \int_0^{u_{n,*}(\tau)} \mathrm{d}u \frac{\mathrm{d}^2\mathsf{P}_n^{\mathrm{mono}}[\xi(\tau), \eta(\tau)]}{\mathrm{d}\tau \mathrm{d}u} , \tag{7}$$

with τ the proper time, related to the phase by $\mathrm{d}\tau/\mathrm{d}\phi = 1/(m\eta_0)$. This is expressed explicitly in terms of the probability, $\mathsf{P}_n^{\mathrm{mono}}$, for nonlinear Compton scattering into the nth harmonic in a *monochromatic background* (see e.g. [1]). However, instead of depending only on the initial intensity parameter, ξ_0, one finds that the intensity takes a local value given by,

$$\xi^2(\tau) = -\frac{a^2[\phi(\tau)]}{m^2} = \xi_0^2 f^2\left(\frac{\phi(\tau)}{\Phi}\right) , \tag{8}$$

which is the component of the potential that varies slowly with respect to wavelength. Furthermore, both the lightfront energy, $\eta(\tau) = \eta_0\omega(\tau)$, where the non-constant frequency $\omega(\tau) = \mathrm{d}b[\phi(\tau)]/\mathrm{d}\phi$ encodes laser pulse chirping, and the harmonic range, $u_{n,*}(\tau) = u_n(\tau)/(1 + u_n(\tau))$ where $u_n(\tau) = 2n\eta(\tau)/(1+\xi^2(\tau))$ is the edge of the classical (nonlinear) harmonic range, are phase (or proper time) dependent. Armed with the LMA rate for a given process, such as eq. (7) for nonlinear Compton scattering or analogous expressions for Breit-Wheeler pair production (see [23, 25]), the next step is to implement this into a numerical simulation.

2.3. Simulations in the transition regime with PTARMIGAN

Simulation frameworks which include strong-field QED effects (e.g. [4, 5, 18, 31]) typically combine classical particle trajectories with probability rates [32]. The information encoded in the rates determines the dynamical quantities that must be obtained from the classical trajectories.

In the LCFA, one defines the classical trajectories of particles in terms of the kinetic momentum eq. (4). Particles propagate between emission events due to the Lorentz force,

$$\frac{\mathrm{d}\pi_\mu}{\mathrm{d}\tau} = -\frac{e}{m} F_{\mu\nu}\pi^\mu , \qquad \frac{\mathrm{d}x^\mu}{\mathrm{d}\tau} = \frac{\pi^\mu}{m} , \tag{9}$$

with $x^\mu(\tau)$ the classical worldline. This is an oscillatory trajectory, and emission occurs at a characteristic angle $\sim 1/\gamma$. Contrastingly, LMA rates already treat oscillations due to the carrier frequency exactly. The classical trajectory must reflect this, and upon averaging the motion over the fast timescale (the laser period) one finds that classical motion is determined by the relativistic ponderomotive force equation for the *quasimomentum* $q_\mu = \langle \pi_\mu \rangle$, i.e. the laser-cycle-average of the instantaneous electron momentum in eq. (4),

$$\frac{\mathrm{d}q_\mu}{\mathrm{d}\tau} = \frac{m}{2}\partial_\mu \xi^2(\tau) , \qquad \frac{\mathrm{d}x^\mu}{\mathrm{d}\tau} = \frac{q^\mu}{m\sqrt{1+\xi^2(\tau)}} . \tag{10}$$

Emission now occurs in an angle $\sim \xi_0/\gamma$. For an illustrative comparison of the particle trajectories in LCFA and LMA based simulations see fig. 4.

PTARMIGAN uses rates derived in the LMA combined with classical motion due to the relativistic ponderomotive force. It uses a Monte Carlo algorithm which:

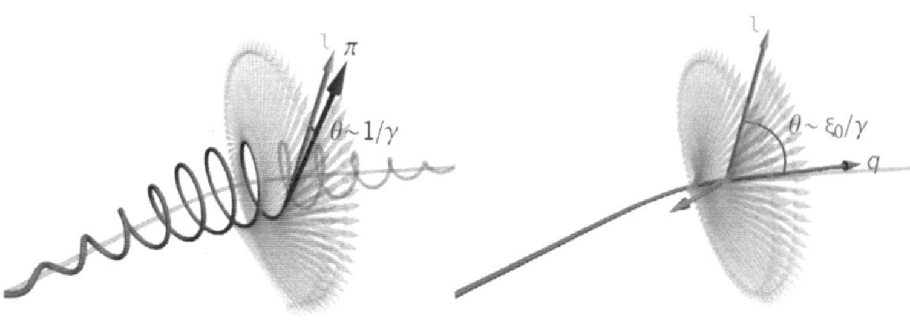

Figure 4. Illustration of photon emission models: LCFA (left), LMA (right). Yellow arrow denotes the emitted photon, momentum l_μ. Laser wavevector, k_μ, shown in red. LCFA trajectory (blue) determined by the Lorentz force, with the momentum given by π_μ (c.f. eq. (4)). In the LMA, fast oscillations due to the carrier frequency are included exactly in the quantum rates, so the electron trajectory (green) is due to the ponderomotive force, with quasimomentum $q_\mu = \langle \pi_\mu \rangle$ the cycle average of π_μ. Figure reproduced from Ref [24].

(i) advances the electron trajectory by solving the ponderomotive force eq. (10),

(ii) evaluates at every time step the probability of emission and pseudorandomly decides whether to emit a photon or not,

(iii) where emission takes place, selects a harmonic index with probability W_n/W, where W_n is the partial rate and $W = \sum_n^\infty W_n$ is the total rate,

(iv) samples u, the lightfront momentum fraction, from the partial spectrum $(\mathrm{d}W_n/\mathrm{d}u)/W_n$,

(v) finally determines the emitted photon momentum, k_μ, given n, u and q, and resets the electron quasimomentum from q to q'.

PTARMIGAN has been benchmarked for both nonlinear Compton scattering [24] and Breit-Wheeler pair production [25], showing excellent agreement to exact result in QED. It is the fundamental tool underpinning analysis of the experimental outcomes at LUXE, and has been used to determine the non-perturbative signatures which will be observed.

3. Signatures of non-perturbative charge-field coupling at LUXE

The flagship experimental campaigns at LUXE will measure non-perturbative charge-field coupling in nonlinear Compton scattering and Breit-Wheeler pair production. PTARMIGAN has been used with realistic parameters for both the laser pulse and electron beam to analyse the non-perturbative signals.

3.1. Non-perturbative signatures in nonlinear Compton scattering

Nonlinear Compton scattering is the emission of a photon by an electron due to the interaction with a strong electromagnetic field, shown in fig. 1(a). The process is distinguished from *linear* Compton scattering, where only a single photon is absorbed from the background. The non-linear interaction with the background when $\xi \gtrsim 1$ causes higher order terms in a perturbative expansion in ξ to dominate, requiring an all-orders, non-perturbative, treatment.

The key non-perturbative signature appears in the energy spectrum. Shown in fig. 5 is the photon energy spectrum versus $u = k \cdot l / k \cdot p$, for a 16.5 GeV electron colliding with a plane wave laser pulse with intensity parameter $\xi_0 = 1$ in: linear QED (red dashed), nonlinear classical electrodynamics (blue dotted) and nonlinear QED (black solid). The first harmonic in the spectrum is clearly defined by the main peak. This is the *Compton edge* [33].

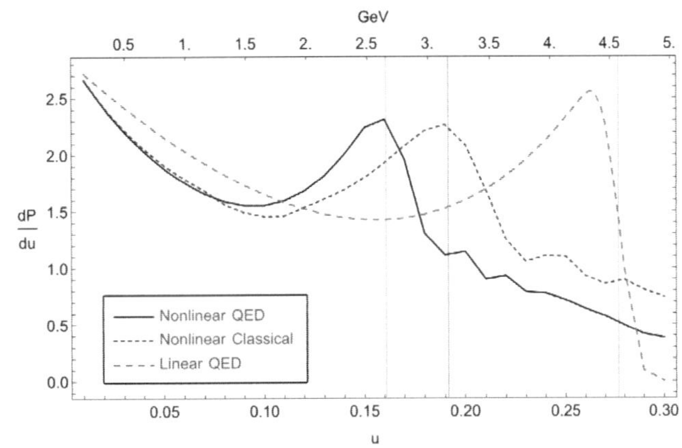

Figure 5. Photon energy spectrum due to the interaction of an electron with energy 16.5 GeV in different physical models: linear QED (red dashed), nonlinear classical (blue dotted), nonlinear QED (black solid) [3]. The position of the $n = 1$ harmonic peak, often called the Compton edge, is determined kinematically and plotted as a grey vertical line in each case.

In linear QED, i.e. perturbative Compton scattering, the position of the Compton edge is $u_{\text{lin.QED}} = \frac{2\eta_0}{1+2\eta_0}$. Multiple harmonic peaks become visible in the energy spectrum when nonlinear effects are included, with their position dependent on ξ. In the nonlinear *classical* theory, Thomson scattering, the position of the nth harmonic peak is $u^n_{\text{nonlin.class}} = \frac{2n\eta_0}{1+\xi^2}$. Comparing the position of the Compton edge ($n = 1$) to the linear QED result, there is a clear red-shifting due to the nonlinear interaction. Recoil effects are included in the transition from nonlinear classical electrodynamics to nonlinear QED leading to a further red-shifting of the harmonics to $u^n_{\text{nonlin.QED}} = \frac{2n\eta_0}{2n\eta_0+1+\xi^2}$. The red-shifting of the nonlinear QED Compton edge ($n = 1$) compared to linear QED is most readily explained by the electron gaining an *effective mass* [33], which is due to the interaction with many background field photons.

The shift of the leading Compton edge ($n = 1$) in the electron energy spectrum is shown in fig. 6. The electron beam has an initial energy of 16.5 GeV, with a finite size of $\sigma_{\text{el}} = 5$ μm. The laser pulse is taken to have a Gaussian profile in both the longitudinal and transverse directions. The red solid line shows the theoretical prediction, with the dashed lines representing a 5% uncertainty on the value of the laser intensity. The black data points show the anticipated data result [3], where the uncertainty in these data points is dominated by an energy scale uncertainty of 2.5%. Here ξ_{nom} is the *nominal* value of the intensity parameter, which takes into account that the highest value ξ is only in the peak of the pulse [3].

3.2. Non-perturbative signatures in Breit-Wheeler pair production

Breit-Wheeler pair creation is the decay of a photon into an electron positron pair due to the interaction with an intense electromagnetic field, see fig. 1(b). As with nonlinear Compton scattering, the nonlinear interaction with the background field leads to an effective mass effect which requires an all-orders non-perturbative treatment of the background field. This manifests as an increase in the threshold harmonic required for the photon to decay into an electron positron pair. Calculated in the LMA, this threshold is $n_\star(\phi) = \frac{2(1+\xi^2(\phi))}{\eta_{\gamma 0}}$, where $\xi(\phi)$ is defined analogously to eq. (8), and $\eta_{\gamma 0}$ is the lightfront energy of the probe photon with momentum l_μ, $\eta_{\gamma 0} = k \cdot l/m^2$ which we use to define the analogue of the quantum nonlinearity parameter for the photon, $\chi_\gamma = \eta_{\gamma 0}\xi$. This dependence on the (local) intensity parameter signifies non-

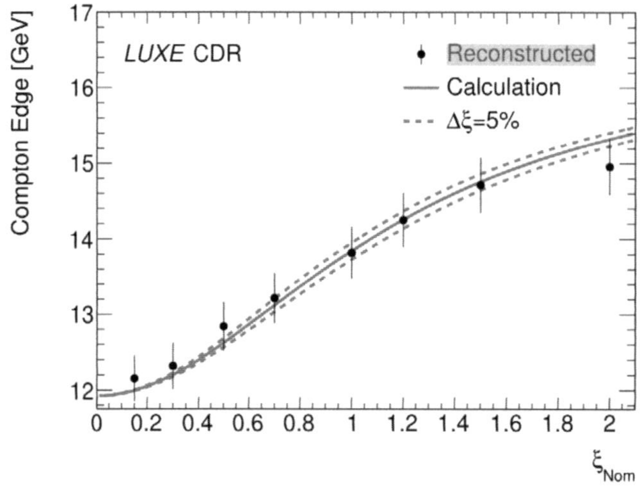

Figure 6. Position of the leading Compton edge in the electron energy spectra as a function of the (nominal) intensity parameter, ξ_{nom}. Red solid line shows theoretical prediction, with dashed lines representing a 5% uncertainty in the intensity parameter value. Black data points calculated using PTARMIGAN for realistic beam and laser parameters (see [3] for details).

perturbativity at small coupling.

For $\xi^2 \ll 1$ and $\chi_\gamma \ll 1$ Breit-Wheeler pair production is perturbative, occurring as a "multiphoton" process in which the probability scales as $\mathsf{P} \sim \xi^{2n_\star}$. However, as the intensity parameter increases to $\xi \gtrsim 1$ with $\chi_\gamma \ll 1$, pair production occurs as a tunnelling-like process. In the limit of a constant background, in this tunnelling regime the probability scales as $\mathsf{P} \sim \chi_\gamma \exp(-8/3\chi_\gamma)$, and since $\chi_\gamma^2 \propto \alpha$, this is non-perturbative in the charge field coupling. This transition from the multiphoton into the tunnelling regime of Breit-Wheeler pair production will be experimentally measured at LUXE. The parameter region accessible to LUXE is shown in the right hand plot of fig. 7.

The key experimental signature is a characteristic "turning of the curve" in the probability (or pair-yield) vs ξ. This is demonstrated in the left hand plot of fig. 7, which shows the probability as a function of ξ for different probe particle energies [3]. The dashed lines correspond to the multiphoton process, when $\xi \ll 1$, while the blue solid lines show the results of simulation based on the LMA. The "turning of the curve" refers to the pivoting of the probability away from the multiphoton result. The data points show the analytical QED plane wave results for a photon energy of 16.5 GeV, which agree with the predictions of the LMA. Also shown is the corresponding probability calculated using the LCFA in red solid lines, which clearly demonstrates the failure of the approach at low ξ.

Turning now to the predicted signal for the experimental parameters of LUXE, fig. 8 is the results from the simulation framework PTARMIGAN of the number of positrons produced per laser shot. This shows the result one could obtain from 10 days of data taking per ξ value (data points, blue: Phase-0, red: Phase-1), assuming no background particles per bunch crossing (left) and 0.01 background particles per bunch crossing (right). The dotted lines show the results of perturbative QED, while the solid line shows the strong-field QED prediction in a plane wave pulse. The errors at low ξ are dominated by the relatively low number of pairs produced pairs, and these can be improved by suppression of the number of background particles. For high intensity parameters, $\xi > 2$, statistical precision falls well below 5%.

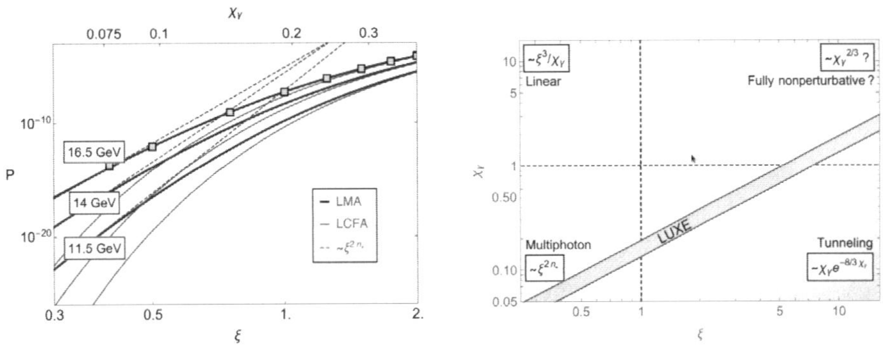

Figure 7. (left) Probability of pair production due to the interaction of a high energy probe with different initial energies in the interaction with the laser pulse. Blue dashed lines correspond to pair production in the multiphoton, perturbative, regime. Green data points are the result of an exact analytical QED calculation using a plane wave background. Blue solid lines show LMA calculation, and excellent agreement is seen in the 16.5 GeV case. Red lines show LCFA result, which becomes more accurate at high ξ, but fails at low values. (right) Pair production ξ, χ_γ parameter space explored by LUXE [3].

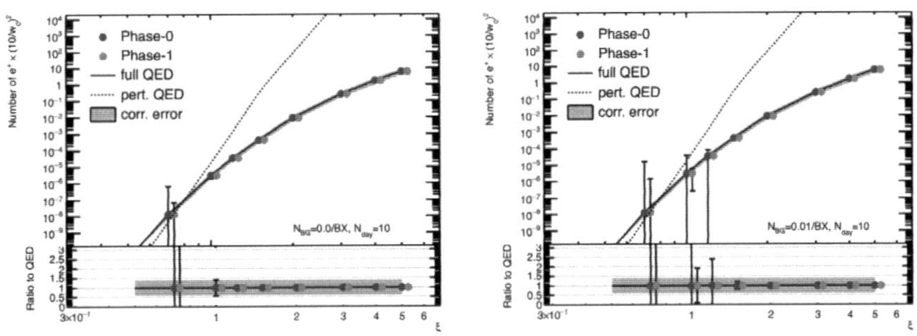

Figure 8. Number of positrons per laser shot, normalised by $10/w_0^2$ versus ξ for phase-0 and phase-1 at LUXE, with w_0 the focal spot size. The left (right) plot shows uncertainties corresponding to no particle background (a particle background of 0.01 particles per beam crossing). Dashed lines correspond to theoretical prediction of perturbative QED, solid lines show theoretical prediction of full strong-field QED. Bottom panels show ratio to QED prediction. More details in [3].

4. Summary

LUXE will explore the transition from perturbative to non-perturbative charge-field coupling. To enable the experimental outcomes of LUXE to be analysed, a new simulation framework, PTARMIGAN, has been developed that models strong-field QED effects using the locally-monochromatic approximation. This framework has been extensively benchmarked, and used to simulate the expected signals of non-perturbative physics which will be measured at LUXE. These are the shifting of the Compton edge in the electron/photon energy spectra in nonlinear Compton scattering, and the "turning of the curve" signifying pair production via a tunnelling mechanism in the Breit-Wheeler process.

4.1. Acknowledgments

AJM is supported by the project High Field Initiative (CZ.02.1.01/0.0/0.0/15_003/0000449) from the European Regional Development Fund. This work has benefited from computing services provided by the German National Analysis Facility (NAF).

References

[1] Ritus V I 1985 *Journal of Soviet Laser Research* **6** 497–617 ISSN 1573-8760 URL `https://doi.org/10.1007/BF01120220`
[2] Heinzl T and Ilderton A 2009 *Opt. Commun.* **282** 1879–1883
[3] Abramowicz H *et al.* 2021 *Eur. Phys. J. ST* **230** 2445–2560 (*Preprint* `2102.02032`)
[4] Ridgers C P, Kirk J G, Duclous R, Blackburn T, Brady C, Bennett K, Arber T and Bell A 2014 *Journal of computational physics* **260** 273–285
[5] Gonoskov A, Bastrakov S, Efimenko E, Ilderton A, Marklund M, Meyerov I, Muraviev A, Sergeev A, Surmin I and Wallin E 2015 *Physical review E* **92** 023305
[6] Lobet M, d'Humières E, Grech M, Ruyer C, Davoine X and Gremillet L 2016 *Journal of Physics: Conference Series* **688** 012058
[7] Harvey C, Ilderton A and King B 2015 *Physical Review A* **91** 013822
[8] Blackburn T G, Seipt D, Bulanov S S and Marklund M 2018 *Physics of Plasmas* **25** 083108
[9] Di Piazza A, Tamburini M, Meuren S and Keitel C 2018 *Physical Review A* **98** 012134
[10] Ilderton A, King B and Seipt D 2019 *Physical Review A* **99** 042121
[11] King B 2020 *Physical Review A* **101** 042508
[12] Nikishov A I and Ritus V I 1964 *Sov. Phys. JETP* **19** 529–541 [Zh. Eksp. Teor. Fiz. **46**, 776 (1964)]
[13] Brown L S and Kibble T 1964 *Phys. Rev.* **133** A705–A719
[14] Ritus V I 1972 *Nucl. Phys. B* **44** 236–252
[15] Breit G and Wheeler J A 1934 *Phys. Rev.* **46** 1087–1091
[16] Reiss H R 1962 *Journal of Mathematical Physics* **3** 59–67 URL `https://doi.org/10.1063/1.1703787`
[17] Narozhny N 1969 *Sov. Phys. JETP* **28** 371–374
[18] Bamber C, Boege S, Koffas T, Kotseroglou T, Melissinos A, Meyerhofer D, Reis D, Ragg W, Bula C, McDonald K *et al.* 1999 *Physical Review D* **60** 092004
[19] Meuren S 2019 Probing strong-field qed at facet-ii (slac e-320) *Third Conference on Extremely High Intensity Laser Physics (ExHILP)*
[20] Cole J, Behm K, Gerstmayr E, Blackburn T, Wood J, Baird C, Duff M J, Harvey C, Ilderton A, Joglekar A *et al.* 2018 *Physical Review X* **8** 011020
[21] Poder K, Tamburini M, Sarri G, Di Piazza A, Kuschel S, Baird C, Behm K, Bohlen S, Cole J, Corvan D *et al.* 2018 *Physical Review X* **8** 031004
[22] Gales S, Tanaka K, Balabanski D, Negoita F, Stutman D, Tesileanu O, Ur C, Ursescu D, Andrei I, Ataman S *et al.* 2018 *Reports on Progress in Physics* **81** 094301
[23] Heinzl T, King B and MacLeod A J 2020 *Phys. Rev. A* **102**(6) 063110 URL `https://link.aps.org/doi/10.1103/PhysRevA.102.063110`
[24] Blackburn T G, MacLeod A J and King B 2021 *New J. Phys.* **23** 085008 (*Preprint* `2103.06673`)
[25] Blackburn T and King B 2021 *arXiv preprint arXiv:2108.10883*
[26] Furry W H 1951 *Phys. Rev.* **81** 115–124
[27] Di Piazza A 2014 *Physical review letters* **113** 040402
[28] Narozhny N and Fofanov M 1996 *Super-Intense Laser-Atom Physics IV* **13** 411
[29] McDonald K T 1997 *order* **2** 0
[30] Seipt D, Kharin V, Rykovanov S, Surzhykov A and Fritzsche S 2016 *Journal of Plasma Physics* **82**
[31] Chen P, Horton-Smith G, Ohgaki T, Weidemann A and Yokoya K 1995 *Nuclear Instruments and Methods in Physics Research Section A: Accelerators, Spectrometers, Detectors and Associated Equipment* **355** 107–110
[32] Blackburn T 2020 *Reviews of Modern Plasma Physics* **4** 1–37
[33] Harvey C, Heinzl T and Ilderton A 2009 *Physical Review A* **79** 063407

AUTHOR INDEX

Institute of Physics
Dirac House, Temple Back
Bristol BS1 6BE UK

ISSN: 1742-6588
ISBN 978-1-7138-5503-3

CURRAN ASSOCIATES INC.
proceedings.com